P9-EEP-478

Applied Atomic Spectroscopy

Volume 2

MODERN ANALYTICAL CHEMISTRY

Series Editor: **David Hercules**
University of Pittsburgh

ANALYTICAL ATOMIC SPECTROSCOPY
By William G. Schrenk

PHOTOELECTRON AND AUGER SPECTROSCOPY
By Thomas A. Carlson

MODERN FLUORESCENCE SPECTROSCOPY, VOLUME 1
Edited by E. L. Wehry

MODERN FLUORESCENCE SPECTROSCOPY, VOLUME 2
Edited by E. L. Wehry

APPLIED ATOMIC SPECTROSCOPY, VOLUME 1
Edited by E. L. Grove

APPLIED ATOMIC SPECTROSCOPY, VOLUME 2
Edited by E. L. Grove

TRANSFORM TECHNIQUES IN CHEMISTRY
Edited by Peter R. Griffiths

Applied Atomic Spectroscopy

Volume 2

Edited by

E. L. Grove

IIT Research Institute
Chicago, Illinois

Plenum Press · New York and London

Library of Congress Cataloging in Publication Data

Main entry under title:

Applied atomic spectroscopy.

(Modern analytical chemistry)
Includes bibliographical references and index.
1. Atomic spectra. I. Grove, E. L., 1913-
QD96.A8A66 543'.085 77-17444
ISBN 0-306-33905-6 (v. 1)
ISBN 0-306-33906-4 (v. 2)

A Division of Plenum Publishing Corporation
227 West 17th Street, New York, N.Y. 10011

Printed in the United States of America

Contributors

Eleanor Berman — Division of Biochemistry, Cook County Hospital, Chicago, Illinois

Bruce E. Buell — Union Oil Company of California, Brea, California

H. Jäger — Institute for Atmospheric Environmental Research, Fraunhofer Society, Garmisch-Partenkirchen, Germany

William Niedermeier — Division of Clinical Immunology and Rheumatology, Department of Medicine, University of Alabama in Birmingham, Birmingham, Alabama

Preface

From the first appearance of the classic *The Spectrum Analysis* in 1885 to the present the field of emission spectroscopy has been evolving and changing. Over the last 20 to 30 years in particular there has been an explosion of new ideas and developments. Of late, the aura of glamour has supposedly been transferred to other techniques, but, nevertheless, it is estimated that 75% or more of the analyses done by the metal industry are accomplished by emission spectroscopy. Further, the excellent sensitivity of plasma sources has created a demand for this technique in such divergent areas as direct trace element analyses in polluted waters.

Developments in the replication process and advances in the art of producing ruled and holographic gratings as well as improvements in the materials from which these gratings are made have made excellent gratings available at reasonable prices. This availability and the development of plane grating mounts have contributed to the increasing popularity of grating spectrometers as compared with the large prism spectrograph and concave grating mounts. Other areas of progress include new and improved methods for excitation, the use of controlled atmospheres and the extension of spectrometry into the vacuum region, the widespread application of the techniques for analysis of nonmetals in metals, the increasing use of polychrometers with concave or echelle gratings and improved readout systems for better reading of spectrographic plates and more efficient data handling.

Many of the far-reaching and on-going changes in industry and environment control would not have been possible without developments in spectroscopy, and committees of ASTM are continuing their work on evaluation and consolidation of procedures.

The available literature dealing with emission spectroscopy has until now been scattered among myriad sources and we in the field have long recognized an urgent need to gather the new ideas and developments together, in a convenient format. However, the enormous amount of work involved in preparing a comprehensive treatise on the subject has been a deterrent. Finally, this major collaborative effort was undertaken: *Applied Atomic Spectroscopy, Volumes 1 and 2* have been written by a group of authors, each of whom has an intimate and expert working knowledge of a special area within the discipline. Individual chapters are treatments in depth of new developments, placed within an historical perspective, in many instances incorporating much of the author's own experience.

I wish to extend my special thanks to all the collaborators for their cooperation and patience. The courtesy of the book and journal publishers who gave permission to reproduce figures and tables is gratefully acknowledged, with special thanks to the U.S. Geological Survey.

We also wish to thank the many practicing spectroscopists for their suggestions and help during the editing process, and last, though not least, Mrs. E. L. Grove and Nancy Robinson for editing, typing, and helping to keep detail in order.

E. L. Grove

Contents

Chapter 2

Petroleum Industry Analytical Applications of Atomic Spectroscopy

Bruce E. Buell

Chapter 3

Analytical Emission Spectroscopy in Biomedical Research

William Niedermeier

Chapter 4

Application of Spectroscopy to Toxicology and Clinical Chemistry

Eleanor Berman

Contents of Volume 1

Applied Atomic Spectroscopy

Volume 2

Precious Metals 1

H. Jäger

1.1 INTRODUCTION

The great importance of gold, silver, and the metals of the platinum group (PGM) is due to their intrinsic qualities and their comparatively rare geological occurrence. Their unique physical and chemical properties, which result in many specialized applications, combined with their high value, justify the large volume of work that has been done on their analysis.

Gold and silver are among the oldest known metals and have been highly esteemed since early times. Platinum, although known to the natives of South America, was first made known in Europe by the Spaniards (Ulloa in 1735). The other metals of the platinum group were discovered later; osmium and iridium by Tennant in 1803 and palladium and rhodium by Wollaston in 1803–1804, all in England, and finally ruthenium by Claus in Russia in 1844.

The precious or noble metals, alloyed and unalloyed, gradually achieved a wide range of uses, owing to their scarcity, their corrosion resistance, their conductivity of heat and electricity, and other properties.

Silver: Currency (historical importance), arts and crafts (jewelry), medicine (dental alloys, artificial bones), photography (silver bromide in emulsions), photochromatic glass (change of transmission by irradiation), engine bearings, electrical engineering and electronic industry (silver solders, silver brazing alloys, connections, relay contacts, silver–magnesium-nickel alloys for high thermal conductivity, printed circuits).

Gold: Monetary system (security of world currencies; at present about 60% of the mined gold is being used for this purpose), arts and crafts (jewelry), medicine (dental alloys, radioactive gold for treatment of cancer), chemical industry (plating against corrosion), electronic industry (contacts, printed

H. Jäger • Institute for Atmospheric Environmental Research, Fraunhofer Society, Garmisch-Partenkirchen, Germany

circuits, connections, semiconductors, precision resistances when alloyed with other metals), space vehicles (coating against infrared and high-intensity solar radiation, coating of bearings, glass-to-metal seals).

Platinum: Jewelry, medicine (dental alloys), chemical industry (crucibles, anodes for electrochemical processes, chemical catalysts), electrical industry (thermocouples, spark plug contacts, resistance thermometers, contacts, potentiometer windings, furnace windings), car exhaust catalysts, glass and glass fiber production.

Palladium: Jewelry (substitute for platinum), medicine (dental alloys), chemical industry (catalysts, hydrogen purification by distillation), electrical industry (thermocouples, contacts, potentiometer windings, brazing alloys).

Remaining platinum group metals: Used primarily as hardening agents, alloyed with the other platinum group metals for the abovementioned purposes.

The precious metals are distributed as traces throughout the earth's crust. According to Goldschmidt,[1] the abundance of the noble metals is as follows: silver, 20 ppb (10^{-9}); palladium, 10 ppb; platinum, 5 ppb; gold, rhodium, iridium, 1 ppb; osmium, ruthenium, $>$0.4 ppb. Owing to their relatively low chemical reactivity, they usually occur in the metallic state, often alloyed with each other. Gold is found as the metal or as the telluride ($AuTe_2$, $AuAgTe_4$). Silver exists as metal deposits, as sulfide ores (argentite Ag_2S, proustite Ag_3AsS_3, pyrargyrite Ag_3SbS_3), or as horn silver (AgCl). The platinum group metals are mostly found as free metals and usually together. The only mineral of importance is sperrylite ($PtAs_2$). They are often mined together with the ores of gold, silver, copper, nickel, and iron. Table 1.1 summarizes the most frequent deposits.

Corrosion resistance was one of the characteristics leading to the term "precious metals." Gold and silver, together with copper, belong to group IB of the periodical system; the metals of the platinum group, together with iron, cobalt, and nickel, belong to group VIIIB. Table 1.2 shows some chemical properties of these metals.

Gold, silver, platinum, and palladium are easily worked. The others are hard and brittle; iridium and rhodium are workable at red heat. It is difficult to fuse or cast the PGM because of their high melting points. Table 1.3 gives some of their physical properties.

Precious metal analysis has two major aspects, which define the analytical requirements: (1) determination of these elements at low and very low concentrations (e.g., prospecting and mining); (2) and analysis of the metals themselves (e.g., purity tests, alloying components).

Smith[2] has given a fine summary of the history of the analysis of the precious metals, gold being the most important. The analysis or assay of gold was regarded as an art for a long time. Assaying is the oldest known form of chemistry and can historically be traced back to early civilizations (e.g., ancient Egypt). In ancient Rome lead was used for the purification of gold and silver. In the Middle Ages cupellation was introduced in France and Germany. Parting assay

Table 1.1 Deposits of Precious Metals

Element	Found as	Where	World deposits
Gold	Native metal, telluride	Quartz veins, alluvial deposits	South Africa (Transvaal, Orange Free State), USSR (Ural, Kasakhstan, Transcaucasia, Central Asia, Siberia, Far East), Canada (British Columbia, Ontario, Quebec, NW-Territories), USA (South Dakota, Utah, Alaska, California, Arizona, Washington, Nevada, Colorado)–together 85% of world production Other: Australia, Mexico, Nicaragua, Colombia, Sweden, India, Korea, Japan, Philippines, Dem. Rep. Congo, Ghana
Silver	Native metal, horn silver, sulfide	Sulfide ores, together with lead, zinc, copper, nickel, and gold ores	Mexico, USA, Canada, Peru–together 50% of world production Other: USSR, Australia, Germany, Spain, Yugoslavia, Sweden, Czechoslovakia, Japan, Burma, South Africa, Dem. Rep. Congo
Metals of the platinum group	Native metal, sperrylite	Alluvial deposits, together with gold, silver, copper, nickel, and iron ores	USSR (Ural), Canada (Ontario), South Africa, (Transvaal), Colombia, USA (California, Oregon, Alaska), South America, Australia

Table 1.2 Chemical Properties of Precious Metals

Element	Symbol	Dissolving acids	Reacts with
Gold	Au	Aqua regia, selenic acid, telluric acid	
Silver	Ag	Nitric acid, hot sulfuric acid	Ozone and hydrogen sulfide on surface
Platinum	Pt	Aqua regia	Chlorine, at high temperature with fused caustic alkaline nitrates, peroxides
Palladium	Pd	Aqua regia, hot nitric acid, hot sulfuric acid	At moderate temperature with halogens, sulfur, arsenic, absorbs 900 times its own volume of hydrogen
Osmium	Os		At room temperature with oxygen to poisonous osmium tetroxide, at heat with fluorine
Iridium	Ir		Oxidation losses over 1000°C; is attacked by chlorine at high temperature
Rhodium	Rh		Oxygen, chlorine at red heat
Ruthenium	Ru		Oxygen, halogens at 800°C

(i.e., the separation of gold and silver) was also known at that time. These methods have not changed essentially, and even in the day of automatic analytical instruments the fire assay procedure is still regarded as the most accurate and reliable method and is widely applied.

Spectral analysis is relatively young by comparison with the fire assay method. Analytical spectroscopy, as used today, had its origin in the beginning of the nineteenth century (Wollaston, Fraunhofer, Kirchhoff, Bunsen), but the first spectrographic analyses of precious metals were done in the early 1920s.

The first spectrochemical papers on the analysis of noble metals seem to be those by Wichers[(3)] and by Meggers *et al.*,[(4)] who in 1921 and 1922 tested the fineness of platinum and 99.99% gold by spectral analysis. During and after 1928, several papers by Gerlach and Schweitzer appeared for the determination

Table 1.3 Physical Properties of Precious Metals

Energy	Atomic number	Atomic weight	Melting point (°C)	Boiling point (°C)	Density (g/cm^3)	Ionization energy (eV)
Silver	47	107.87	960.8	2212	10.50	7.57
Gold	79	196.97	1064.4	2966	19.28	9.22
Ruthenium	44	101.07	2250	3900	12.30	7.36
Rhodium	45	102.91	1960	3727	12.44	7.46
Palladium	46	106.40	1552	2927	11.97	8.33
Osmium	76	190.20	3000	~5000	22.48	8.5
Iridium	77	192.20	2443	4527	22.42	9.2
Platinum	78	195.09	1769	3827	21.45	8.96

of the lead content in gold[5] and in gold-silver-copper alloys.[6] They used the metals as electrodes with spark excitation. Even then, tests were done using inert gas atmospheres, and the technique of selective evaporation from the melted sample was applied to increase intensities. The determination of the contents was done by means of calibration curves or homologous line pairs (lines of matrix and contamination which show similar intensities at certain concentrations). At that time, examinations concerning the homogeneity of gold samples were also carried out spectroscopically.[7] In another paper the iridium, rhodium, and palladium contents of platinum were determined.[8]

During the following decades the spectral analysis of the noble metals remained in the emission field. Atomic absorption picked up quickly after the basic research work by Walsh[9] and Alkemade and Milatz[10] in 1955. In fact, today most papers on precious metal analysis describe AAS applications. X-ray spectroscopy was used in 1930 by Ida and Walter Noddak for noble metal determination,[11] but only during the past decade have an increasing number of X-ray analyses been reported. The latter also applies to neutron activation analysis. All these modern instrumental methods have to compete with the accepted traditional fire assay and other chemical methods.

Much has been published describing analytical methods for precious metals. Details of the spectroscopic research work, from its inception until 1950, can be found in Twyman's *Metal Spectroscopy*.[12] Further developments in the spectral analysis of precious metals and other analytical techniques are collected in review papers by Beamish and coworkers[13-22] and by Ivanov and Busev.[23]

The aim of this work is not to offer analytical recipes but to give a review of the current methods, employing atomic emission and absorption spectroscopy as well as nuclear methods. Readers who are interested in the particulars of certain techniques should refer to the original papers.

1.2 EMISSION SPECTROSCOPY

1.2.1 Analysis of Noble Metals

When reviewing the emission spectral methods of analysis developed thus far for the noble metals, two remarkable things should be mentioned: (1) although one deals with metals, i.e., good electrical conductors, most of the methods employ graphite electrodes or sample-graphite mixtures; and (2) with a few exceptions, only photographic recording instruments have been used. The reasons for (1) are: In noble metal analysis, in contrast to the analysis of other types of metals (e.g., steel), one tries or, more often, is forced to use sample quantities that are as small as possible. Here the spectral analytical standard method, the excitation of a sample-graphite system, offers a solution. This method, originally developed for the analysis of nonconductive materials, is also suitable for metal analysis. Another, yet stronger reason for (1) is the lack of commercially available and internationally recognized noble metal standards. There are noble metals of highest purity available (99.999 to 99.9999%) with sufficiently accurate total analyses, but alloys, which are essential for establishing calibration curves, have to be produced and tested by the analyst himself. Exact composition, homogeneity, structure, and so on, of such standard samples are not always guaranteed, and such tests are costly and tedious. Again one can solve this problem by mixing metal powders, salts, or solutions. This leads almost inevitably to the use of graphite as electrode material or as sample admixture.

For (2), in the past, the noble metal industry did not feel the need for high-speed direct-reading methods; and analyses could take hours. In comparison with the time taken for sample and standard preparations, the evaluation of a photographic plate is done relatively quickly. However, two large producers, Englehard Minerals and Chemical Corp. and Johnson, Matthey Chemicals Ltd. use direct readers. The U.S. Geological Survey group at Denver developed techniques and have used direct readers for geological materials of value for some time.[23a]

In most cases air was chosen as the discharge atmosphere, thus disregarding the advantages of a controlled and inert atmosphere.

The spectral analysis of the noble metals can be divided into two groups. In the first group the sample is introduced into the plasma with the aid of graphite. These techniques are categorized by the methods of sample preparation, whereas the plasmas and excitation processes are very similar. In the second group, use is made of the metal character of the sample. Here the methods are characterized by different evaporation and excitation mechanisms, whereas the sample remains in the original state.

1.2.1.1 Sample-Graphite Techniques

Conventional spectrochemical techniques are used for this group. In all cases an arc or a spark discharge takes place between graphite electrodes. These procedures do not utilize the properties specific to precious metals. On the other hand they yield advantages like versatility, little matrix effects, and the choice of any internal standard. Due to the use of air as discharge atmosphere, parts of the spectra are influenced or obscured by band emission, which prevents the evaluation of some of the most sensitive spectral lines. Apart from these drawbacks many of the published methods that have been developed give excellent analytical results.

There are two general methods for the preparation of standards, viz. the mixing of dry substances or the mixing of solutions, the latter being used more frequently. Dry mixing is mostly done with pure metal powders or oxides. The preparation of solutions of some of the noble metals is difficult, as osmium, iridium, rhodium, and ruthenium are not soluble in acids. Several authors give recipes for dissolving these elements: osmium,[24] iridium, [24–27] rhodium,[25,26,28] and ruthenium.[24,25] Admixtures of impurities are dissolved as pure metals, oxides, or nitrates. With few exceptions, almost all the elements of interest can be dissolved.

1.2.1.1a Powder DC Arc Techniques. The powder dc arc technique, employing cupped graphite electrodes, is most commonly used in spectrochemistry. The published methods show only slight differences in such details as sample and standard preparation. Usually, the sample is mixed with graphite which is either added to the sample solution[29,30] or added after the sample has been evaporated to dryness.[31] Regarding the analysis of silver, the metal can be converted to its nitrate, which is then packed into the electrode crater without the addition of graphite.[32] Spectrographic buffers are also used to increase the sensitivity of silver nitrate analyses.[33]

There is little information given concerning the time spent on sample preparation, but dissolution, evaporation, grinding, mixing, and so on, may take several hours. In all cases the sample is excited in the anode of a free-burning dc arc. Exposure times run up to several minutes, because in some cases the sample is completely evaporated. Table 1.4 summarizes the methods described.

1.2.1.1b Pellet Techniques. To prepare pellet standards, solutions[25,26,35–37,39] as well as dry substances[25,38] are used. Graphite again can be added during the different stages of preparation. Preparation times of several hours and even days are reported. Pellets 4–6 mm in diameter are produced and fitted onto graphite electrodes. Lewis and Ott employ brass sample holders.[26,38] Lincoln and Kohler use a controlled atmosphere (Stallwood jet, Ar/O_2 70:30) to increase sensitivity and precision.[35,36] Either dc arc, intermittent ac arc, or a high-voltage spark are used for excitation. Table 1.5 summarizes the pellet methods.

1.2.1.1c Solution Residue. Using this technique, standard and sample preparation time is relatively short. The sample solutions are evaporated onto the flat tops of carbon or copper electrodes. In some cases both electrodes are treated in this way. Carbon electrodes usually require pretreatment either to seal the pores, to prevent the sample from being partially absorbed by the electrode,[40–43] or to allow the sample to penetrate to a predetermined

Table 1.4 Powder DC Arc Technique

Sample	Impurities	Excitation	Internal standard	Reference
Au	31 elements	dc arc		34
Au	Ga	dc arc	Au	30
Ag as nitrate	10 elements	dc arc	Ag	32
Ag nitrate	25 elements	dc arc	Ag	33
Ag	31 elements	dc arc		34

Table 1.5 Pellet Technique

Sample	Impurities	Excitation	Internal standard	Reference
Au	26 elements	dc arc	Au	35
Ag, Pd, Pt	9 elements	Unipolar intermittent ac arc	Ag, Pd, Pt	39
Pt	27 elements	dc arc	Pt	36
Pd	28 elements	dc arc	Pd	37
Pd	14 elements	dc arc, high-voltage spark	Pd	38
Rh	12 elements	dc arc, high-voltage spark	Rh	26
Ir, Rh	12 elements	ac arc	Rh, Ir	25

depth.[44] In the first case a few drops of apiezon solution or oleic acid are used; in the second case the electrodes are immersed in a polystyrene solution and then dried.

The solution residue technique is not restricted to graphite electrodes; high-purity copper, which is comparatively inexpensive, can also be used. Further advantages of copper electrodes are the absence of porosity, the relatively simple spectrum, and the suppression of CN bands due to the lack of carbon from the electrodes.[46] Up to 0.1 ml of solution can be placed on the flat surface of the electrode. Dipping the electrode in apiezon solution prevents the sample solution from trickling down the sides.

If both electrodes serve as sample electrodes, spark or ac arc excitation may be applied (copper electrodes, with spark only). The use of a dc arc allows only one sample electrode. As a result of sample consumption, line intensities decrease exponentially with time. Since the line/background ratios also decrease, exposure times should not exceed half the total burn period. Residue techniques are summarized in Table 1.6.

Table 1.6 Solution Residue Technique

Sample	Impurities	Excitation	Internal standard	Reference
Au	23 elements	ac arc	Au, background	44
Au	17 elements	dc arc		40
Au	11 elements	Spark		45
Au	Pt, Pd, Rh	Intermittent dc arc	Mo	43
Au	P	dc arc		41
Au	Sb	Spark	Mo	42

1.2.1.1d Solution Techniques. The shortest possible periods of sample preparation can be expected from techniques where the sample solution is introduced directly into the source. The spectral analysis of solutions has been known for several decades. Detailed investigations in this field have been done since 1945. The methods in use—the porous cup, the vacuum cup, or the rotating disc—have many advantages: simple and quick sample preparation, simple introduction of any internal standard, homogeneity of the sample, decomposition of the matrix with the resulting elimination of metallurgical history effects, a large sample volume, and relatively good precision, accuracy, and sensitivity. The method also has some disadvantages: considerable heating of the solution occurs in some cases, which can lead to boiling of the solution and result in inconsistent line ratios; low line/background ratios due to long exposure times, and fractionation because of absorption (porous cup). Solution techniques such as the plasma jet will be dealt with in Section 1.2.4.

Porous Cup. The sample electrode, used as the upper electrode, consists of a graphite rod with a very deep cavity bored into it, leaving only a relatively thin base. The solution diffuses through the base, which has been made porous by preburning. Thorough investigations of this method were published by Feldman,[47] who also reviewed the solution techniques known at that time. The energy of the high-voltage spark should not heat up the solution, but should be used for the evaporation of the diffused solution. All solutions which wet carbon may be analyzed. Interelement effects cannot be excluded.

Vacuum Cup. The vacuum cup electrode has a long concentric hole bored into it. At the lower end a hole through the side allows the sample solution to enter from a Teflon receptacle which is fixed on the shoulder of the electrode. The underpressure resulting from the gas expansion during the discharge sucks the solution up into the analytical gap and it is distributed into the plasma. According to Zink,[48] the transport rate is between 1 ml/100 s and 1 ml/40 s, depending on the size of the holes. With a still larger bore, the droplet size becomes too large, which hampers volatilization. The size of the bore is limited by the viscosity of the solution.

Rotating Disc. The rotating disc or rotrode method is probably the most popular solution technique. The various parameters of the system have to be optimized for each application. The wheel diameter and width, rotational speed, immersion depth, and preburn and exposure time have to be adjusted. Eardley and Clarke[50] suggested a $\frac{1}{2}$-inch wheel and 12–16 rpm at a constant immersion depth. Hyman could not detect selective evaporation when analyzing gold alloys.[51] The high-voltage spark used does not cause overheating of the wheel and the solution. A controlled argon–oxygen atmosphere improves the sensitivity of some elements. Dixon and Steele observed mutual and base metal interferences,[52] which could not be suppressed completely by the addition of a buffer.

Table 1.7 Solution Technique

Sample	Impurities	Excitation	Internal standard	Reference
Noble metal solution	Pd, Pt, Ir, Rh	Porous cup + high-voltage spark	Co	27
Au	8 elements	Vacuum cup + ac arc	Au	49
Au	8 elements	Rotating disc + high-voltage spark	Au, As, Sb	51
Noble metal solution	Au, Pt, Pd, Rh, Ru, Ir, Os	Rotating disc + controlled condensed discharge	Ni	52
Pt	Rh	Rotating pin + low-voltage condensed spark	Pt	28

Rotating Pins. After studies on the behavior of solutions with heating when using the rotrode method, Bardocz and Varsanyi introduced the rotating pin electrode for the investigation of platinum alloys.[28] The discharge–a low-voltage condensed spark–is ignited only when the slowly rotating electrodes (3.75 rpm) pass the counterelectrode.

Table 1.7 summarizes the applications of the solution techniques for the analysis of the noble metals. Figure 1.1 shows the electrode systems of the various techniques. Precision data of the methods mentioned in Section 1.2.1.1 vary from a few percent relative standard deviation up to more than 25%, depending on the method, the element, and the absolute concentration. Interference effects have been reported only by a few authors. Data on accuracy are given with great reservation since independently analyzed standards or comparative studies are usually not available, but estimated values are presumed to be of the same order as the precision data. Detection limits characteristic of the different methods are given in Table 1.9.

1.2.1.2 Metallic Samples

1.2.1.2a Globule Arc. The globule arc has a long history for the analysis of copper. The method was first suggested in 1919.[53] Milbourn,[54] Werner,[55,56] Publicover,[57] and Tölle[58] continued with this technique. In noble metal analysis the globule arc has been applied mainly for analyzing gold.[59-63] The sample (up to several 100 mg), in the form of shavings, filings, cuttings, wire pieces, or the like, is placed in the shallow cavity of the graphite electrode, where, immediately after ignition of the dc or ac arc, the molten globule forms. Gold does not oxidize but forms a homogeneous globule with high thermal and electrical conductivity where the anode burning spot is concentrated. The

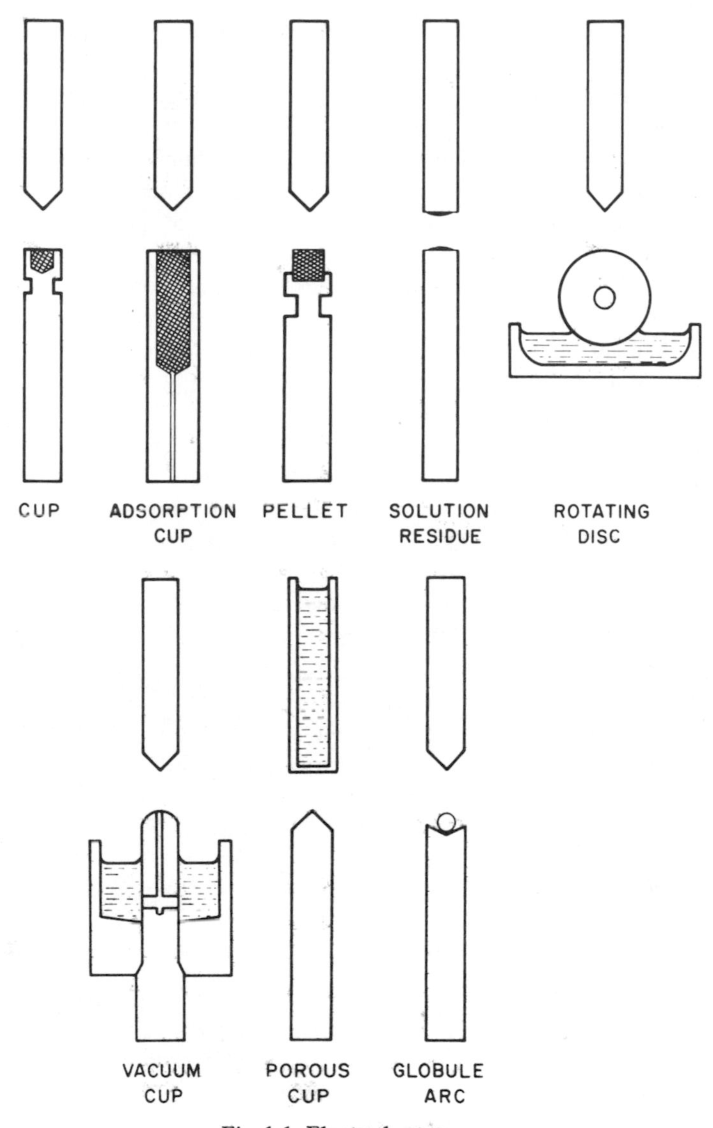

Fig. 1.1 Electrode systems.

sample evaporates and jumps out of the cavity after reaching a critical diameter. Depending on their boiling points, the impurities distill off preferentially. Elements of high boiling points can also evaporate at an early stage if they form volatile oxides on the surface of the globule. Copper and silver evaporate slowly because of the formation of intermetallic compounds of the gold–copper and gold–silver systems.[63] Gerbatsch and Artus observed a distinct cathode layer enrichment (factor of enrichment up to 6).[63] They also mentioned that the globule arc can be applied for the analysis of silver. In this case, however, there is no cathode enrichment. Tölle had some difficulties when analyzing samples of palladium powder in a globule arc.[64] He describes the formation of a silicate phase which hampers the volatilization of silicon, aluminum, magnesium, and iron. Only silver evaporates preferentially, whereas gold and even elements like lead and zinc distill off continuously. An addition of 2.5% graphite improves the evaporation of silicon, magnesium, and iron.

The globule arc offers a suitable method for analyzing beads produced by the fire assay enrichment of geologic samples (see Section 1.2.3.3a). Dorrzapf and Brown analyzed gold beads,[65] Tomingas and Cooper,[66] Whitehead and Heady,[67] Sim,[68] and Cooley *et al.*[69] silver beads, and Broadhead *et al.* platinum beads.[24] Tomingas and Cooper observed interelement effects; copper, and to a lesser extent, gold, at higher concentrations enhances the palladium and silver lines. The production of good standards is not easy and rather tedious. Gerbatsch and Artus produced a master alloy containing all the impurities of interest,[63] which was then diluted with highest-purity gold wire in an induction furnace.

Published precision data are of the order of a few % relative standard deviation, but can reach values of 20% and more near the detection limit. Only a few independent comparative measurements for accuracy tests are available. They are, however, very good, considering the low concentrations.[63,68] The globule arc is one of the most sensitive analytical sources. Gerbatsch and Artus could report guarantee limits (2 times the detection limit) for silver and copper of 0.005 ppm, tin 0.02 ppm, iron 0.07 ppm, gallium 0.08 ppm, and lead and antimony 0.3 ppm.[63] The detection limit depends, of course, on the size of the analyzed bead. Globule arc methods are summarized in Table 1.8.

Table 1.8 Globule Arc

Sample	Impurities	Excitation	Internal standard	Reference
Au	Cu, Ag	dc arc	Au	62
Au	Cu, Ag, Pd, Fe, Pb, Zn	dc arc	Au	61
Au	Cu, Ag, Fe, Bi, Sb, Pb		Background	60
Au	Cu, Ag, Fe, Ga, Pb, Sb, Sn			63
Au	Pt, Pd, Rh	dc arc		65
Ag	Pd	ac arc	Ag	66
Ag	Pt, Pd, Rh, Au	dc arc	Ag, Pt, Rh	67
Ag	Au, all PGM	dc arc	Ag	68
Ag	Au, Pt, Pd, Rh, Ru, Ir	dc arc		69
Pt	It, Os, Ru	dc arc	Pt	24
Pd	Au, Ag, Pb, Zn, Si, Mg, Al, Fe	dc arc		64

1.2.1.2b Carrier Distillation. The carrier distillation technique takes advantage of the fact that halogens form volatile compounds. If one adds metal halides to the sample metal in the crater of a carbon electrode, these halides become a good flux after ignition of the dc arc. It is not known precisely whether the halides fuse with the metal or only envelop the melt. At high crater temperatures metal halides are no longer stable and will decompose. The halogens thus produced will react with the sample impurities on the surface of the globule. These new halides, because of their volatility, will immediately be transported into the plasma. This reaction continues until either the halogen or the metal is consumed. Tymchuk *et al.* determined iron traces in copper and silver with an efficiency of 96% using this technique.(70) This value was independent of the iron concentration and the sample size. Matrix elements such as copper, silver, gold, and the platinum group metals do not react with the halogens, and therefore remain in the electrode until the carrier is consumed.

Tymchuk *et al.* also tested various types of halide carriers: fluorides, chlorides, bromides, and iodides of copper and silver.(71) Not all affect the impurities in the same way; e.g., for each trace element there was an optimum carrier, and mixing of the carriers did not appear to offer any advantages. In general, fluorides show the best enhancement. These halides have to be produced with the greatest possible care, because traces of the analyte element in the carrier cause nonlinear calibration curves.

Tymchuk *et al.*(73) [based on previous research(71,72)] recommend the use of copper fluoride as the most suitable carrier for the analysis of silver, gold, and the PGM. Burnoff curves show that the common impurities in all precious metals are preferentially evaporated. The PGM must be mixed with high-purity copper (1 : 2.5) to reduce their melting point. During the first 10 s after igniting the arc, almost all the impurities are removed. Meanwhile the matrix remains practically unchanged, as the flux completely envelopes the sample and the arc cannot concentrate on the sample itself.

This fact is most important in analysis. As the matrix is not consumed during the exposure time, the line intensities of the impurities are independent of sample size and type and depend only on the absolute amount of impurities present in the sample. Therefore, calibration curves for impurities in noble metals can be established with another metal of known composition, e.g., copper, by varying the copper quantity. This type of analysis is, of course, confined to those impurities which react with the carrier. Copper and the noble metals therefore are not enhanced. After the carrier has been consumed, normal globule arc conditions take over. Precision data vary between 4 and 10%, depending on the element and its concentration. Accuracy data (compared with AAS values and added amounts) are of the same order. Detection limits for impurities in silver and gold are about the same as in copper (from 0.02 ppm upward), whereas in PGM they are about 4 times higher, because of the necessity of having to dilute the sample with copper.

1.2.1.2c Glow Discharge. The glow discharge source is a relatively new development in the field of analytical spectroscopy. It was described first by Grimm in 1968.[74] The "abnormal" glow discharge takes place at pressures of 2–20 Torr. Metal samples used as the cathode are atomized by cathodic sputtering.[75] The sputtering rate amounts to only a few mg/min. On a commercial model of the source (RSV, West Germany) burning spots have diameters of 8 mm. The minimum sample size allowable is 15 mm diameter. The sample has to be plane to seal off the lamp for evacuation of the discharge chamber (Fig. 1.2).

The emitted spectra consist chiefly of atomic lines. The spectra show negligible background radiation and are free from any interfering molecular bands because spectroscopically pure argon is used as the discharge atmosphere. The characteristic occurrence and intensities of the spectra differ so markedly from the well-known arc and spark spectra that Salpeter has compiled special spectral atlases.[76]

Because the sample material is removed by ionic bombardment, atomic layer by atomic layer, no fractional distillation occurs. Line intensity ratios therefore remain constant during the exposure. Surface contamination is removed during a preburn period. A recent report deals with some aspects of the sputtering of gold and other metallic samples in the Grimm discharge source.[75] The sample surface should be at least 40 mm in diameter in order to have more than one burning spot. Standards, in the form of alloys, have to be prepared.

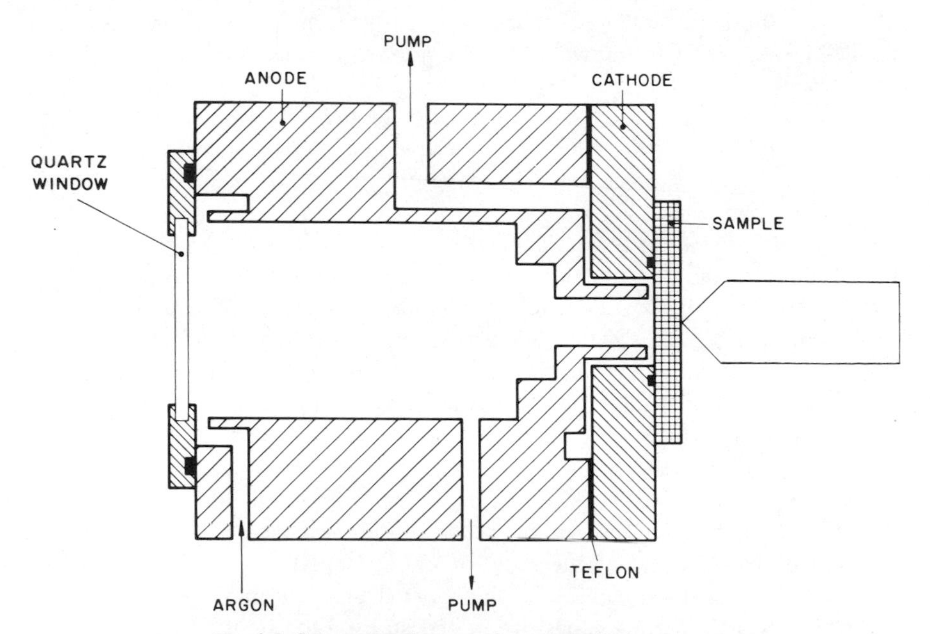

Fig. 1.2 Schematic diagram of the glow-discharge lamp.

Standards and samples should have the same metallurgical history to ensure measurements which lie on a common calibration curve.

For the analysis of high-concentration alloys such as raw gold (80–90% gold, up to 15% silver, and 7% copper, and having lead sometimes in the percent range),[77,78] the reference to one of these elements as an internal standard is not practicable. The background emission, since it is negligible, also cannot be used. For these reasons the sputtering rate of the sample has been introduced as the internal standard. This is possible because the sample removal by sputtering is steady and controllable. The relative sputtering rate S for each sample is related to the intensities I of the alloying elements by the relation

$$S = a \cdot I_{\mathrm{Au}} + b \cdot I_{\mathrm{Ag}} + c \cdot I_{\mathrm{Cu}} + \cdots \tag{1.1}$$

where a, b, c are factors determined by measurements of a reference sample containing known concentrations C of Au, Ag, Cu, $\cdots$:

$$a = (C_{\mathrm{Au}}/I_{\mathrm{Au}})_{\mathrm{standard}} \qquad b = \cdots \tag{1.2}$$

The sum of the normalized intensity values J is then constant for all samples:

$$\frac{a \cdot I_{\mathrm{Au}}}{S} + \frac{b \cdot I_{\mathrm{Ag}}}{S} + \frac{c \cdot I_{\mathrm{Cu}}}{S} + \cdots = J_{\mathrm{Au}} + J_{\mathrm{Ag}} + J_{\mathrm{Cu}} + \cdots = \text{const.} \tag{1.3}$$

The advantage of this method is an increase in precision, which allows the establishment of a calibration curve up to very high concentration of the matrix element (99.995% Au) (Fig. 1.3). There is practically no upper limit of detection for all elements.

For the analysis of fine gold (99.5% plus) the calibration curves are almost straight.[79] In the case of raw gold analysis, self-absorption when using resonance lines and interferences of the main elements cause nonlinear and split calibration curves.[78] The nonlinearity can be overcome by utilizing nonresonance lines. The interferences are caused by elements which hamper the sputtering of the sample (copper in the case of gold). These changes in the sputtering rate cannot be completely eliminated by referring the intensities to the sputtering rate, because such a change in sputtering influences the total excitation mechanism in the plasma. The curvature and splitting of the calibration curves has been expressed by second- and third-order polynomial equations, the coefficients of which have been computed by an iteration program.[78] This enables on-line data calculations to be made in connection with a direct-reading spectrometer to attain high-speed analyses, which are required in the gold refining industry.

Relative standard deviations for the determination of impurity concentrations in raw and in fine gold are 1–2%. The accuracy of the determination of the fineness of refined gold has been determined by analyzing 100 samples. The average deviation from the Au fire assay values was 0.009% at a level of 99.6% gold. Compared with other methods, the glow discharge is not a very sensitive

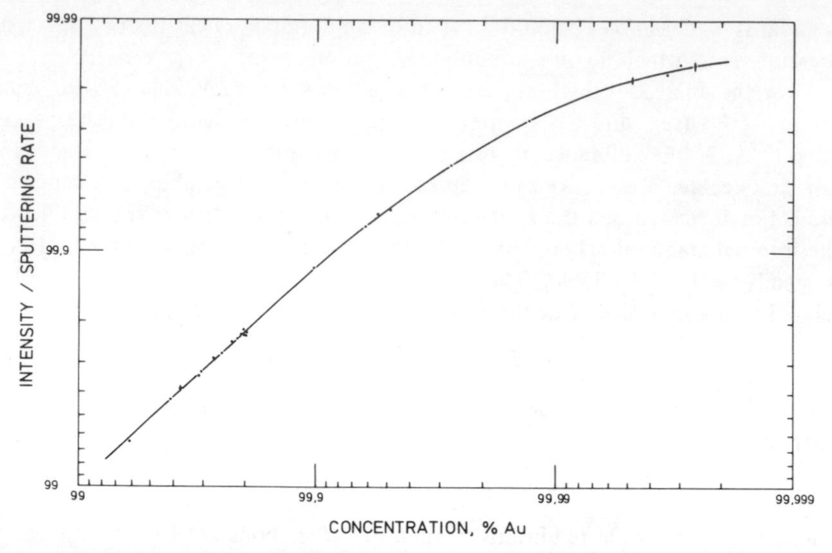

Fig. 1.3 Calibration curve for gold in gold samples. The calibration is not linear at high concentrations because minor constituents in the ppm range have been included in the sputtering rate without background correction.

source. Detection limits are: copper, <1 ppm; silver, 1 ppm; other contaminants, 10–20 ppm. On the other hand, the fineness of the matrix element itself can be determined with an accuracy which is comparable only to fire assay results.

Technical improvements such as the incorporation of a secondary high-density current discharge to increase line intensities[80] might make this excitation source even more attractive in the future. The application of the glow discharge lamp to the analysis of other precious metals is only a question of time.

1.2.1.2d Point-to-Point Technique. In point-to-point or point-to-plane techniques the metals to be analyzed are used directly as the electrode or electrodes. But since the work by Gerlach and Schweitzer,[5–8] these methods have rarely been applied in noble metal analyses. In the recent literature only one publication, by Manrique, who analyzed 99.99% silver with the point-to-point technique, could be traced.[81] Detection limits are of the order of 1 ppm for bismuth, lead, and tellurium and 10 ppm for copper.

1.2.1.3 Comparison of Methods

Some of the technical details of the described methods have been listed in Tables 1.4 to 1.8. An objective comparison of the methods is rather difficult. Particulars concerning precision, accuracy, and detection limits depend on many factors, such as instrumental facilities, concentration levels, and mathematical treatment of data, and consequently methods and results cannot always be directly compared. In most cases no independent test is given for ascertaining the accuracy. By comparison with other methods of excitation, the glow discharge lamp gives excellent precision and accuracy. Table 1.9 shows detection limits (as published by the authors) as only one criterion for comparing the methods. Only a few characteristic publications have been referred to and only elements which are comparable have been selected.

1.2.2 Determination of Noble Metals in Other Metals and Materials

Relatively few publications are available describing the determination of noble metals in other metals and materials. A reference will be made only to those methods which are currently in use. In addition to the graphite cup and the globule arc, which have been described previously, the popular point-to-plane technique and a special application of the hollow cathode lamp will be reported.

1.2.2.1 Graphite Electrode

Mitchell described a quick method for analyzing 14 high-purity metals.[34] Of the precious metals, only silver was determined, but the method could be extended easily to include the other precious metals as well. Standards were produced by mixing oxides or other suitable compounds with graphite and highly refined metal powders. The samples were burned in the positive crater electrode of a dc arc at between 10 and 15 A in air (aluminum was arced in a 70:30 Ar/O_2 atmosphere). The photographic plate was moved at certain time intervals, which depended on the evaporation periods of the elements. Jackwerth *et al.* determined palladium in high-purity metals (nickel, copper, silver).[82] Palladium was coprecipitated with silver cyanide and the precipitate burned in a dc arc. Talalev determined palladium and ruthenium in aluminum oxide using a dc arc.[83] Since the most sensitive palladium and ruthenium lines lie in the CN-band range, an Ar/O_2 atmosphere (1:1) was used. Larina *et al.* determined gold in cathodic copper.[84] The sample was dissolved and the solution was mixed with activated charcoal. Gold was completely adsorbed. The solution then was filtered and the residue containing gold was calcined, mixed with graphite, and arced at 14 A dc. An interesting possibility for analyzing gold and silver in cyanide solutions was described by Savichev and Shugurov.[85] Activated charcoal is placed in the crater of a graphite electrode from which a small channel extends through to the other end of the electrode. When the solution is sucked through, gold and silver are adsorbed from the solution. Russo determined the amount of gold in coatings on brass.[86] Samples and standards were prepared in the form of solutions. Chalkov and Ustimov analyzed small concentrations of gold in lead production materials.[87] Fire assay enrichment was applied, resulting in silver beads. Sample and standard preparation was again done by means of solutions.

Table 1.9 Detection of Impurities in Noble Metals, Various Methods

Matrix	Impurities	Graphite Cup		Pellet		Solution residue		Solution		Globule arc		Carrier distillation		Glow discharge	
		Det. lim. (ppm)	Line (nm)	Det. lim. (ppm)	Line (nm)	Det. lim. (ppm)	Line (nm)	Det. lim. (ppm)	Line (nm)	Det. lim. (ppm)	Line (nm)	Det. lim. (ppm)	Line (nm)	Det. lim. (ppm)	Line (nm)
Au	Ag	0.1		5	338.3	1	328.1	10	328.1	0.005	328.1			1	338.3
	Pt	1		5	265.9	1	306.5								
	Pd			0.1	340.5	1	324.3								
	Cu	0.1		0.5	327.4	0.5	327.4	10	324.8	0.005	327.4			<1	324.8
	Fe	0.1		5	302.1	0.5	302.1	10	302.1	0.03	302.1	0.05	302.1	19	260.0
	Sb	1								0.2	287.8	0.23	259.8		
	Pb	1		1	368.4	30	280.2	10	283.3	0.15	261.4	0.10	283.3	9	405.8
	Ref.	34		35		40		49		63		73		79	
Ag	Au			2	267.6					5	267.6				
	Pt			10	265.9					20	306.5				
	Pd			10	363.5					1	342.1				
	Cu	0.1		0.4	324.8							0.05	302.1		
	Fe	0.1		0.5	248.3					(Fire assay bead)		0.23	259.8		
	Sb	0.5										0.10	283.3		
	Pb	0.1		1	283.3										
	Ref.	34		39						68		73			

Pt	Au		0.5	267.6				
	Ag		0.1	338.3				
	Pd		1	340.5	25	324.3		
	Cu		0.1	324.8				
	Fe	5	2	302.1	(Fire assay bead)		0.2	302.1
	Sb		20	231.2			0.9	259.8
	Pb		5	280.2			0.4	283.3
	Ref.	31	36		24		73	
Pd	Au		3					
	Ag		6					
	Pt		40					
	Cu		2					
	Fe		10				0.2	302.1
	Sb		50				0.9	259.8
	Pb		3				0.4	283.3
	Ref.		37				73	

Table 1.10 Determination of Noble Metals in Various Materials, Graphite Electrode Technique

Matrix	Noble metal	Enrichment	Excitation	Internal standard	Detection limit	Reference
14 high-purity metals	Ag		dc arc		<0.1–0.3 ppm	34
High-purity Ni, Cu, Ag	Pd	Coprecipitation	dc arc	Ag, background	9×10^{-4} ppm in nickel	82
Al_2O_3	Pd, Ru		dc arc	Al	5 ppm, 25 ppm	83
Cathode copper	Au	Adsorption by charcoal	dc arc	Ge	0.01 ppm in copper	84
Cyanide solution	Au, Ag	Adsorption by charcoal	ac arc		0.01 ppm in solution	85
Brass	Au		dc arc	Bi	100 ppm	86
Lead production materials	Au	Fire assay	ac arc	Co	0.002 ppm in matrix	87
Metallurgical materials	Au, Ag	Coprecipitation	dc arc	Cu	0.1 ppm, 5 ppm	88

Balfour *et al.* described a method for analyzing silver, gold, and other elements in a variety of metal matrices.[88] The method is based on enrichment of the impurities by coprecipitation with copper as the sufide. Advantages of this method are the simultaneous determination of several elements in complex matrices (e.g., alloys with aluminum, chromium, cobalt, iron, and titanium) and the formation of a uniform matrix for excitation in a dc arc. Table 1.10 summarizes the various methods.

1.2.2.2 Globule Arc

The globule arc was primarily a method for analyzing copper and, as reported, it has also been used for the analysis of precious metals. The sample is usually excited by dc arc in the crater of the positive carbon electrode. The evaporation mechanism has already been described. The sensitivity of the method may be increased by using copper oxide instead of copper or by converting copper into its oxide.[55] Tölle used a special chamber to arc the sample in an oxygen atmosphere.[58]

Larina *et al.* applied the globule arc to determine silver in cathodic copper.[84] Kreimer *et al.* determined gold, platinum, and palladium in copper after enrichment by coprecipitation.[89] Copper is converted into copper oxide by roasting. Sample pellets are melted into buttons in the crater of the positive electrode of a dc arc and then excited after changing the polarity. Table 1.11 summarizes these methods.

1.2.2.3 Other Methods

Goleb, in two publications, described the determination of ruthenium, rhodium, palladium, and other elements in uranium fission alloys.[90,91] The first of the references deals with the point-to-plane technique, and the second with a water-cooled hollow cathode lamp. An advantage that spectral analysis has over chemical analysis of radioactive substances is the low sample consumption resulting in small quantities of radioactive waste. The point-to-plane technique

Table 1.11 Copper Globule Arc

Matrix	Noble metal	Enrichment	Excitation	Internal standard	Reference
Cu	Ag		dc arc	Cu	56
Cu	Au, Ag		dc arc	Cu	57
Cu	Ag		dc arc in cuvette	Cu	58
Cu	Ag		ac arc	Cu	84
Cu	Au, Pt, Pd	Coprecipitation with copper	dc arc		89

utilizes a Petry stand and a high-voltage condensed spark. In the hollow cathode discharge [Schüler-Gollnow hollow cathode[92]] oxides are sputtered in argon (3 Torr, 15 min), which is then replaced by neon (3 Torr). Constant conditions are reached during a 20-min discharge (constant current 80 mA, voltage adjusted to between 150 and 300 V). The analytical results obtained from the point-to-plane, hollow cathode, and wet chemical methods agree very well.

1.2.3 Determination of Noble Metals in Geologic Samples, Minerals, and Ores

The determination of noble metals in geological samples, minerals, and ores requires the utmost sensitivity of detection. Determinations in the lower ppm range are usually not sufficient; detection limits in the ppb range or even sub-ppb range are necessary. In spite of their high sensitivity, direct emission methods of analysis as well as AAS, AFS, FES, and x-ray fluorescence methods cannot normally be used for the direct analysis of these samples. Only gamma and mass spectrometry offer sufficient sensitivity. For routine investigations, prospecting, and so on, however, this costly equipment is usually not available, and measuring techniques are often too time consuming. The historical method for this type of determination of the noble metals is fire assay enrichment and parting assay.

The spectral analysis of geologic and other samples does not differ essentially from the methods already described. Only sample preparation differs markedly, because elaborate enrichment techniques usually have to be employed to bring concentrations to a detectable level. The enrichment steps finally present a sample ready for analysis by the well-known methods. The methods can be classified according to sample preparation as follows:

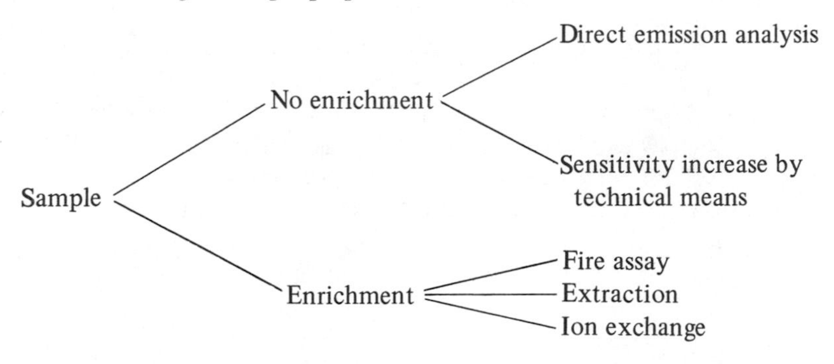

1.2.3.1 Normal Geochemical Analysis

Standard methods of spectral analysis of geologic samples are generally not suitable for the determination of the noble metals, owing to their inadequate sensitivity. The methods have, however, been developed to such a degree that they can be performed even in mobile laboratories.[93] A rapid semiquantitative method which includes all the noble metals is given by Myers *et al.*[94]: the sample-graphite mixture is burned to completion in a dc arc and 68 elements are estimated by visual observation of their lines on the photographic plate and by comparison with standard spectra. The detection limits for the noble metals are: osmium, iridium, ruthenium, 100 ppm; rhodium, 50 ppm; platinum, 30 ppm; gold, 20 ppm; palladium, 3 ppm; and silver, 1 ppm.

1.2.3.2 Methods with Increased Sensitivity

It is possible to improve the sensitivity of spectral analysis without tedious sample preconcentration by several means. To begin with the spectrograph: it is possible to optimize the recording technique (e.g., by the choice of the correct slit width), the use of a spectrograph with higher resolving power, or to increase the blackening of the photographic plate by the use of a cylindrical lens as long as the width of the plate and placed in front of the plate. The latter reduces the image of the full height of the entrance slit, which increases the blackening.[95] In practice, when determining palladium in ores Terekhovich improved the line/background ratio significantly by using a higher resolution (1 Å/mm instead of 19 Å/mm,[96] enabling as little as 0.1 ppm to be determined in a dc arc.

An improvement can also be achieved by imaging only that part of the plasma which gives preferential emission. Avni and Boukobza used the cathode region to determine 50 elements in rock samples.[97] The choice of the cathode region and of the correct buffer were based on measurements of the temperature and particle density distributions along the arc axis. The use of a buffer reduces the differences between standards and samples and increases the concentration of traces in the plasma by extending their residence time in the arc by slowing down the axial velocity. Only silver was determined by 0.1 ppm, but good sensitivity can also be expected for the other noble metals.

Carrier distillation was tested as another method to increase sensitivity. From the various carriers tested (gallium oxide, sodium chloride, sodium flouride, and silver chloride), Strasheim and Van Wamelen selected sodium chloride, which transported gold from the investigated ore sample into the plasma.[98] Pavlenko *et al.* used gallium oxide to determine palladium, rhodium, and ruthenium in ores with arc excitation.[99] The carrier did not affect platinum and iridium. They explained the effect of the carrier by the increased concentration of the particles in the plasma (change of diffusion, convection, and transport rates affect concentration near the anode and in the middle of the analytical gap). Palladium, rhodium, and ruthenium could be detected at the 1-, 1-, and 10-ppm level, respectively.

Grimes and Marranzino[100] developed an ac spark-rotating disc-solution technique for geological samples. The samples were first treated with hydrogen fluoride, taken to dryness, then treated with aqua regia. The clear liquid was used for analysis. Detection limits were 0.2, 0.2, 1, and 5 ppm for gold, silver, palladium, and platinum, respectively. A modification of this technique was adapted for use with the U.S. Geological Survey mobile spectrographic laboratories. These laboratories were designed to go into remote areas and quickly do the analyses for survey crews. A laboratory and the interior of one of the mobile units are shown in Figs. 1.4 and 1.5.

1.2.3.3 Enrichment Techniques

One has to employ enrichment procedures when even the most sensitive techniques show "blank readings." The oldest enrichment technique—fire assay—was introduced early in applied spectrochemistry.[101-103] Chemical methods were later employed. Today the noble metals are usually preconcentrated by one of the following techniques: fire assay, precipitation, extraction, or ion exchange.

The methods are described in various articles, some of which we include.[104-110]

1.2.3.3a Fire Assay Enrichment. This method is complex and shall be described only briefly, following Smith.[2] The description is also given because fire assay is the oldest method used for determining noble metals in ores as well as for the testing of fineness of refined noble metals.

Fig. 1.4 U.S. Geological Survey mobile spectrographic laboratory. The motor-driven generator, power cords, and other heavy gear are mounted on the trailer. (U.S. Geological Survey.)

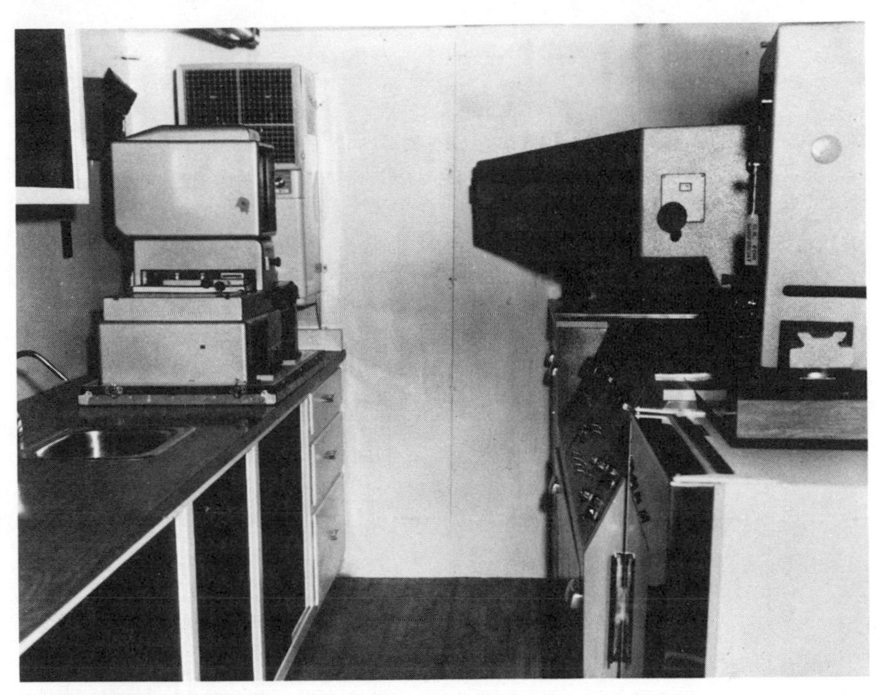

Fig. 1.5 Interior view of one of the mobile spectrographic laboratories. (U.S. Geological Survey.)

The normal fire assay consists of melting the sample together with materials which release metallic lead. Therefore, a suitable flux has to be added to enable the sample to melt at moderate temperature as well as a reducing agent, to release the noble metals. The noble metals are collected by dissolution in the lead, which sinks to the bottom of the melt because of its higher density. After cooling, the lead is separated from the slag. The noble metals are then, in turn, separated from the lead and the base metals by cupellation; i.e., the lead button is exposed to an oxidizing air stream in a porous crucible (cupel) at red heat where litharge (PbO) is formed. As soon as this happens, the litharge is absorbed by the walls of the crucible, taking with it all the simultaneously formed base metal oxides. The noble metals remain as an alloyed bead. Separation of the noble metals is performed by parting assay (selective acid dissolution), which will not be dealt with in this chapter.

When used as a preconcentration technique for spectral analysis, the fire assay process can be stopped at either of two stages. First, the lead button obtained after the fusion can be analyzed. Second, a noble metal which is added during fusion together with the lead, or later, during cupellation, can be used as a collector to form a bead suitable for spectral analysis.

Table 1.12 Fire Assay Spectrochemical Methods

Original material	Analytical sample	Excitation	Internal standard	Element	Line (nm)	Detection limit (μg)	Reference
Ores, minerals, rocks	50-mg Pb bead	High-voltage spark	Pb	Ag		9	
				Pd		0.6	
				Pt		0.6	117
				Au		0.6	
				Rh		0.6	
Geologic samples	Residue of 0.8- to 1.2-mg Au bead dissolved in aqua regia	Intermittent dc arc	Mo	Pt	265.9	0.1	
				Pd	340.5	0.06	43
				Rh	339.7	0.05	
Geologic samples	4-mg Au bead + 2-mg $(NH_4)_2OsCl_6$	dc arc		Pt	265.9	0.1	
				Pd	340.5	0.02	
				Rh	343.5	0.02	65
				Ir	322.1	1.5	
				Ru	343.7	1.5	
Geologic samples	9.3-mg Ag bead + 3-mg $(NH_4)_2OsCl_6$	dc arc		Au	267.6	0.015	
				Pt	265.9	0.03	
				Pd	324.3	0.015	69
				Rh	343.5	0.03	
				Ir	322.1	1.5	
				Ru	343.7	1.5	
Precious-metal bearing materials	10-mg Ag bead + 10-mg Pt or Rh	dc arc	Ag	Au	312.3	1	
			Rh	Pt	306.5	0.2	67
			Rh	Pd	340.5	0.05	
			Pt	Rh	322.3	0.05	
Ores, concentrates, heavy sands	10-mg Pt bead	dc arc	Pt	Pd	324.3	0.25	
				Rh	332.3	0.25	24
				Ru	287.5	0.25	
				Ir	285.0	0.5	
				Os	263.7	0.5	
Ores, concentrates, metallurgical products	100-mg Ag bead	ac arc	Ag	Pd	342.1	4.4	66

During recent years several other collection materials besides lead have proved useful, e.g., copper-nickel-iron collection,[111,112] tin collection,[113] and nickel matte collection.[114] These methods have, however, not yet been employed in spectral analysis, notwithstanding the good recovery values. Only Danilova *et al.* are using copper and nickel present in precious metal ores as collectors.[115]

Lewis used an "arrested" cupellation for the determination of gold, silver, platinum, palladium, and rhodium.[116,117] Since the noble metal beads resulting from cupellation would be too small for spectral analysis, Lewis interrupted cupellation as soon as the lead bead attained a weight of 50-100 mg. Tomingas and Cooper (palladium)[66] and Whitehead and Heady (gold, platinum, palladium, rhodium)[67] collected the noble metals in silver beads, the silver having been added during fusion (100 and 10 mg, respectively). Cooley *et al.* fused the samples with 10 mg of Ag_2O to collect gold and the PGM except osmium in a silver bead.[69] Haffty and Riley (platinum, palladium, rhodium)[43] and Dorrzapf and Brown (platinum, palladium, rhodium)[65] collected noble metals in gold beads, the gold (1 and 4 mg, respectively) also being added during fusion. Broadhead *et al.* investigated the use of 10 mg of platinum as collector for iridium, osmium, and ruthenium,[24] because platinum turned out to be the best carrier for these elements.

Haffty and Riley[43] and Danilova *et al.*[115] used solution-graphite techniques for spectral analysis. But the most common method is the excitation of the samples in the craters of graphite electrodes (globule arc). Tomingas and Cooper reported an interelement effect[66]: increasing copper content caused a rise in the palladium intensity. Separate calibration curves for various copper/silver ratios have been established. Whitehead and Heady added a buffer of 10 mg of platinum or 10 mg of rhodium (depending on the element to be determined) to obtain a stable discharge (8-A dc arc) and to prevent boiling of the silver bead,[67] which would otherwise jump out of the electrode. Dorrzapf and Brown added 2 mg of ammonium chlorosmate [$(NH_4)_2OsCl_6$] to the gold bead to stabilize the platinum emission in a 15-A dc arc.[65] To increase accuracy and sensitivity they arced in an Ar/O_2 atmosphere (70:30). The same technique was applied by Cooley *et al.* to arc their silver bead.[69] Broadhead *et al.* did not need any addition because the platinum bead could be acred in a 20-A dc arc without difficulty.[24] The methods are summarized in Table 1.12.

1.2.3.3b Chemical Enrichment. Several authors coprecipitated gold and silver with tellurium or iron. Voskresenskaya *et al.* first removed silica with a gelatine; subsequently, gold was precipitated with tellurium.[118] Novikov and Bondarenko also used tellurium to precipitate silver[119] and Polikarpochkin *et al.* coprecipitated gold with iron.[120] Hahn-Weinheimer tested the organic precipitant chinolinselenol for platinum, palladium, osmium, iridium, rhodium, and ruthenium.[121] Tarasova *et al.* precipitated gold and the PGM (except osmium) with thiourea.[122] Rubinovich *et al.* enriched gold, platinum, and

Table 1.13 Spectrometric Determination of Noble Metals in Geochemical Samples after Chemical Enrichment

Element	Lines (nm)		Method of enrichment	Detection limit (μg)	Reference
Au	267.6	242.8	Coprecipitation	0.01	118, 120
			Extraction	0.15	123, 124
			Ion exchange	0.02	123, 125
Ag	328.1		Coprecipitation	0.003	119
Pt	306.5	265.9	Coprecipitation	0.01	121
			Precipitation		122
			Extraction	0.5	123
			Ion exchange	0.1	123, 125
Pd	340.5	342.1	Coprecipitation	0.01	121
	324.3	302.8	Precipitation		122
			Extraction	0.5	123
			Ion exchange	0.05	123, 125
Rh	343.5	332.3	Coprecipitation	0.05	121
	339.7		Precipitation		122
			Ion exchange		125
Ru	349.9	343.7	Coprecipitation	0.05	121
	291.6	272.6	Precipitation		122
			Ion exchange		125
Os	290.9	283.9	Coprecipitation	2	121
Ir	322.1	266.5	Coprecipitation	0.05	121
	254.4		Precipitation		122
			Ion exchange		125

palladium in geologic samples by ion exchange and adsorption by activated charcoal and compared both methods.[123] The ion exchange method yielded a detection limit which was lower by a factor of 10. Kreimer extracted gold as gold diethyldithiocarbamate.[124] Brooks and Ahrens had previously suggested ion exchange enrichment for the spectrochemical analysis of noble metals.[125] They claimed an enrichment factor of 20,000.

For spectral analysis, the residues are frequently calcined and then mixed with graphite powder; sometimes the graphite is added directly to the residues. The samples are then vaporized in the crater of a carbon electrode (possibly with the aid of a carrier) and excited in the analytical gap of a dc arc. Table 1.13 summarizes these methods.

1.2.4 New Developments in Spectral Analysis

In addition to the glow discharge lamp, which has been discussed (Section 1.2.1.2c), some highly sophisticated light sources have been developed as excitation sources for solution analysis: plasma jet,[126-130] capacitively coupled microwave plasma (CMP),[131-134] and inductively coupled high-frequency plasma (ICP).[135-140] Detailed reviews and comparative studies are given by Fassel,[141] Greenfield *et al.*,[142-144] and Boumans *et al.*[145] Besides an investigation on a very sensitive analysis of high-purity gold,[146] these techniques have not been directly applied to typical samples for precious metal analysis, nor have any possible occurring interferences, which might be typical of these samples, been investigated. Both Boumans *et al.*[145] and Greenfield *et al.*[144] regard the ICP superior to the other two light sources due to better sensitivity and less tendency to interference effects. Interesting applications of the ICP in precious metal analysis can be expected in the near future, especially in trace analysis, because ICP detection limits are in most cases equal or better than those of AAS and AFS.[138,145] Some details are given in Table 1.14.

Table 1.14 Analysis of Solutions by Plasma Jet, CMP, and ICP

Element	Technique	Line (nm)	Detection limit (ppm in solution)	Reference
Au	Plasma jet	242.8	0.2	142
	CMP	267.6	0.02	132
	CMP	267.6	0.07	134
	ICP		0.04	138
Ag	Plasma jet	338.3	0.2	142
	CMP	328.1	0.001	143
	ICP		0.004	138
Pt	Plasma jet	265.9	0.5	142
	Plasma jet	265.9	0.6	146
	CMP	265.9	0.1	132
	CMP	265.9	0.3	134
	ICP		0.08	138
Pd	Plasma jet	361.0	0.5	142
	Plasma jet	340.5	0.011	146
	CMP	340.5	0.4	134
	ICP	363.5	0.002	137
	ICP		0.007	138
Os	Plasma jet	225.6	0.7	142
Ir	Plasma jet	254.4	1.5	142
Rh	Plasma jet	343.5	0.5	142
	Plasma jet	343.5	0.001	146
	CMP	369.2	0.05	134
	ICP		0.003	138
Ru	Plasma jet	372.8	0.6	142

1.3 FLAME TECHNIQUES

1.3.1 FES, AAS, and AFS

FES, AAS, and AFS are dealt with together, in spite of their different physical principles. Because similar techniques of measurement instrumentation, sample preparation, and nebulization are used, a discussion of each method is not justified.

A paper by Fassel and Golightly revealed higher sensitivity for osmium, palladium, and ruthenium by using FES, and for silver, gold, iridium, platinum, and rhodium by AAS,[147] whereas Christian and Feldman found that AAS was more sensitive than FES for all noble metals except rhodium.[148] Sychra *et al.* compared FES, AAS, and AFS (and various flames) for the determination of palladium.[149] With optimized flame conditions and proper line selection, all three methods showed approximately the same sensitivity. In recent papers Sen Gupta[150] and Fishkova[151] reviewed AAS methods for the determination of noble metals, whereas Heinemann reports optimum AAS measuring conditions.[152-155]

1.3.1.1 Interferences

Because many of the published investigations deal with interference problems, they shall be discussed briefly. Chemical interferences occur mainly as cationic interferences (different acids and acidities). They influence the critical number of free atoms in the flame plasma. This reduction in the numbers of free atoms is caused by compound formation in the solution, which can only be partially decomposed by the flame or by chemical reactions at flame temperature which form refractory-type compounds (oxides, nitrates, chlorides, etc.).

These difficulties must be solved by adding buffers which either convert interfering elements into stable compounds, thus eliminating their influence on the noble metals, or which form compounds with the noble metals which can easily be destroyed by the flame but at the same time prevent the formation of unwanted compounds.

By 1959-1960 no mention had yet been made of interferences in noble metal analysis.[156,157] During the following years, however, an increasing number of interference observations were published. Many buffers have since been tested by numerous authors. For the relatively cold air-propane and air-acetylene flames, mainly copper, lanthanum, and uranium additions have been recommended for interference suppression.

Strasheim and Wessels were the first to report a systematic study of cation and anion interferences on gold, platinum, palladium, and rhodium.[158] To eliminate the interferences, they suggested the separation of the noble metals from the base metals and the suppression of mutual interferences (except for rhodium) by adding 2% copper at constant hydrochloric acid concentration.

Slavin also recommended the addition of 2% copper for the determination of platinum in the presence of palladium and gold.[159] Schnepfe and Grimaldi determined palladium and platinum in a gold bead and suppressed interferences by adding 0.5% cadmium and 0.5% copper.[160] Van Loon added copper and sodium (500 and 1000 ppm, respectively) when determining iridium.[161] Copper and sodium addition had also been used by Janssen and Umland to avoid mutual influences of platinum, palladium, rhodium, and iridium (1–1.5% copper and sodium),[162] and by Grimaldi and Schnepfe to determine iridium in gold beads (0.7% copper and 0.3% sodium).[163]

Van Loon suppressed interferences when determining palladium, platinum, and gold in silver beads by adding 1% lanthanum,[164] which had also been used by Yudelevich *et al.* to avoid palladium influences on gold determination (3% lanthanum).[165] Macquet *et al.* also tested the use of lanthanum (2%), and in addition potassium phosphate (3%) to prevent interferences when analyzing platinum in platinum complexes.[166,167] Makarov *et al.* recommended lanthanum (0.5%) as a buffer for platinum, ruthenium, and iridium determination.[168]

Scarborough added uranium to suppress mutual interferences of palladium, rhodium, ruthenium, and molybdenum.[169] Uranium had also been used by Montford and Cribbs for ruthenium determination,[170] and Heinemann tested uranium and lanthanum for palladium and platinum determination.[154,155]

Ginzburg *et al.* recommended the addition of 1% magnesium for analyzing rhodium.[171] Belcher *et al.* suggested extraction to avoid interferences on silver.[172] Eckelmans *et al.* reported that mutual interferences among gold, platinum, and palladium can be suppressed by using the dithizonates of these elements.[173] They also mentioned that complex formation with dithizone enhances the atomization of gold. Khrapai *et al.* buffered with cadmium and vanadium for the analysis of platinum and rhodium in silver.[174]

Atwell and Hebert were the first to try to eliminate interferences on rhodium by employing the hotter nitrous oxide–acetylene flame.[175] The remaining interferences by iridium and ruthenium could be suppressed by adding 0.5% zinc. The hotter flame, on the one hand, hampers the formation of refractory compounds but, on the other hand, reduces sensitivity. Pitts *et al.* also investigated the air–acetylene and nitrous oxide–acetylene flames for platinum determination.[176,177] In the air–acetylene flame anionic and cationic (mutual and base metal) interferences were suppressed by adding 0.2% lanthanum. In the hotter flame all the interferences were eliminated, but sensitivity was reduced. In a careful study Mallett *et al.* investigated mutual and base metal interferences on gold, platinum, palladium, rhodium, and ruthenium in air–acetylene and nitrous oxide–acetylene flames and in various acids.[178,179] The hotter flame strongly reduced mutual interferences of platinum, palladium, gold, and rhodium but not ruthenium. It did, however, strongly reduce the sensitivity, again with the exception of ruthenium. The ruthenium intensity is enhanced by gold. Of the various additions tested for the air–acetylene flame, uranium and vanadium turned out to be the best. Uranium (1%) showed the best result, and when the base metals

Table 1.15 AAS, AFS, and FES Determination of Single Noble Metals in Solution

Element	Lines (nm)	Method	Best published detection limit (ppm)		Reference
			In solution	In sample	
Au	242.8	AAS	0.009		148, 151, 189
	242.8	AAS + extraction	0.1	0.004	185, 187, 188
	267.6	AFS	0.05		191
	242.8	AFS + extraction	0.005	0.00007	181
	267.6	FES	0.5		148, 190
Ag	328.1	AAS	0.001		148, 151
	328.1	AAS + extraction	0.05	0.01	172, 185
	328.1	AFS	0.001		191, 192
	328.1	AFS + extraction	0.017	0.00004	193
	328.1	FES	0.002		137, 148, 190
Pt	265.9	AAS	0.05		148, 151, 176, 189
	265.9	FES	2		148, 190
Pd	244.8/ 247.6	AAS	0.01		148, 149, 151, 189, 195
	244.8	AAS + extraction		0.0015	165, 194
	340.5	AFS	0.04		149, 191
	363.5	FES	0.04		148, 149, 190
	340.5	FES + extraction		0.5	196

Rh	343.5	AAS	0.02		148, 151, 171, 189, 197, 198
	343.5	FES	0.03		199
	369.2	FES + extraction		0.5	196
Ru	349.9	AAS	0.3		170, 151, 199
	372.8	FES	0.06		148
Os	290.9/ 305.9	AAS	0.4		148, 151, 201
	301.8	AAS + extraction	1		200
	442.1	FES	2		199
Ir	285.0/ 264.0	AAS	1		148, 151, 202, 203 204, 205
	380.0/ band emission at 550	FES	0.4		199

had been removed, vanadium (1%) suppressed both mutual and base metal interferences.

The investigations above deal exclusively with AAS. Eshelman *et al.* described anion and cation interferences on palladium and rhodium when using FES[180] and applied chemical separation. Matoušek and Sychra determined gold by AFS in various flames.[181] The oxygen-hydrogen-argon flame proved to be free of interferences. In a comparison investigation on palladium determination, Sychra *et al.* suppressed interferences in AAS by adding lanthanum or EDTA, in AFS by adding strontium or EDTA.[149] In FES interferences appeared to be only of a spectral nature.

1.3.2 Single Elements in Solution

Besides the methods mentioned, which deal mainly with interferences, there are a large number of publications on the determination of single noble metals in solution. Some of them apply extraction techniques to increase sensitivity or add enhancing compounds. Table 1.15 gives a summary of those papers dealing with single-element determinations.

1.3.3 Analysis of Noble Metals

Only a few publications are available on the analysis of noble metals. The reason may be the time required for analysis if more than one element has to be determined. Five publications on silver analysis should be mentioned. Hickey analyzed silver up to a fineness of 99.99% by determining the remaining silver in solution after precipitation of the main portion of silver.[182] Mathieu and Guiot determined the main impurities copper and iron in high-purity silver (99.99%) after extracting silver from the solution.[183] Makarov *et al.* analyzed artificial silver beads for platinum, palladium, rhodium, and gold having precipitated silver from the solution.[184] Sen Gupta determined gold, PGM, and some common metals in native silver after separation of silver as silver chloride by double precipitation.[185] Khrapai *et al.* analyzed pure silver for 16 elements, including some PGM, after organic extraction of the trace elements.[174] Ashy and Headridge developed a sensitive method to determine iridium and ruthenium in rhodium sponge applying solvent extraction.[186]

1.3.4 Determination of Noble Metals in Geologic Samples, Minerals, and Ores

Flame techniques (only AAS applications have been reported) are, like emission techniques, in general not sufficiently sensitive for noble metal determination in geologic samples. Fire assay and chemical enrichment have to be employed. These methods are summarized in Table 1.16.

Table 1.16 Determination of Noble Metals in Geologic Samples, AAS

Sample	Element	Preconcentration	Line (nm)	Detection limit (ppm) In solution	Detection limit (ppm) In sample	Reference
Carbonate rocks	Au	Sodium cyanide extraction				212
Low grade ores	Au	Methyl isobutyl ketone extraction				213
Copper ores geochemical samples	Au	Methyl isobutyl ketone extraction				206, 207
Sulfide ores, concentrates	Au	Methyl isobutyl ketone extraction			0.1	210
Copper-based materials	Au Ag Pd	Dichloroethane extraction	242.8 328.1 244.8			211
Copper ores, geochemical samples	Ag		328.1			206, 207
Oxide and sulfide ores	Ag		328.1			209
Sulfide materials	Ag		328.1			208
Ores	Ag	Lead assay button	328.1	0.6		214
	Au	Silver assay bead	242.8	0.5		
	Pd	Silver assay bead	247.6	1.5		
	Au	Copper assay button	242.8	0.15		
	Pd	Copper assay button	247.6	0.5		
Ores and concentrates	Au Pt Pd	Tin-collection fire assay				215
	Au	Silver assay bead	242.8	1		164
	Pt	Silver assay bead	265.9	5		
	Pd	Silver assay bead	247.6	1		
Geochemical samples	Pt	Gold assay beads	265.9	0.04	0.06	160
	Pd	Gold assay beads	340.5	0.04	0.06	
Rocks	Au Pt Pd	Silver assay bead + extraction in diethyldithio-carbamate	242.8 265.9 247.6		0.4 0.01 0.02	216
Mafic rocks	Ir	Gold assay bead	264.0	0.25	2.5	163

Table 1.17 Solid Sample Evaporation

Element	Line (nm)	Method	Technique	Sample quantity (μl)	Detection limit: Relative (ppm)	Detection limit: Absolute (g)	Reference
Au	242.8	AAS	Carbon cuvette + burner		0.01 (1% abs.)		222
	242.8	AAS	Solid fuel		0.1	3×10^{-7}	217
	242.8	AAS	Carbon cuvette			7×10^{-11}	219
		AAS	Extraction + carbon cuvette	50			231
	242.8	AAS	Carbon filament + burner		0.03		232
	242.8	AAS	Carbon filament	5	0.04	2×10^{-10}	224
		AAS	Carbon filament			4×10^{-11}	225
		AAS	Carbon filament	1	0.02	2×10^{-11}	226
		AAS	Carbon filament	50	0.0005		228
	267.6	AFS	Carbon filament	5	0.0008	4×10^{-12}	224
		AFS	Carbon filament			1×10^{-11}	225
Ag	328.1	AAS	Carbon cuvette			8×10^{-13}	220
	328.1	AAS	Carbon cuvette			5×10^{-13}	219
		AAS	Carbon cuvette				233
		AAS	Carbon filament			1×10^{-12}	225
	328.1	AAS	Carbon filament	0.5	0.004	2×10^{-12}	226
	328.1	AAS	Carbon filament			5×10^{-13}	219
	328.1	AAS	Carbon filament	1			229
	328.1	AFS	Carbon cuvette			1.5×10^{-12}	220
	328.1	AFS	Carbon filament	1	0.001	1×10^{-12}	223
		AFS	Carbon filament			5×10^{-13}	225
	328.1	AFS	Carbon filament	0.5	0.0008	4×10^{-13}	230

Pt	265.9	AAS	Carbon cuvette + burner		0.5 (1% abs.)		222
	265.9	AAS	Carbon cuvette	20		1.1×10^{-9}	221
	265.9	AAS	Carbon cuvette			2.5×10^{-10}	219
		AAS	Carbon filament	50	0.005		228
Pd	247.6	AAS	Carbon cuvette + burner		0.01 (1% abs.)		222
	247.6	AAS	Carbon cuvette			5×10^{-11}	219
		AAS	Extraction + carbon cuvette	50			231
	247.6	AAS	Carbon filament	50	0.0005		228
Rh	343.5	AAS	Carbon cuvette + burner		0.02 (1% abs.)		222
	343.5	AAS	Carbon cuvette			6.3×10^{-11}	219
	343.5	AAS	Carbon cuvette	20		2.5×10^{-10}	221
		AAS	Carbon filament	50	0.001		228
Ru		AAS	Carbon filament	50	0.004		228
Ir	264.0	AAS	Carbon cuvette	20		7.5×10^{-9}	221
	264.0	AAS	Carbon filament	50	0.01		228

1.3.5 Flameless Techniques

AAS and AFS are not limited to the analysis of solutions. The dissolution of samples is often difficult and tedious. Moreover, dissolving a sample implies a dilution. The sensitivity, if related to the actual sample quantity, is no better than that obtained by emission methods. In cases of samples which dissolve with difficulty, it is an advantage to avoid the strong acid solutions, with their corrosive effects and their addition of impurities. Evaporation by other means can also produce the required number of ground-level atoms for AAS and AFS.

1.3.5.1 Solid Fuel

The first move in this direction was the work of Venghiattis.[217] He used a powder mixture of solid fuel and sample which was burned in a suitable container below the light path of the hollow cathode lamp. Gold ores were investigated with this method (Table 1.17).

1.3.5.2 Graphite Cuvette

In the electrically heated graphite cuvette there is no dilution by any type of fuel. L'vov described the use of the graphite cuvette for AAS.[218,219] Massmann applied the cuvette for AFS and compared the results with those of AAS.[220] In L'vov's system the sample is introduced into the preheated cuvette on top of an additional electrode. In Massmann's system for AAS the sample is placed in the cuvette through a hole in its center by means of a micropipet. For AFS a cup-like cuvette is used, the light being incident from above and the fluorescence radiation being observed through a slit from the side of the cuvette. All systems work in an inert argon atmosphere. Adriaenssens and Knoop discuss interferences and the optimal working conditions for the analysis of platinum, iridium, and rhodium with a Massmann cuvette.[221] A paper by Rubeska and Stubar should be mentioned in this connection because flame and cuvette techniques are combined.[222] The flame gases are directed into a preheated absorption tube and are blown away by an air stream when leaving the tube (Table 1.17).

1.3.5.3 Graphite Rod

Anderson *et al.*[223] and Aggett and West[224] described the development of a carbon filament atom reservoir which was used for the determination of silver and gold. The latter work also deals with interference investigations. The sample may be placed on a flat part in the middle of the filament,[224] in a cavity or hole,[225,226] or in a long groove.[227,228] The filament is electrically heated in an inert argon atmosphere. In addition, Amos *et al.*[225] and Molnar *et al.*[226] employed an oxyhydrogen argon flame to increase sensitivity and stability. Guerin

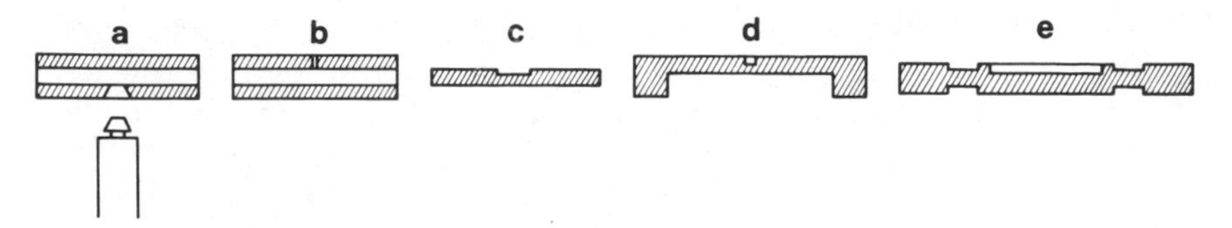

Fig. 1.6 Cuvettes (a, b) and filaments (c–e) for flameless AAS. [From: (a) L'vov,[219] (b) Massmann,[220] (c) Aggett and West,[224] (d) Amos *et al.*,[225] and (e) Guerin.[228]]

discusses interference effects.[228] Chuang and Winefordner analyze engine oil for silver and other elements with the graphite filament technique.[229] Patel *et al.* report a very sensitive method which combines multiple-element electrodeless discharge lamps with the graphite filament and AFS.[230]

Considering the detection limits, solid sample evaporation shows much improvement over flame methods. Figure 1.6 shows some types of cuvettes and filaments used. Data on the cuvette and filament techniques are summarized in Table 1.17.

1.4 X-RAY FLUORESCENCE

Most of the papers on x-ray fluorescence applications in precious metal analysis have been published during the last 15 years. This indicates how very new this analytical technique still is. Further progress is to be expected. Aspects such as the original form of the material to be analyzed, the sample type, and interferences will be discussed briefly.

1.4.1 Original Form of the Material and Sample Type

A wide range of materials has been examined: alloys (dental alloys), ores, mattes, glasses, catalysts, fission alloys, noble metals in solution, and powdered materials. A variety of methods may be used, depending on the type of sample obtained from the original material. There are three main methods: analysis of solutions, analysis of solids (metal discs, powder, powder pellets), and analysis of paper filters or reagent papers.

1.4.2 Interferences

Three types of interference are present in x-ray fluorescence: an intensity reduction by partial absorption by other matrix elements, an intensity enhancement due to the presence of other elements (secondary radiation), and line interferences. The proper choice of lines eliminates some of the interference effects.

This results in a distinct division of the noble metals into two groups: silver, palladium, ruthenium, and rhodium are determined using the K_α lines, and gold, platinum, iridium, and osmium are determined using the L_α or L_β lines. Line interferences, such as Cu $K_{\beta 1}$ on Os L_α,[234-236] W $L_{\beta 1}$ on Au $L_{\alpha 1}$,[237] or Sn K_α on Ag K_β,[238] have been reported. Some authors avoid the interference problems by an enrichment of the noble metals, while others use mathematical correction.[238-240] The correction method of Lucas-Tooth and Price[241] has been applied successfully. The application of different internal standards is also discussed.[242-244,235]

1.4.3 Enrichment Techniques

In many materials, e.g., ores, fission alloys, or glass, the concentrations of noble metals fall below the detection limit. In these cases enrichment techniques are applied: ion exchange,[235,237,245,246] coprecipitation,[234,244,247] extraction,[236,246] and fire assay.[234,235,248] In many cases the concentrate is analyzed directly on the filter, either as the filtrate or as a chemical reaction product (reagent paper, anion exchange paper).

1.4.4 Precision, Accuracy, and Detection Limits

X-ray fluorescence is not a very sensitive technique. The best detection limits for noble metals reported lie between 4 and 24 ppm.[235] On the other hand, there is no upper limit.

Concentrations of up to 100% may be determined. Because of the excellent precision of single measurements, it is comparatively easy to correct interferences mathematically, thus yielding very good accuracy, for instance in the analysis of alloys.[240] A survey on x-ray methods is given in Table 1.18.

1.4.5 Other Techniques

Other x-ray techniques, such as electron probe x-ray microanalyses and x-ray photoelectron spectroscopy, have thus far been applied to the analysis of noble metals in very few cases. With the aid of the former method, a new mineral (Os, IR)S was discovered,[254] and with the latter, gold-silver alloys (0-100% Au) have been analyzed.[255]

Radioisotope x-ray analysis was used by Burkhalter and Marr for the determination of gold in ores; gamma radiation of ^{57}Co served as the source. Concentrations down to 20-30 ppm gold could be determined.[256] Snyman used ^{241}Am as the source for the rapid determination of the fineness of refined gold; he determined silver as the main impurity, assuming that silver was a constant 84% of the total amount of impurities. He also determined gold directly.[257]

Table 1.18 X-Ray Methods

Original material	Enrichment	Sample	Determined elements	Reference
Uranium-base fission alloys	Ion exchange, extraction	Solution	Pd, Rh, Ru, Mo, Zr	246
Matte solution		Solution	Au, Pt, Ir	243
Matte solution		Solution	Pd, Rh, Ru	242
Au solution		Solution	Au	249
Matte-leach residues		Solution	Pt, Pd, Au	235
Alumina catalysts		Powder	Pt	253
Complex materials		Powder	Ag, Pd, Ru, Rh	250
Platinum alumina catalysts		Powder pellet	Pt	239
Sludges, platinum concentrates		Powder pellet	Au, Ag, Pt, Pd, Rh, Ru	248
Ores, nickel sulfide mattes	Nickel–sulfide collection	Pellet	Pd, Rh, Ru, Ag, Pt, Ir, Os	235
Solution	Ion exchange	Resin pellet	Pd, Rh, Ru, Pt, Ir, Au	235
Noble metal dental alloys		Metal disc	Au, Ag, Cu, Pt, Pd, Zn,	251
Noble metal alloys		Metal disc	Au, Ag, Pt, Pd, Cu, Zn	240
Silver dental alloys		Metal pellet	Ag, Cu, Zn, Sn	219
Platinum group metal solution		Absorbent paper	Pt, Pd, Rh, Ir	252
Solution	Precipitation on reagent paper	Paper disc	Os, (Pt, Pd, Ir)	236
Sulfide ores	Coprecipitation	Diaphr. filter	Au	244
Base metal solution	Ion exchange	Ion exchange paper	Os, Ru	245
Solution	Ion exchange	Ion exchange paper	Au	237
Glass	Coprecipitation	Filter	Pt	247
Sludges, platinum concentrates	Fire assay silver bead	Filter	Ir	248
Various materials	Precipitation, fire assay		Au, Pt, Pd, Rh, Ru, Os	234

1.5 ACTIVATION ANALYSIS

A record of emission analytical techniques would not be complete without briefly mentioning neutron and proton[258,259] activation analysis. Gamma and beta spectrometry do not replace the methods described so far but extend the concentration range to the sub-ppb level. Because gamma spectrometry has exceptional sensitivity, adequate precision, and no laboratory contaminations, the applications include the analysis of high-purity noble metals and the determination of noble metals in samples of the earth's crust and extraterrestrial materials.

1.5.1 Technique

Usually, gamma spectra, emitted by (n, γ) reactions, are counted, but in some cases the β decay is evaluated. The irradiation and the necessary cooling period are normally followed by a radiochemical separation of the analytical element to avoid any interfering emission from other radionuclides. Separation is usually done with the aid of a carrier; a sufficiently large quantity of the (inactive) analytical element is added to the sample. Since the active analytical element and the inactive carrier do not differ in their chemical properties, the chemical yield of the separation can be determined from the recovery of the carrier.

The introduction of high-resolution Ge(Li) detectors allows the nondestructive multielement analysis of geochemical samples.[260,261] Irradiation with epithermal neutrons suppresses the activities of the major elements (e.g., in ores), which yields a relative sensitivity increase for the trace elements.[262]

Systematic errors such as self-shielding and absorption can be avoided by preparing standards which are as identical to the sample and as homogeneous as possible. The addition method, i.e., addition of known amounts of the analytical element to the matrix, has proved advantageous.

Some disadvantages of activation analysis should also be mentioned. In many cases long irradiation and cooling times and lengthy chemical separations render the technique unsuitable for quick routine analyses. In addition, reactor facilities are not available to all laboratories.

1.5.2 Interferences

Several possible interferences can limit the sensitivity. (n, p), (n, α), and (n, f) reactions and β decay can produce additional amounts of the measured radionuclide and must be taken into account. Moreover, other radionuclides of the analytical element may form which cannot be removed by chemical separation and have to be discriminated against. The interferences are dealt with by Gijbels.[263]

Table 1.19 Activation Analyses

Sample	Determined elements and references
Au	Pt (282)
Ag	30 elements (283)
Pt	Ag (266); Au (267); Pd (268); Os, Ir (269); Ru (270); Au, Ru, Os, Ir (271); Ir (272); As, Sb (273); Co (274); Fe (275); Se, Te (276); 8 elements (284)
Pd	Os, Ir (269); Co, Fe (277)
Os	Pt, Pd, Ir, Au, Fe (277); Ru, Ir (278); Au (279)
Ru	Os (280)
Rh	Ir (281); Au, Pt, Pd, Ir, Cu (285); 35 elements (258)
Rocks and ores	Au (286–288, 298–300); Au, Pd (289); Pd (290); Os (291); Au, Pt, Pd, Os, Ir, Ru (292, 294); Au, Pt, Ir (301); Os, Ru (295); Ag (296); Au, Pt, Pd, Ir (297)
Extraterrestrial materials	Au (293, 306); Au, Pt, Ir (301); Pt, Ir (302); Pd (303); Ag, Rh (304); Ru, Os (305)
Cu	8 elements (308)
Catalysts	Pt (309)
Lead foam	Pt, Pd, Rh (310)
Filter media	Au (237); Rh (311); Au, Ag (312)
Silicon	Au, Ag, Pt (313)
Glass	Pt (314)
Aerosols	Ag (315, 316)
Biological materials	Au (317)

1.5.3 Applications

Three articles by Gijbels,[263] Beamish *et al.*,[264] and Rakovskii *et al.*[265] survey the applications of activation analysis. A large number of investigations deal with trace determinations in noble metals.[266–285] The other important field of application is the determination of the noble metals in samples of the earth crust and in samples of extraterrestrial origin.[260–262, 286–307] Various other materials have also been analyzed for noble metals.[237, 308–317] Table 1.19 summarizes the references.

ACKNOWLEDGMENTS

The author is indebted to Mrs. D. B. de Villiers and Dr. L. R. P. Butler, National Physical Research Laboratory, C.S.I.R., Pretoria, South Africa, for reviewing the manuscript.

1.6 REFERENCES

1. V. M. Goldschmidt, *Geochemistry*, Oxford University Press, London (1954).
2. E. A. Smith, *The Sampling and Assay of the Precious Metals*, Charles Griffin & Company Ltd., London (1947).
3. E. Wichers, *J. Amer. Chem. Soc.* 43 (1921).
4. W. F. Meggers, C. C. Kiess, and F. J. Stimson, *U.S. Bur. Std. Sci. Papers 18*, 235 (1922).
5. W. Gerlach and E. Schweitzer, *Z. Anorg. Allgem. Chem. 173*, 92 (1928).
6. W. Gerlach and E. Schweitzer, *Z. Anorg. Allgem. Chem. 181*, 101 (1929).
7. W. Gerlach and E. Schweitzer, *Z. Anorg. Allgem. Chem. 173*, 104 (1928).
8. W. Gerlach and E. Schweitzer, *Z. Anorg. Allgem. Chem. 181*, 108 (1929).
9. A. Walsh, *Spectrochim. Acta 7*, 108 (1955).
10. C. T. J. Alkemade and J. M. W. Milatz, *J. Opt. Soc. Amer. 45*, 583 (1955).
11. I. Noddak and W. Noddak, *Naturwissenschaften 18*, 757 (1930).
12. F. Twyman, *Metal Spectroscopy*, Charles Griffin & Company Ltd., London (1951).
13. F. E. Beamish, *The Analytical Chemistry of the Noble Metals*, Pergamon Press, Oxford (1966).
14. F. E. Beamish and J. C. van Loon, *Recent Advances in the Analytical Chemistry of the Noble Metals*, Pergamon Press, Oxford (1972).
15. F. E. Beamish, *Talanta 1*, 3 (1958).
16. F. E. Beamish, *Talanta 2*, 244 (1959).
17. F. E. Beamish, *Talanta 8*, 85 (1961).
18. A. Chow and F. E. Beamish, *Talanta 10*, 883 (1963).
19. A. Chow, C. L. Lewis, D. A. Moddle, and F. E. Beamish, *Talanta 12*, 277 (1965).
20. F. E. Beamish, *Talanta 12*, 743 (1965).
21. F. E. Beamish, *Talanta 13*, 1053 (1966).
22. F. E. Beamish, C. L. Lewis, and J. C. van Loon, *Talanta 16*, 1 (1969).
23. V. M. Ivanov and A. I. Busev, *Zh. Anal. Khim. 26*, 1606 (1971).

23a. A. T. Meyers and R. G. Havens, personal communication.

24. K. G. Broadhead, B. C. Piper, and H. H. Heady, *Appl. Spectrosc. 26*, 461 (1972).
25. A. A. Kuranov, V. D. Ponomareva, and N. I. Chentsova, *Zh. Anal. Khim 15*, 476 (1960).
26. C. L. Lewis and W. L. Ott, *Methods for Emission Spectrochemical Analysis*, American Society for Testing and Materials, Philadelphia (1971), p. 594.
27. G. H. Ayres and E. W. Berg, *Anal. Chem. 24*, 465 (1952).
28. Á. Bardócz and F. Varsányi, *Anal. Chem. 28*, 989 (1956).
29. R. H. Bell and C. A. Jedlicka, *Methods for Emission Spectrochemical Analysis*, American Society for Testing and Materials, Philadelphia (1968), p. 578.
30. R. F. Doll, *Methods for Emission Spectrochemical Analysis*, American Society for Testing and Materials, Philadelphia (1971), p. 599.
31. C. Tosi, *Mikrochim. Acta 6*, 1162 (1966).
32. *Methods for Emission Spectrochemical Analysis*, *Amer. Soc. Testing Mater. Designation E378-68* (1971), p. 260.
33. V. le Roy and A. J. Lincoln, in *Developments in Applied Spectroscopy* (E. L. Grove, ed.), Vol. 8, Plenum Press, New York (1970), p. 199.
34. C. J. Mitchell, *Can. J. Spectrosc. 12*, 90 (1967).
35. A. J. Lincoln and J. C. Kohler, *Engelhard Ind. Tech. Bull. 4*, 81 (1963).
36. A. J. Lincoln and J. C. Kohler, *Anal. Chem. 34*, 1247 (1962).
37. J. C. Kohler and A. J. Lincoln, *Methods for Emission Spectrochemical Analysis*, American Society for Testing and Materials, Philadelphia (1971), p. 667.

38. C. L. Lewis and W. L. Ott, *Methods for Emission Spectrochemical Analysis*, American Society for Testing and Materials (1971), p. 589.
39. W. Diehl, *Metall. 19*, 712 (1965).
40. G. E. Crawley, *Methods for Emission Spectrochemical Analysis*, American Society for Testing and Materials, Philadelphia (1971), p. 628.
41. G. E. Crawley, *Methods for Emission Spectrochemical Analysis*, American Society for Testing and Materials, Philadelphia (1971), p. 633.
42. E. M. Murt and I. Corson, *Methods for Emission Spectrochemical Analysis*, American Society for Testing and Materials, Philadelphia (1971), p. 560.
43. J. Haffty and L. B. Riley, *Talanta 15*, 111 (1968).
44. O. F. Degtyareva and M. F. Ostrovskaya, *Zavodsk. Lab. 26*, 564 (1960).
45. A. Zelle and J. Fijakowski, *Chem. Anal. (Warsaw)* 7, 317 (1962).
46. M. Fred, N. H. Nachtrieb, and F. S. Tomkins, *J. Opt. Soc. Amer. 37*, 279 (1947).
47. C. Feldman, *Anal. Chem. 21*, 1041 (1949).
48. T. H. Zink, *Appl. Spectrosc. 13*, 94 (1959).
49. L. D. Shubin and J. H. Chaudet, *Appl. Spectrosc. 18*, 137 (1964).
50. R. P. Eardley and H. S. Clarke, *Appl. Spectrosc. 19*, 69 (1965).
51. H. M. Hyman, *Appl. Spectrosc. 16*, 129 (1962).
52. K. Dixon and T. W. Steele, *J. S. African Chem. Inst. 25*, 275 (1972).
53. C. W. Hill and G. P. Luckey, *Trans. Amer. Inst. Mining Met. Engrs. 60*, 342 (1919) (cited in Ref. 54).
54. M. Milbourn, *J. Inst. Metals 55*, 275 (1934).
55. O. Werner, *Metallwiss. Tech. 16*, 1062 (1962).
56. O. Werner, *Proc. IX Colloq. Spectrosc. Int. Lyon 2*, 461 (1961).
57. W. E. Publicover, *Anal. Chem. 37*, 1680 (1965).
58. H. Tölle, *Z. Anal. Chem. 240*, 162 (1968).
59. A. A. Kuranov and N. P. Ruksha, *Fiz. Sb. L'vovsk. Gos.* Univ. *4*, 421 (1958).
60. A. A. Kuranov, N. P. Ruksha, and M. M. Sviridova, *Analiz Blagorodnych Metallov*, Akad. Nauk SSSR, Moscow (1959), p. 139.
61. A. Strasheim, D. B. de Villiers, and D. Brink, *J. S. African Inst. Mining Met. 62*, 728 (1962).
62. A. L. Esterhuizen, *J. S. African Inst. Mining Met. 62*, 739 (1962).
63. R. Gerbatsch and G. Artus, *Z. Anal. Chem. 223*, 81 (1966).
64. H. Tölle, *Mikrochim. Acta*, 771 (1973).
65. A. F. Dorrzapf and F. W. Brown, *Appl. Spectrosc. 24*, 415 (1970).
66. N. Tomingas and W. C. Cooper, *Appl. Spectrosc. 11*, 164 (1957).
67. A. B. Whitehead and H. H. Heady, *Appl. Spectrosc. 24*, 225 (1970).
68. P. G. Sim, *Appl. Spectrosc. 28*, 23 (1974).
69. E. F. Cooley, K. J. Curry, and R. R. Carlson, *Appl. Spectrosc. 30*, 52 (1976).
70. P. Tymchuk, J. A. H. Desaulniers, D. S. Russell, and S. S. Berman, *Appl. Spectrosc. 21*, 151 (1967).
71. P. Tymchuk, D. S. Russell, and S. S. Berman, *Spectrochim. Acta 21*, 2051 (1965).
72. P. Tymchuk, A. Mykytiuk, and D. S. Russell, *Appl. Spectrosc. 22*, 268 (1968).
73. P. Tymchuk, F. C. Goodhue, and D. S. Russell, *Can. J. Spectrosc. 14*, 77 (1969).
74. W. Grimm, *Spectrochim. Acta 23B*, 443 (1968).
75. H. Jäger and F. Blum, *Spectrochim. Acta 29B*, 73 (1974).
76. E. W. Salpeter, *Spektren in der Glimmentladung*, Specola Vaticana, Città del Vaticano (1971).
77. H. Jäger, *Anal. Chim. Acta 58*, 57 (1972).
78. H. Jäger, *Anal. Chim. Acta 71*, 43 (1974).
79. H. Jäger, *Anal. Chim. Acta 60*, 303 (1972).

80. R. M. Lowe, *Spectrochim. Acta 31B*, 257 (1976).
81. E. Manrique, *Methods for Emission Spectrochemical Analysis*, American Society for Testing and Materials, Philadelphia (1971), p. 639.
82. E. Jackwerth, G. Graffmann, and J. Lohmar, *Z. Anal. Chem. 247*, 149 (1969).
83. B. M. Talalev, *Zh. Anal. Khim. 12*, 1443 (1966).
84. L. K. Larina, N. S. Belenkova, and N. F. Sachkova, *Zh. Anal. Khim. 22*, 808 (1967).
85. E. I. Savichev and E. V. Shugurov, *Zh. Anal. Khim. 22*, 1320 (1967).
86. C. F. Russo, *Appl. Spectrosc. 22*, 790 (1968).
87. N. Y. Chalkov and A. M. Ustimov, *Zavodsk. Lab. 33*, 448 (1967).
88. B. E. Balfour, D. Jukes, and K. Thornton, *Appl. Spectrosc. 20*, 168 (1966).
89. S. E. Kreimer, N. V. Tuzhilina, P. M. Mikhailov, and A. V. Andreeva, *Zavodsk. Lab. 36*, 1063 (1970).
90. J. A. Goleb, *Appl. Spectrosc. 15*, 57 (1961).
91. J. A. Goleb and J. K. Brody, *Appl. Spectrosc. 15*, 166 (1961).
92. H. Schüler and H. Gollnow, *Z. Physik 93*, 611 (1935).
93. F. C. Canney, A. T. Myers, and F. M. Ward, *Econ. Geol. 52*, 289 (1957).
94. A. T. Myers, R. G. Havens, and P. J. Dunton, *U.S. Geol. Surv. Bull. 1084-I* (1961).
95. K. Laqua and H. Waechter, *Proc. XIV Colloq. Spectrosc. Int. Debrecen* (1967), p. 775.
96. S. L. Terekhovich, *Zavodsk. Lab. 34*, 426 (1968).
97. R. Avni and A. Boukobza, *Appl. Spectrosc. 23*, 483 (1969).
98. A. Strasheim and J. Van Wamelen, *J. S. African Chem. Inst. 15*, 60 (1962).
99. L. I. Pavlenko, A. V. Karyakin, and L. V. Simonova, *Zh. Anal. Khim. 26*, 934 (1971).
100. D. J. Grimes and A. P. Marranzino, *U.S. Geol. Surv. Circ. 591* (1968).
101. H. de Lazlo, *Ind. Eng. Chem. 19*, 1366 (1927).
102. J. M. Lopez de Azcona, *Spectrochim. Acta 2*, 185 (1942).
103. A. G. Scobie, *Trans. Can. Inst. Mining Met. 48*, 309 (1945).
104. H. Zachariasen and F. E. Beamish, *Talanta 4*, 44 (1960).
105. F. E. Beamish, *Talanta 5*, 1 (1960).
106. G. H. Faye and W. R. Inman, *Anal. Chem. 35*, 985 (1963).
107. G. G. Tertipis and F. E. Beamish, *Talanta 10*, 1139 (1963).
108. K. C. Agrawal and F. E. Beamish, *Talanta 11*, 1449 (1964).
109. F. E. Beamish, *Talanta 14*, 991 (1967).
110. F. E. Beamish, *Talanta 14*, 1133 (1967).
111. G. H. Faye, *Anal. Chem. 37*, 696 (1965).
112. J. G. Sen Gupta and F. E. Beamish, *Anal. Chem. 34*, 1761 (1962).
113. G. H. Faye and P. E. Moloughney, *Talanta 19*, 269 (1972).
114. A. P. Kuznetsov, Y. N. Kukushkin, and D. F. Makarov, *Zh. Anal. Khim. 29*, 2155 (1974).
115. F. I. Danilova, V. A. Orobinskaya, V. S. Parfenova, R. M. Nazarenko, V. G. Khitrov, and G. E. Belousov, *Zh. Anal. Khim. 29*, 2142 (1974).
116. C. L. Lewis, *Can. Spectrosc. 9*, 81 (1964).
117. C. L. Lewis, *Trans. Can. Inst. Mining Met. 50*, 163 (1957).
118. N. T. Voskresenskaya, N. F. Zvereva, and L. L. Rivkina, *Zh. Anal. Khim. 20*, 1288 (1965).
119. V. M. Novikov and V. K. Bondarenko, *Zavodsk. Lab. 34*, 1080 (1968).
120. V. V. Polikarpochkin, I. Y. Korotaeva, and V. N. Sarapulova, *Zavodsk. Lab. 33*, 441 (1967).
121. P. Hahn-Weinheimer, *Z. Anal. Chem. 162*, 161 (1958).
122. I. I. Tarasova, L. S. Dudenkova, V. G. Khitrov, and G. E. Belousov, *Zh. Anal. Khim. 29*, 2147 (1974).
123. R. S. Rubinovich, R. Y. Epshtein, and O. N. Soshal'skaya, *Zh. Anal. Khim. 18*, 216 (1963).

124. S. E. Kreimer, P. M. Mikhailov, and A. S. Lomekhov, *Zh. Anal. Khim. 22*, 1105 (1967).
125. R. R. Brooks and L. H. Ahrens, *Spectrochim. Acta 16*, 783 (1960).
126. M. Margoshes and B. F. Scribner, *Spectrochim. Acta 15*, 138 (1959).
127. M. Riemann, *Z. Anal. Chem. 215*, 407 (1966).
128. H. Gotô and I. Atsuya, *Z. Anal. Chem. 225*, 121 (1967).
129. M. Marinković and B. Dimitrijević, *Spectrochim. Acta 23B*, 257 (1968).
130. H. Raab, *J. Physics E*, *5*, 779 (1972).
131. W. Tappe and J. van Calker, *Z. Anal. Chem. 198*, 13 (1963).
132. C. D. West and D. N. Hume, *Anal. Chem. 36*, 412 (1964).
133. H. Gotô, K. Hirokawa, and M. Suzuki, *Z. Anal. Chem. 225*, 130 (1967).
134. S. Murayama, H. Matsuno, and M. Yamamota, *Spectrochim. Acta 33B*, 513 (1968).
135. R. H. Wendt and V. A. Fassel, *Anal. Chem. 37*, 920 (1965).
136. G. W. Dickinson and V. A. Fassel, *Anal. Chem. 41*, 1021 (1969).
137. P. W. J. M. Boumans and F. J. de Boer, *Spectrochim. Acta 27B*, 391 (1972).
138. V. A. Fassel and R. N. Kniseley, *Anal. Chem. 46*, 1110A (1974).
139. S. Greenfield, I. Ll. Jones, H. McD. McGeachin, and P. B. Smith, *Anal. Chim. Acta 74*, 225 (1975).
140. P. W. J. M. Boumans and F. J. de Boer, *Spectrochim. Acta 30B*, 309 (1975).
141. V. A. Fassel, *Proc. XVI Colloq. Spectrosc. Int. Heidelberg*, Adam Hilger Ltd., London (1972), p. 63.
142. S. Greenfield, H. McD. McGeachin, and P. B. Smith, *Talanta 22*, 1 (1975).
143. S. Greenfield, H. McD. McGeachin, and P. B. Smith, *Talanta 22*, 553 (1975).
144. S. Greenfield, H. McD. McGeachin, and P. B. Smith, *Talanta 23*, 1 (1976).
145. P. W. J. M. Boumans, F. J. de Boer, F. J. Dahmen, H. Hoelzel, and A. Meier, *Spectrochim. Acta 30B*, 449 (1975).
146. B. Zmbova and M. Marinković, *Talanta 20*, 647 (1973).
147. V. A. Fassel and D. W. Golightly, *Anal. Chem. 39*, 466 (1967).
148. G. D. Christian and F. J. Feldman, *Appl. Spectrosc. 25*, 660 (1971).
149. V. Sychra, P. J. Slevin, J. Matoušek, and F. Bek, *Anal. Chim. Acta 52*, 259 (1970).
150. J. G. Sen Gupta, *Mineral Sci. Eng. 5*, 207 (1973).
151. N. L. Fishkova, *Zh. Anal. Khim. 29*, 2121 (1974).
152. W. Heinemann, *Z. Anal. Chem. 279*, 351 (1976).
153. W. Heinemann, *Z. Anal. Chem. 280*, 127 (1976).
154. W. Heinemann, *Z. Anal. Chem. 280*, 359 (1976).
155. W. Heinemann, *Z. Anal. Chem. 281*, 103 (1976).
156. R. Lockyer and G. E. Hames, *Analyst 84*, 385 (1959).
157. A. C. Menzies, *Anal. Chem. 32*, 898 (1960).
158. A. Strasheim and G. J. Wessels, *Appl. Spectrosc. 17*, 65 (1963).
159. W. Slavin, *Atomic Absorption Spectroscopy*, Wiley-Interscience, New York (1968).
160. M. M. Schnepfe and F. S. Grimaldi, *Talanta 16*, 591 (1969).
161. J. C. van Loon, *At. Absorption Newsletter 8*, 6 (1969).
162. A. Janssen and F. Umland, *Z. Anal. Chem. 251*, 101 (1970).
163. F. S. Grimaldi and M. M. Schnepfe, *Talanta 17*, 617 (1970).
164. J. C. van Loon, *Z. Anal. Chem. 246*, 122 (1969).
165. I. G. Yudelevich, G. A. Vall, V. G. Torgov, and T. M. Korda, *Zh. Anal. Khim. 26*, 1550 (1971).
166. J. P. Macquet, J. Hubert, and T. Theophanides, *Anal. Chim. Acta 72*, 251 (1974).
167. J. P. Macquet and T. Theophanides, *Anal. Chim. Acta 72*, 261 (1974).
168. D. F. Makarov, Y. N. Kukushkin, and T. A. Eroshevich, *Zh. Anal. Khim. 29*, 2128 (1974).
169. J. M. Scarborough, *Anal. Chem. 41*, 250 (1969).

170. B. Montford and S. C. Cribbs, *Anal. Chim. Acta 53*, 101 (1971).
171. V. L. Ginzburg, D. F. Makarov, and G. I. Satarina, *Zh. Anal. Khim. 24*, 264 (1969).
172. R. Belcher, R. M. Dagnall, and T. S. West, *Talanta 11*, 1257 (1964).
173. V. Eckelmans, E. Graauwmans, and S. de Jaegere, *Talanta 21*, 715 (1974).
174. B. P. Khrapai, E. A. Startseva, L. E. Zhuravleva, and N. M. Popova, *Zh. Anal. Khim. 29*, 2137 (1974).
175. M. G. Atwell and J. Y. Hebert, *Appl. Spectrosc. 23*, 480 (1969).
176. A. E. Pitts, J. C. van Loon, and F. E. Beamish, *Anal. Chim. Acta, 50*, 181 (1970).
177. A. E. Pitts, J. C. van Loon, and F. E. Beamish, *Anal. Chim. Acta, 50*, 195 (1970).
178. R. C. Mallett, D. C. G. Pearton, E. J. Ring, and T. W. Steele, *Talanta 19*, 181 (1972).
179. R. C. Mallett, D. C. G. Pearton, E. J. Ring, and T. W. Steele, *J. S. African Chem. Inst. 25*, 219 (1972).
180. H. C. Eshelman, J. Dyer, and J. Armentor, *Anal. Chim. Acta 32*, 411 (1965).
181. J. Matoušek and V. Sychra, *Anal. Chim. Acta 49*, 175 (1970).
182. L. G. Hickey, *Anal. Chim. Acta 41*, 546 (1968).
183. G. Mathieu and S. Guiot, *Anal. Chim. Acta 52*, 335 (1970).
184. D. F. Makarov, Y. N. Kukushkin, and T. A. Eroshevich, *Zh. Prikl. Khim. 47*, 1215 (1974).
185. J. G. Sen Gupta, *Anal. Chim. Acta 63*, 19 (1973).
186. M. A. Ashy and J. B. Headridge, *Analyst 99*, 285 (1974).
187. L. R. P. Butler, A. Strasheim, and F. W. E. Strelow, *Proc. XII Colloq. Spectrosc. Int. Exeter* (1965), p. 284.
188. T. Groenewald, *Anal. Chem. 40*, 863 (1968).
189. H. C. van Rensburg and P. B. Zeeman, *Anal. Chim. Acta 43*, 173 (1968).
190. E. E. Pickett and S. R. Koirtyohann, *Spectrochim. Acta 23B*, 235 (1968).
191. J. I. Dinnin, *Anal. Chem. 39*, 1491 (1967).
192. D. W. Ellis and D. R. Demers, *Anal. Chem. 38*, 1943 (1966).
193. T. S. West and X. K. Williams, *Anal. Chem. 40*, 335 (1968).
194. G. Erinc and R. J. Magee, *Anal. Chim. Acta 31*, 197 (1964).
195. D. E. Harrington, *At. Absorption Newsletter 9*, 106 (1970).
196. M. H. Campbell, *Anal. Chem. 40*, 6 (1968).
197. P. Heneage, *At. Absorption Newsletter 5*, 64 (1966).
198. J. R. Deily, *At. Absorption Newsletter 6*, 65 (1967).
199. G. D. Christian and F. J. Feldman, *Anal. Chem. 43*, 611 (1971).
200. T. W. Olinski and N. H. Knight, *Appl. Spectrosc. 22*, 532 (1968).
201. F. Fernandez, *At. Absorption Newsletter 8*, 90 (1969).
202. C. E. Mulford, *At. Absorption Newsletter 5*, 63 (1966).
203. D. C. Manning and F. Fernandez, *At. Absorption Newsletter 6*, 15 (1967).
204. D. L. Fuhrman, *At. Absorption Newsletter 8*, 105 (1969).
205. D. F. Makarov and Y. N. Kukushkin, *Zh. Anal. Khim. 24*, 1118 (1969).
206. F. M. Tindall, *At. Absorption Newsletter* 4, 339 (1965).
207. F. M. Tindall, *At. Absorption Newsletter 5*, 140 (1966).
208. I. Rubeška, Z. Šulcek, and B. Moldan, *Anal. Chim. Acta 37*, 27 (1967).
209. G. Walton, *Analyst 98*, 335 (1973).
210. B. Strong and R. Murray-Smith, *Talanta 21*, 1253 (1974).
211. F. I. Danilova, V. A. Orobinskaya, V. S. Parfenova, R. F. Propistsova, and S. B. Savvin, *Zh. Anal. Khim. 29*, 2150 (1974).
212. E. C. Simmons, *At. Absorption Newsletter 4*, 281 (1965).
213. S. L. Law and T. E. Green, *Anal. Chem. 41*, 1009 (1969).
214. V. L. Ginzburg, D. M. Livshits, and G. I. Satarina, *Zh. Anal. Khim. 19*, 1089 (1964).
215. P. E. Moloughney and G. H. Faye, *Talanta 23*, 377 (1976).

216. N. L. Fishkova, *Zavodsk. Lab. 36*, 1461 (1970).
217. A. A. Venghiattis, *Spectrochim. Acta 23B*, 67 (1967).
218. B. V. L'vov, *Spectrochim. Acta 17*, 761 (1961).
219. B. V. L'vov, *Spectrochim. Acta 24B*, 53 (1969).
220. H. Massmann, *Spectrochim. Acta 23B*, 215 (1968).
221. E. Andriaenssens and P. Knoop, *Anal. Chim. Acta 68*, 37 (1973).
222. I. Rubeška and J. Štubar, *At. Absorption Newsletter 5*, 69 (1966).
223. R. G. Anderson, I. S. Maines, and T. S. West, *Anal. Chim. Acta 51*, 355 (1970).
224. J. Aggett and T. S. West, *Anal. Chim. Acta 55*, 349 (1971).
225. M. D. Amos, P. A. Bennett, K. G. Brodie, P. W. Y. Lung, and J. P. Matousek, *Anal. Chem. 43*, 211 (1971).
226. C. J. Molnar, R. D. Reeves, J. D. Winefordner, M. T. Glenn, J. R. Ahlstrom, and J. Savory, *Appl. Spectrosc. 26*, 606 (1972).
227. T. W. Steele and B. D. Guerin, *Analyst 97*, 77 (1972).
228. B. D. Guerin, *J. S. African Chem. Inst. 25*, 230 (1972).
229. F. S. Chuang and J. D. Winefordner, *Appl. Spectrosc. 28*, 215 (1974).
230. B. M. Patel, R. D. Reeves, R. F. Browner, C. J. Molnar, and J. D. Winefordner, *Appl. Spectrosc. 27*, 171 (1973).
231. E. Jackwerth and P. G. Willmer, *Talanta 23*, 197 (1976).
232. V. I. Men'shikov, V. D. Malykh, and T. D. Shestakova, *Zh. Anal. Khim. 29*, 2132 (1974).
233. F. J. Langmyhr, J. R. Stuhergh, Y. Thomassen, J. E. Hanssen, and J. Dolezal, *Anal. Chim. Acta 71*, 35 (1974).
234. P. N. Gerrard and N. Westwood, *J. S. African Chem. Inst. 25*, 285 (1972).
235. B. G. Russell, B. T. Eddy, G. Beckmann, and T. W. Steele, *J. S. African Chem. Inst. 25*, 297 (1972).
236. R. Neeb, *Z. Anal. Chem. 179*, 21 (1961).
237. T. E. Green, S. L. Law and W. J. Campbell, *Anal. Chem. 42*, 1749 (1970).
238. S. D. Rasberry, H. J. Caul, and A. Yezer, *Spectrochim. Acta 23B*, 345 (1968).
239. P. A. Verkhovodov, *Zh. Anal. Khim. 8*, 1214 (1967).
240. J. D. Eick, H. J. Caul, D. L. Smith, and S. D. Rasberry, *Appl. Spectrosc. 21*, 324 (1967).
241. H. J. Lucas-Tooth and B. J. Price, *Metallurgia, 64*, 149 (1961).
242. F. T. Wybenga and A. Strasheim, *Appl. Spectrosc. 20*, 247 (1966).
243. A. Strasheim and F. T. Wybenga, *Appl. Spectrosc. 18*, 16 (1964).
244. H. Pietzner and H. Werner, *Z. Anal. Chem., 221*, 186 (1966).
245. H. Taylor and F. E. Beamish, *Talanta, 15*, 497 (1968).
246. J. O. Karttunen, *Anal. Chem. 35*, 1044 (1963).
247. C. W. Fuller, G. Himsworth, and J. Whitehead, *Analyst 96*, 177 (1971).
248. V. A. Shestakov, N. A. Arkhipov, D. F. Makarov, and Y. N. Kukushkin, *Zh. Anal. Khim. 29*, 2176 (1974).
249. A. Chow and F. E. Beamish, *Talanta 13*, 539 (1966).
250. R. A. van Nordstrand, A. J. Lincoln, and A. Carnevale, *Anal. Chem. 36*, 819 (1964).
251. B. W. Mulligan, H. J. Caul, S. D. Rasberry, and B. F. Scribner, *J. Res. Natl. Bur. Std. 68A*, 5 (1964).
252. W. M. MacNevin and E. A. Hakkila, *Anal. Chem. 29*, 1019 (1957).
253. A. J. Lincoln and E. N. Davis, *Anal. Chem. 31*, 1317 (1959).
254. J. Ottemann and S. S. Augustithis, *Mineralium Deposita 1*, 269 (1967).
255. P. E. Larson, *Anal. Chem. 44*, 1678 (1972).
256. P. G. Burkhalter and H. E. Marr, *Int. J. Appl. Radiation Isotopes 21*, 395 (1970).
257. G. C. Snyman, *Int. J. Appl. Radiation Isotopes 18*, 243 (1967).

258. J. L. Debrun, J. N. Barrandon, P. Benaben, and Ch. Rouxel, *Anal. Chem. 47*, 637 (1975).
259. G. Deconninck and G. Demortier, *J. Radioanal. Chem. 24*, 437 (1975).
260. P. W. de Lange, W. J. de Wet, J. Turkstra, and J. H. Venter, *Anal. Chem. 40*, 451 (1968).
261. J. Turkstra, P. J. Pretorius, and W. J. de Wet, *Anal. Chem. 42*, 835 (1970).
262. J. Turkstra, H. J. Smit, and W. J. de Wet, *J. S. African Chem. Inst. 25*, 254 (1972).
263. R. Gijbels, *Talanta 18*, 587 (1971).
264. F. E. Beamish, K. S. Chung, and A. Chow, *Talanta 14*, 1 (1967).
265. E. E. Rakovskii, T. D. Krylova, and G. M. Baevskaya, *Zh. Anal. Khim. 29*, 2166 (1974).
266. D. F. C. Morris and R. A. Killick, *Talanta 3*, 34 (1959).
267. D. F. C. Morris and R. A. Killick, *Talanta 8*, 793 (1961).
268. R. A. Killick and D. F. C. Morris, *Talanta 8*, 601 (1961).
269. D. F. C. Morris and R. A. Killick, *Talanta 8*, 129 (1961).
270. R. A. Killick and D. F. C. Morris, *Talanta 9*, 349 (1962).
271. R. Gijbels and J. Hoste, *Anal. Chim. Acta 41*, 419 (1968).
272. L. S. Jawanovitz, F. B. McNatt, R. E. McCarley, and D. S. Martin Jr. *Anal. Chem. 32*, 1270 (1960).
273. R. A. Killick and D. F. C. Morris, *Talanta 9*, 879 (1962).
274. D. F. C. Morris and R. A. Killick, *Talanta 10*, 987 (1963).
275. D. F. C. Morris and R. A. Killick, *Talanta, 10*, 1153 (1963).
276. D. F. C. Morris and R. A. Killick, *Talanta 10*, 279 (1963).
277. A. G. Ganiev, V. V. Rybnov, and L. M. Yakusheva, *Zh. Anal. Khim. 23*, 918 (1968).
278. R. Gijbels and J. Hoste, *Anal. Chim. Acta 32*, 17 (1965).
279. R. Gijbels and J. Hoste, *Anal. Chim. Acta 39*, 132 (1967).
280. R. Gijbels and J. Hoste, *Anal. Chim. Acta 29*, 289 (1963).
281. D. F. C. Morris, D. N. Slater, and R. A. Killick, *Talanta 8*, 373 (1961).
282. H. Förster and W. Görner, *J. Radioanal. Chem. 24*, 369 (1975).
283. E. N. Gilbert, G. V. Veriovkin, B. N. Botchkaryov, A. A. Godovikov, V. A. Mikhailov, and V. Y. Zhavoronkov, *J. Radioanal. Chem. 26*, 253 (1975).
284. H. Maleszewska and R. Dybczyński, *J. Radioanal. Chem. 31*, 177 (1976).
285. S. Sterliński, H. Maleszewska, and R. Dybczyński, *J. Radioanal. Chem. 31*, 61 (1976).
286. A. R. De Grazia and L. Haskin, *Geochim. Cosmochim. Acta 28*, 559 (1964).
287. F. O. Simon and H. T. Millard, *Anal. Chim. 40*, 1150 (1968).
288. E. N. Gilbert, G. V. Glukhova, G. G. Glukhov, V. A. Mikhailov, and V. G. Trogov, *J. Radioanal. Chem. 8*, 39 (1971).
289. E. A. Vincent and A. A. Smales, *Geochim. Cosmochim. Acta 9*, 154 (1956).
290. J. H. Crocket and G. B. Skippen, *Geochim. Cosmochim. Acta 30*, 129 (1966).
291. J. W. Morgan, *Anal. Chim. Acta 32*, 8 (1965).
292. J. H. Croket, R. R. Keays, and S. Hsieh, *J. Radioanal. Chem. 1*, 487 (1968).
293. K. S. Chung and F. E. Beamish, *Anal. Chim. Acta 43*, 357 (1968).
294. R. R. Keays and J. H. Crocket, *Econ. Geol. 65*, 438 (1970).
295. V. P. Razhdaev and V. N. Nikitin, *Zh. Anal. Khim. 29*, 2172 (1974).
296. R. A. Nadkarni and G. H. Morrison, *Anal. Chem. 47*, 2285 (1975).
297. E. N. Gilbert, G. V. Veriovkin, and V. A. Mikhailov *J. Radioanal. Chem. 31*, 365 (1976).
298. R. K. Iyer and K. R. Krishnamoorthy, *J. Radioanal. Chem. 33*, 243 (1976).
299. E. A. Vincent and J. H. Crocket, *Geochim. Cosmochim. Acta 18*, 130 (1960).
300. E. A. Vincent and J. H. Crocket, *Geochim. Cosmochim. Acta 18*, 143 (1960).
301. P. A. Baedecker and W. D. Ehmann, *Geochim. Cosmochim. Acta 29*, 329 (1965).

302. W. Nichiporuk and H. Brown, *Phys. Rev. Letters 9*, 245 (1962).
303. H. Brown and E. Goldberg, *Science 109*, 347 (1949).
304. E. D. Goldberg and H. Brown, *Anal. Chem. 22*, 308 (1950).
305. U. Schindewolf and M. Wahlgren, *Geochim. Cosmochim. Acta 18*, 36, (1960).
306. G. L. Bate and J. R. Huizenga, *Geochim. Cosmochim. Acta 27*, 345 (1963).
307. T. K. Choy, *Anal. Chim. Acta 34*, 372 (1966).
308. A. Govaerts, R. Gijbels, and J. Hoste, *Anal. Chim. Acta 79*, 139 (1975).
309. J. W. Butler and D. E. Becknell, *J. Radioanal. Chem. 10*, 47 (1972).
310. K. S. Park, R. Gijbels, and J. Hoste, *J. Radioanal. Chem. 5*, 31 (1970).
311. E. L. Steele and W. W. Meinke, *Anal. Chim. Acta 26*, 269 (1962).
312. C. E. Gleit, P. A. Benson, W. D. Holland, and I. J. Russell, *Anal. Chem. 36*, 2067 (1964).
313. B. A. Thompson, B. M. Strause, and M. B. Leboeuf, *Anal. Chem. 30*, 1023 (1958).
314. H. Menke and M. Weber, *Z. Anal. Chem. 267*, 122 (1973).
315. L. A. Rancitelli and R. W. Perkins, *J. Geophys. Res. 75*, 3055 (1970).
316. A. Albini and A. Cesana, *J. Radioanal. Chem. 34*, 185 (1976).
317. D. Gibbons, *Int. J. Appl. Radiation Isotopes 4*, 45 (1958).

Petroleum Industry Analytical Applications of Atomic Spectroscopy

2

Bruce E. Buell

2.1 INTRODUCTION

This presentation is intended as a critical evaluation of analytical applications of emission spectrographic (ES) and direct-reading emission spectrometric (ESD) methods, as well as atomic absorption (AA), and flame emission (FE) spectrometric methods. References to the use of atomic fluorescence (AF) are limited. Much of the literature from 1949 on has been reviewed, and many references are given. Coverage is relatively complete beginning with 1965. The history and the need and importance of applications for each type of product are discussed.

The presentation is not limited to a review, but also discusses application trends and merits of techniques as evaluated by the author from personal experience and observations. In cases of special importance, and where there is much literature coverage, details are presented and the merits of debatable areas are discussed.

Coverage and details of instrumental techniques have been limited because of the excellent coverage in other books. In this section, however, detection limits, plasma torches for emission, AA and FE burners, and nonflame AA methods are discussed in more detail.

Because crude oil is "the vital fluid of the petroleum industry" and has become of critical importance to the nation and the world, a background section on "Crude Oil Importance, Composition, and Origin" is presented before the discussion on analytical applications.

Bruce E. Buell • Union Oil Company of California, Brea, California

2.2 DEFINITION AND SCOPE OF PETROLEUM INDUSTRY

Petroleum industries in this modern age have diversified interests. In addition to the usual oil- and gasoline-derived products, subsidiaries and affiliates may include petrochemicals, mining companies, and energy sources other than petroleum-derived (nuclear power, geothermal, etc.). For the purposes of this chapter emphasis will be placed on direct petroleum products and necessary substances involved in petroleum industry operation and research. Substances requiring analysis include crude oils and related fractions (distillates and residuums), partially refined fractions (feedstocks, gasoline- and oil-base stocks, residuals), finished products (lubricating oils, gasolines, jet fuels, greases, waxes, etc.), by-products and wastes (oil well brines, cooling waters, oil well cores, etc.). Research-oriented samples, of course, may be widely diversified; thus, the need for special analyses and a research laboratory with a methods development program.

Subsidiaries may need to analyze samples such as plants, soils, fertilizers, petrochemicals (insecticides, solvents), minerals, and others.

2.3 CRUDE OIL IMPORTANCE, COMPOSITION, AND ORIGIN

2.3.1 Importance

Without an adequate supply of crude oil, for which the annual requirements are increasing, there would be no petroleum industry. The increasing demand for energy is speeding the day when our known crude oil reserves will be depleted; in fact, we have entered an era of energy shortage. Eventually the large deposits of shale oil in the United States, and perhaps the Athabasca tar sands will be needed with an attendant increase in the cost of petroleum products. In Colorado, Utah, and Wyoming it is estimated there are 2 trillion barrels[1] of recoverable shale oil, and the recoverable oil from the Athabasca tar sands is estimated at 300 million barrels.[2,3] It is estimated that energy demands will increase 93% for the period from 1970 to 1985.[4] It is also estimated[4] that oil and gas must continue to provide most of our energy requirements until probably 1990. As of 1970, the use of oil is shown as approaching 15 million barrels daily and natural gas use as 10 million barrels daily. The daily production of U.S. oil is at about 8 million barrels. For additional data on world production and reserves and the economics involved, refer to Refs. 5–7. This demonstrates the extreme importance of the petroleum industry and crude oil and indicates why even such a relatively small detail as metal analyses in crude oils might be very important.

2.3.2 Composition

Crude oil is a mobile, complex mixture of organic material produced by drilling wells into underground oil reservoirs. Wells may vary from shallow to very deep (30,000 ft) and may produce a mixture of oil, water, and natural gas. Crude oils contain compounds varying from low-boiling, straight-chain hydrocarbons to high-molecular-weight, complex heterocyclic compounds. Much research has been done on the composition of crude oils, including trace metals. The book *Fundamental Aspects of Petroleum Geochemistry* contains chapters on petroleum hydrocarbons and nonhydrocarbons.[8] The metals in crudes are concentrated in the highest-boiling, nonhydrocarbon fractions. At Union Oil, extensive research has led to the development of methods for the complete characterization of heterocyclic compound types in these fractions[9-13] except for metalloorganic compounds. American Petroleum Institute Project 60 on Heavy Ends, and the U.S. Bureau of Mines research have also included much study on selected crude oils using similar and additional techniques.[14-16]

Crude oil characteristics vary widely and have been reviewed for U.S. crudes by Smith.[17] Viscosity of crudes vary widely, and API gravity varies from less than 15 to over 45. Sulfur content varies from less than 0.1% to over 2% (an exceptional value of 7.5% is found in an Oxnard, California, crude[18] and correlates with the V content. Nitrogen content varies from less than 0.05% to 1% at the most, and there appears to be some correlation of high nitrogen with high Ni content. Metal content varies from less than 10 ppm to a high of about 2000 ppm. More recently a project has been completed for the *Oil and Gas Journal* and published as a series during 1976 as a guide to world crude oil export streams.[19] Data for gravity, sulfur content, and so on, are included.

2.3.3 Crude Oil Fractions and Refining

Upon distillation of a crude oil the fractions obtained can be classified as light petroleum gases (C_3/C_4), gasoline (boiling up to 400°C), light gas oils or petroleum distillate (LPD, 400-700°), heavy gas oils (HPD, 700-1100°), and residuum fractions. The metal-containing compounds are concentrated in the heavier fractions. Modern refinery operations are designed to convert more of the heavy fractions by cracking operations[20,21] into gasoline, since that remains the premium fraction in the United States. Unicracking-JHC and GOfining and RESIDfining[21] are modern processes developed by Esso and Union Oil to achieve this objective.

In addition to LPG and gasoline, refined products include petroleum naphthas, jet fuel, lubricating oil, residual fuel oil, grease, asphalt, wax, coke, and petrochemicals.

2.3.4 Origin of Crude Oils and Metal Content

The metal compounds in crude oil, particular Ni and V and their porphyrin (Ph) compounds, have played a significant part in theories concerning the origin of crude oils. There have been many studies of Ph compounds and because of their importance and association with Ni and V, detailed data on analytical procedures is given in a separate section (Section 2.9.3).

Because of the importance of crude oil and the relation of its origin to the studies of metals in crudes, this section will review references as to crude oil origin.

2.3.4.1 Theories of Origin

Through the years many theories have been proposed as to the origin of crude oils. The theory most accepted currently is biogenic and considers that organic matter from once-living organisms, such as phytoplankton, zooplankton, and bacteria, accumulated in the sediments of aquatic basins and gradually was converted to crude oil. Literature concerning the biogenic origin of crude oil is prolific and is reviewed in Section 2.3.4.2. Other theories concern abiogenic[2,22,23] and dual roles.[24-28] Wolsky and Chapman[29] found evidence for the presence of inorganic forms in Boscan asphaltenes and mentions vanadium sulfide deposits as a possible origin (cf. Refs. 16 and 17 in Ref. 28). One theory proposes extraterrestrial or cosmic contributions, which is reviewed by Lind.[30] Another "pedogenic" origin (concerning organic matter in soils) is discussed by Cate.[31] The history of various theories, as well as a good review of abiogenic theories, is thoroughly covered by Dott and Reynolds.[2]

2.3.4.2 Literature on Biogenic Origins of Crude Oil

Crude oil origin is currently believed to have involved accumulation of organic material from plants and organisms in basins of marine and possibly non-marine sediments. Aquatic organisms and vegetable matter such as phytoplankton, zooplankton, and bacteria can contribute tremendous amounts of organic material,[32,33] especially in the oceans, and have been studied extensively as possible crude oil precursors. Phytoplankton produces the largest mass of organic material (more than 10 times than that from zooplankton), which amounts to 42 million tons (dry weight) annually in the ocean off the coast of southern California[32]

McNab *et al.*[34] discussed three phases for organic material in sediments to form oil at an ACS symposium on the origin of petroleum, and presented an extensive review of crude oil assay data for U.S. oil fields. It appears that this era, beginning in the 1950s, was an age when technology was developing rapidly and studies such as those concerning organic material in sediments advanced markedly shedding much light on the probable origin of crude oil. The 1954 work by Smith[35] in which the presence of hydrocarbons in recent sediments was established gave impetus to this era of enlightenment. By 1956, Stevens[36] stated: "There is now no reasonable doubt concerning the organic origin of crude oil." He also concluded that new techniques were "facilitating, to a high degree, the study of the origin of oil."

Dott and Reynolds[2] in their reference to the excellent 1964 study of Mair[37] which discussed the abundance of certain compound types in plants and animals as source materials for petroleum hydrocarbons, state "confirming beyond reasonable doubt the organic origin of petroleum." There have been many good studies of sediments, including metals and chlorophyll-chlorin-porphyrin compounds.[37-51] The cruises of the *Glomar Challenger* and results of the Deep Sea Drilling Project[52] are of extreme interest (see Baker, Ref. 42, p. 498, and Refs. 38 and 39). Among sediment studies, that by Bradley[41] is unique in finding organic ooze that apparently has been preserved in a fully oxidizing environment containing only a few bacteria (in contrast to the high bacteria count most often found and the requirement of a reducing environment thought by many to be needed for preserving organic material).

There have been many reviews in books[8,28,53-62] and journals[63-73] on the origin of crude oil. One of the most thorough reviews is that by Hedberg[68] on different and varied phases of petroleum origin. Among the books cited, those by Breger,[53] Dott and Reynolds,[55] Eglinton and Murphy,[56] Manskaya and Drozdova,[60] and Nagy and Colombo[8] contain many excellent chapters on different phases of petroleum geochemistry and origin. The Russian book[60] also has some excellent chapters, especially good sections on the organic chemistry of metals. The book by Mason,[61] in addition to the chapter on the origin of petroleum, has excellent coverage on the biosphere and hydrosphere and the geochemistry of metals.

Books by Davis[74] and by Beerstecher[75] have excellent coverage of petroleum microbiology, including a chapter in[75] on the role of microorganisms in the synthesis of petroleum and a table (p. 30) on the classification of the major groups of microorganisms. A book by Allen[76] contains excellent contributions by various experts on marine biogeochemistry, as does a two-volume *Treatise on Marine Ecology and Paleoecology*,[77] which contains extensive annotated bibliographies. A student of the origin of crude oils should also be aware of the Russian works by Radchenko,[78] which contains extensive data and discussions on the oil fields of the world. Also of extreme importance in the origin of crudes are the studies of hydrocarbons, of which papers by Mair[37] and Martin *et al.*[79] and a chapter by Meinschein in Ref. 56 are good examples. Prior to printing, additional references were published as selected papers reprinted from the AAPG publication on the origin of petroleum.[80] These include many of the references cited and others.

Since porphyrins and other fossil pigments are of such importance to the origin of crude oils and the metals contained therein, it should be emphasized that in the books cited the subject is treated in chapters by Baker in Ref. 56, Dunning in Ref. 53, Hodgson in Ref. 62, Krauskopf[58] Rimington and Kennedy[81] (see Part A, p. 25), and Vallentyne in Ref. 76. Falk authored an extensive work on porphyrins and metal porphyrins,[82] and Inhoffen *et al.*[83] presented an extensive review on Ph reactions, including 221 references. Because Ph compounds occur in crudes only as Ni and V compounds, as opposed to Mg in chlorophyl and Fe in hemoglobin (possible Ph precursors), the relations of these metals in plants, animals, the ocean, and sediments are important to studies on the origin of metals in crudes. Works by Bowen[84] and Vinogradov,[85] and studies by Goldberg,[86] Krauskopf,[87] and Schutz and Turekian[88] on the distribution of trace metals in the oceans may provide clues. For example, Krauskopf reviews enrichment factors for the concentration of trace metals from the ocean into plant and animal organisms (also discussed by Goldberg). Since V and V-Ph are prevalent in crudes, it is interesting to note that the highest enrichment factor (in the range 10,000-300,000) is for V, while that for Ni is also quite high.

2.3.4.3 Characteristics of Crude Oil That May Relate to Metal Content Origin

In the study of metal and Ph compounds in crudes, certain specialized areas are of interest. Because they are associated with the heavier oils and the heavy, asphalt fraction, variations according to the gravity of oils are of interest. Since variation of gravity and metal content vary with age, depth, and basin geometry and geochemistry, these subjects will be reviewed. Of extreme value also are C and S stable isotope studies. Since crudes with high V content appear to have S content and crudes with high Ni appear to have high N content, S and N compounds are also important.

2.3.4.3a Ages of Crude Oils. Knebel and Rodriguez-Eraso[89] in 1956 listed 53% of crude oils as Mesozoic, 38% as Tertiary, and only 9% as older. Since then, however, additional oil fields may have been discovered, with extremes in age from very young[90] to very old. Also, evidence of Precambrian fossils, organic material, and oil-like material has been identified.[91–101] In regard to the latter area, due consideration must be given to the possibilities of contamination from younger sources; nevertheless, the Precambrian evidence seems positive. It is interesting to note that there appears to be an abundance of Paleozoic oil fields formed before the advent of abundant land plants and animals (Ref. 68, p. 1768). The Rocky Mountain area is interesting because of the occurrence of both young oils and the older, Cambrian oil fields, such as the Lost Soldier and the Wertz in Wyoming (Ref. 78, p. 26). The many studies try-ing to correlate V/Ni ratios with age have been briefly reviewed by Erdman and Harju,[102] who aptly sum up such studies as contradictory. There appears to be correlation with age or depth only within certain oil fields, and there is probably more correlation with geochemical history.

2.3.4.3b API Gravity of Crude Oils versus Depth. Crudes with lower API gravity are higher in metal content and generally appear to correlate with younger oils and shallow depths. There are, however, many exceptions, such as heavy oils at greater depths in the Greater Oficina (Venezuela), Burma, Apsheron (Russia), and Kettleman Hills (California) fields[68] and lighter shallow oils in the Rocky Mountain region as recorded by Dobbin.[103] In reference to this latter area, Hunt[104] states that environmental deposition has a greater influence on gravity, and Todd[105] believes that changes are caused by contact with artesian ground waters. Murray[71] states that gravities change basinward and may correlate with the higher Cl or CO_3 content of oil-field waters.

2.3.4.3c Basin Horizontal Geometry and Geochemistry. Oil fields are found worldwide, and most are associated with marine basins, although some are believed to have been derived from nonmarine basins (Refs. 106 and 107; Section 2.3.4.3d). They are found most often in sandstones, 59% of the world's total known oil according to Knebel and Rodriguez-Eraso (Ref. 89, p. 313), and 40% are found in carbonate reservoirs.

Bonham,[108] in a study of 66 crude oils mostly from Oklahoma, found higher metal contents toward basin edges. Higher salinity and sulfate content of associated oil-field waters occurs toward the center of basins.[109] Also see Murray (Ref. 71; Section 2.3.4.3b) and Weeks.[7]

2.3.4.3d Carbon, Sulfur, and Oxygen Isotopes. Ratios of C^{13}/C^{12} and S^{34}/S^{32} have been determined in crude oil and sediment studies to furnish clues as to their origin and mode of formation. Lower ratios of C^{13}/C^{12} are found for marine origins and higher for nonmarine. Eckelmann *et al.*[110] determined C-isotope abundance for 128 crude oils, and the preponderance of oils showed enrichment compared to organic extracts from ocean sediments. Silverman and Epstein[111] also studied C isotopes and discussed the striking enrichments in the Uinta Basin attributed to nonmarine origins.

Thode *et al.*[112] and Harrison and Thode[107] studied S-isotope ratios for many different oil fields in Canada and the United States. They discussed the extremes found and also cited the enrichments found in the Uinta Basin as confirmation of a nonmarine origin and the action of bacterial reduction of seawater sulfate to produce sulfide enriched in ^{34}S.

Oxygen isotopes have not been studied as much, but according to Epstein and Mayeda,[113] fresh water is depleted in ^{18}O.

2.3.4.3e Nitrogen and Sulfur Compounds and Their Relation to Vanadium and Nickel. In discussing the nature of Ni and V compounds in crude oil, Sugihara *et al.*[114] pointed out a possible correlation of V with crudes high in S and Ni with crudes high in N content. An examination of the literature shows that there are many data (summarized in Section 2.9.3) to support these observations. Since this is closely related to metal content, further discussion will be reserved for Section 2.9.3.

Because of this trend, additional clues as to the origin and nature of metal compounds might possibly be obtained from studies of metal concentrations in various high-boiling N and S compounds. Further knowledge of N compounds and the concentrations in crudes can be obtained from studies of the API and U.S. Bureau of Mines (Ref. 115, pp. 27 and 33). The same sources can furnish similar data for S (Ref. 115, pp. 33, 17). More recently, additional data for S have been published for world-wide export streams(19) and U.S. crude oils.(116)

2.4 HISTORY OF ATOMIC SPECTROSCOPY APPLICATIONS

Applications of emission spectroscopic techniques in the American petroleum industry (P.I.) began in about 1945, near the close of World War II. Since such techniques are used primarily for determining the inorganic elements in oil (minor constituents), they had not received the attention given to the organic composition of oil. Through the years and with the diversification of the P.I. and their affiliates, emission spectroscopy has become an accepted, highly used, and valuable technique.

2.4.1 Emission Spectrographic and Spectrometric Methods

The first application of ES methods for the analysis of metals *directly* in oils was by Calkins and White in 1946.(117) ES methods were not used much in the P.I. until a sudden surge of interest in 1947–1950. This slow development is rather surprising, since Thomas in 1924 summarized the metal contents in petroleum ash(118) Shirey in 1931 gave some qualitative ES data for seven American crude oils(119); Triebs in 1934 discovered the presence of Ni and V porphyrin compounds in crudes(120); and Thomas in 1938 summarized data for the metal contents of crude oils and cited spectrographic analysis as a useful technique.(121) Chemical and colorimetric methods were used at first and were studied by the American Society for Testing and Materials (ASTM) and others, 1942–1949.(122,123) To this day chemical methods are preferred in some laboratories. In 1948 Russell described emission spectroscopy in an oil laboratory.(124) By 1949, proceedings of the American Petroleum Institute (API) included three papers on spectrographic methods.(125–127) One of these papers, by Hughes *et al.*,(127) presented a general review of rapid analyses by means of emission spectroscopy. They used an ashing procedure and cited three years of use. This included the qualitative and quantitative analysis of 1674 samples in 1946, and 3400 samples in 1948, of which about 48% were used oils.

Such use of ES methods is typical of the high value they are to the P.I. and also is indicative of the large use for wear metal analysis in used lubricating oils. This latter application has received more attention than any other use of ES in the P.I. and is rivaled only by studies of the metal contents and type of metal compounds in crude oils (Section 2.9.3). From about 1953 to 1956, the ASTM conducted an extensive cooperative program for application to diesel locomotive lubricating oils.[128] The analysis of wear metals gradually became used more and more by many industries and also the Army and Navy in preventive maintenance programs. Such programs can be applied to any type of engine using lubricating oils and have resulted in large monetary savings and the prevention of possible deaths through impending aircraft failures (Section 2.9.4).

In the 1950s the use of direct-reading (ESD) methods using multichannel spectrometers began replacing ES methods using photographic recording of spectra. This resulted in great savings of time and money and increased precision. The use of ESD is now an accepted, valuable technique in the P.I. and is applied to many types of samples.

2.4.2 Flame Spectrometric Methods

At about the time ES applications were increasing in 1950, the first P.I. applications of FE were published and included determinations of Ba, Ca, Li, and Sr by Conrad and Johnson,[129] and Pb in gasoline by Gilbert.[130] The excellent stability of flame sources, compared to the arc and spark sources used for ES, and the high sensitivity for alkali and alkaline earth elements made FE a natural for a number of P.I. applications. Despite considerable use of FE, publication of P.I. applications was limited, although some cooperative studies were conducted in the 1950s by the ASTM. This relatively late start and slow development of applications is rather surprising, since Lundegardh,[131] the father of modern flame spectroscopy, had developed FE to a remarkable degree in 1929 for agricultural applications. As instrumentation and techniques became better, more applications developed. Now FE is a valuable method for use in the P.I., and with the development of better flame sources and the use of modulation techniques (Section 2.6.3), it now rivals the newer technique of AA.

2.4.3 Atomic Absorption Methods

Although AA was introduced as an analytical technique in 1955, the first P.I. applications were not published until 1961 for determining Pb in gasoline,[132] and 1962 and 1963 for metals in oil.[133,134] As with FE, although no publications appeared, AA had seen considerable use in the P.I. in the late 1950s.

Instrumentation and techniques developed very rapidly for AA, and Walsh shows a fantastic increase in instrument sales after 1964.[135] This coupled with some excellent advantages provided by AA (especially the excellent sensitivity for certain elements and a possible greater freedom from spectral interferences) led to a large boom in AA applications. One of the odd phases of this development was the fact that early proponents of AA made highly exaggerated claims for performance, even to stating that the technique was almost free of interferences. Although this may be nearly true for elements which are outstandingly better by AA (such as Zn), most workers agree today that many interferences are possible for AA as well as FE and ES methods (Section 2.7).

Today, AA rivals the use of ESD for analyzing wear metals in used oil and has many other applications. With the development of better flame sources and more recently nonflame methods to increase sensitivity, the scope of AA applications should become wider yet.

2.4.4 Atomic Fluorescence Methods

Atomic fluorescence spectroscopy (AF) is a more recent technique which may be considered to be in its infancy. It has seen little use in the P.I. but may develop more applications as instrumentation and techniques advance. Cresser and Keliher[136] in 1970 reviewed AF as a practical analytical technique and in their conclusions state: "At the present time the method has only been proven superior to AAS and atomic emission spectroscopy for a few elements, and its role at the present stage must be regarded as complementary rather than competitive." Sychra *et al.*[137] applied AF to the determination of Ni in petroleum fractions as low as 0.01 ppm, finding it to be more sensitive than AA.

2.4.5 Current Use of All Techniques

Well-equipped modern emission spectroscopic laboratories use ES for qualitative determinations and ESD for applications where spark or even plasma jet excitation sources are more suitable and/or multichannel measurements are faster and more economical. Well-designed combination instrumentation for AA and FE (perhaps also AF) are used for specialized and routine determinations where accurate and sensitive determinations are required.

2.5 REVIEW OF LITERATURE

2.5.1 Reviews

In the Annual Reviews section of *Analytical Chemistry* a good review of petroleum is published in the odd-numbered years[138] (the even years include reviews on ES and FE, including AA and AF). In 1971 the Society for Analytical

Chemistry in England published its first *Annual Reports on Analytical Atomic Spectroscopy*,[139] which includes a section on petroleum. Other reviews and bibliographies include a survey of literature from 1946 to 1960 on spectrographic techniques for determining metals in lubricating oils by Il'ina,[140] a review by Buell[141] on FE applications in the P.I. through 1959, and a review on uses of ES in the P.I. by Noar.[142] Bibliographies on AA, FE, and AF are published yearly since 1962 by the Perkin-Elmer Corporation,[143] including indexing of entries under petroleum since 1963. A bibliography covering the period 1800 to 1966 has been prepared by Mavrodineanu.[144] Chronologically, other reviews applying to the P.I. vary from one on rapid analyses by means of emission spectroscopy by Hughes *et al.* in 1949[145] to one on atomic absorption in the petroleum industry by Barras and Smith in 1967.[146] Reviews applied specifically to lubricating oils are included in Section 2.9.4.

Books include those by Kyuregyan,[147,148] McCoy,[149] and Milner,[150] with the latter two containing much on chemical methods and very little on ES. Aleksandrov *et al.*[151] prepared a compilation of methods used in the P.I. with sections on ES and FE methods. Various other books, including those on AA and FE, contain sections devoted to P.I. applications. A book by Pinta, *Detection and Determination of Trace Elements*, contains an extensive bibliography on ES applications.[152] The book is a recent translation of a book originally published in French in 1962.

2.5.2 Growth of Literature

The number of literature references on ES application in the P.I. was reviewed by Gamble in 1951.[153] He cites only four references from 1920 through 1945, and 12 references from 1945 through 1950. The 1950 petroleum review in *Analytical Chemistry*[138] included 10 references. The 1971 review[138] included 39 references, of which less than half were in English. In the 1966 ES review in *Analytical Chemistry*, Margoshes and Scribner[154] reviewed the growth of all ES literature. They reviewed 8000 references for the 2-year period 1964–1965 and cite a growth rate of twofold about every 5 years. On this basis the ES application in the P.I. have not kept pace.

2.6 INSTRUMENTAL TECHNIQUES

Because of the many good books and chapters in books on emission spectroscopic instrumentation and techniques (see Section 2.6.8), only a brief treatment will be given here. New advances, relative merits, sensitivities, and detection limits will be presented. Since interferences and methods of eliminating and compensating for them are so complex and important to accuracy, they will be discussed separately in Section 2.7.

2.6.1 Emission Spectrographs

An emission spectrograph (ES) consists of an excitation source, instrumentation to separate light rays emanating from excited samples (in this modern age usually a grating monochromator equipped with a narrow entrance slit), and a means of recording the resulting spectrum photographically.

2.6.1.1 Advantages

The spectrograph has the advantage of being able to simultaneously record spectra for all elements present in a given sample and provides a permanent, photographic record. It is unsurpassed for the qualitative analysis of metallic elements and some nonmetallic elements and is also excellent for semiquantitative determinations. It also has the advantage of making determinations directly on solid or powder materials. The spectrograph can also be used for quantitative work but generally does not provide precision as good as the faster ESD methods.

2.6.1.2 Excitation and Analysis Techniques

A low-voltage dc arc or high-voltage spark is generally used for spectrographic determinations, although other excitation sources, such as plasma jets or flames, can be used. For qualitative and semiquantitative determinations, a shallow graphite cup electrode is most often used. The cup electrode is packed with powdered sample, and the sample is excited with a dc arc. Because the dc arc fluctuates widely, a given amount of an element not present in samples is generally added as an internal standard. Even with this technique the dc arc usually does not provide precision as good as spark methods. Precision is generally in the range 5-20%, although Mitchell(155) shows that better precision can be obtained with a dc arc and cited relative percent standard deviations of 1.35, 2.03, 3.49, 7.88, and 5.47 for Co, V, Mo, Sn, and Zn. Various other ES techniques are given under appropriate sections.

2.6.2 Emission Spectrometers (ESD)

2.6.2.1 Instrumentation

By simply replacing the photographic recording with narrow exit slits and photodetectors, a spectrograph becomes a spectrometer. Spectrometers may also be designed so that a large number of exit channels can be used for analyzing a wide variety of elements simultaneously. Thus we have the modern, direct-reading, multichannel emission spectrometer. Readout systems may vary and can even be connected directly to computerized calibration and calculating systems. Some new types of instrumentation that may offer some advantages include an echelle grating spectrometer(156,157) and an optical multichannel analyzer.[158,159] The former provides greater dispersion (to 0.4 Å/mm) and is provided with a special plasma jet excitation source. The latter eliminates the need for exit slits and individual detectors through the use of a TV-type detector but may be limited in range. Time-resolved spectroscopy, reviewed by Laqua and Hagenah,[160] is a technique likely to improve detection limits but does not yet seem to have been utilized much for practical determinations. Repetitive wavelength scanning used so effectively for FE and plasma torches (Section 2.6.3.2) has seen little, if any use, but should be quite effective.

2.6.2.2 Advantages and Disadvantages

The greatest advantage provided by multichannel spectrometers is the precise, simultaneous determination of many elements. Faster determinations and great savings in time and money that result, especially when using computerized calculations, are quite significant. ESD methods are considered to be more precise than spectrographic methods and provide good sensitivity, which is better for certain elements, such as B, Gd, Lu, Pr, Si, and Ta, than AA or FE methods provide. ESD methods can also be applied directly to solids and has especially helped the metal and alloy industries.

2.6.2.3 Excitation and Analysis Techniques

ESD methods most often use high-voltage spark excitation and various types of electrodes for the analysis of liquid samples. Over types of excitation sources may also be used, such as plasma jets and even flames. The use of such sources leads to a question of nomenclature definition. At what point does one make a distinction between ESD and FE methods? The use of a flame source could be classified as FE no matter what spectrometer or spectrograph is used. But how about a plasma torch, which is not a flame but is flamelike? The segregation of FE as a separate technique is involved in past history but perhaps should be called ESD using a flame source. Thus, we would also have ESD using arc, spark, plasma torch, and enclosed discharges (hollow cathodes, rf furnace as used in Ref. 208, etc.).

2.6.2.3a Spark Excitation. Using conventional spark excitation, metal samples can be excited directly, but inorganic powdered samples (which can be excited directly with a dc arc at some sacrifice to precision) require special handling, such as mixing with graphite and pressing into pellets or electrode shapes. For analyzing liquids, various techniques and types of electrodes can be used (Sections 2.9.1 and 2.9.4). The most popular type of electrode appears to be a disc electrode (Rotrode) rotating directly in the sample. The author finds that a modified vacuum cup electrode[161] gives better precision and is easier to use for aqueous solutions. Others[162,163] prefer the porous cup electrode and find it easier to use than the Rotrode. All these techniques produce large signal fluctuations; thus, internal standards and time integration are used.

Ultimately, the more stable plasma and/or flame excitations and better techniques of sampling (ultrasonic nebulization, etc.) may provide better methods.

2.6.2.3b Flame Excitation. Buell[164] and Fassel and Golightly[165] have shown that with modern spectrometers flame excitation provides better detection limits than spark excitation for many elements. The technique can be readily adapted to ESD but requires a high-speed spectrometer. Detection limits for FE are improved by a factor of 3- to 30-fold by using ac FE techniques (see Section 2.6.3.2) as compared to the best previously reported which makes the technique highly competitive with AA (see Section 2.6.7.9 and Table 2.4).

2.6.2.3c Plasma Excitation. Plasma excitation sources provide some of the best detection limits reported for emission methods (Section 2.6.7 and Table 2.5). These sources include inductively coupled plasma (ICP), microwave excited plasma (MP), and dc plasma arc (PA). These sources are not as simple to use as flames, and some problems are involved. Sample introduction methods have improved, but limitations are imposed by their background radiation. See Sections 2.6.3.1 and 2.7.1.1 and Refs. 166 and 167 for discussions of problems and solutions to stray-light limitations. Skogerboe and coworkers[166,167] used repetitive wavelength scanning in the derivative mode (Section 2.6.3.2) to minimize this problem and obtained excellent results with MP. Skogerboe and Coleman[168,169] reviewed MP sources and applications to multielement spectrometric analyses. Spectrometer–MP systems have also been described by the EPA,[170] Wooten,[171] and Greenfield and Smith.[172] Layman and Hieftje[173] developed a novel microarc atomizer coupled to a MP that provided excellent detection limits and may be of considerable merit.

Fassel and Knisely[174–176] reviewed ICP sources and with their coworkers have done much toward their analytical development. Multielement ICP spectrometers are now commercially available from Applied Research Laboratories and Jarrell-Ash. Schleicher and Barnes[177] developed a remote coupling unit for ICP which provides greater flexibility in positioning this source. Freedom from interferences for plasmas, cited in Ref. 178, may prove to be an outstanding advantage for such sources.

Skogerboe *et al.*[179] reviewed PA sources and described the use of a commercially available unit (Spectrometrics) with a multielement spectrometer equipped with simultaneous background correction. Detection limits were obtained that were comparable to AA–flame methods. Rippetoe and Vickers[180] modified a PA which utilized excess KCl for improved results and provided lower background, which was better in certain areas than a nitrous oxide–acetylene flame.

2.6.3 Flame Emission Spectrometers (FE)

2.6.3.1 Instrumentation

Currently most instrumentation for FE spectroscopy is designed for AA and includes the ability to use both AA and FE. Since these are companion techniques with each providing some advantages, good instrumentation should be designed for optimum use of both techniques. To provide the best detection limits, a spectrometer for FE should have high dispersion, high speed, and good quality of design and optics to minimize stray light. Stray light from bright flames such as nitrous oxide–acetylene, plasma torches, and excesses of sample matrix elements can be a major factor influencing detection limits (Section 2.7.1.1 and Ref. 167). Unfortunately, many AA-FE spectrometers in use have poor characteristics for FE. Their wide use has been detrimental to the full exploitation of FE and has led to the false impressions that AA is much better than FE. In the author's laboratory a Perkin-Elmer Model 306 was found to have up to 1000-fold greater background emission caused by excess Ca (2000 ppm) and nitrous oxide–acetylene flame bands than the spectrometer described in Section 2.6.3.2.

In addition to the spectrometer, a burner–nebulizer system is required for sample excitation. A given burner system can be used for both AA and FE, and their design is of major importance in providing the best detection limits (Section 2.6.6).

For the detector a high-grade multiplier phototube (with high efficiency and low noise) should be used to provide the best detection limits. The output of the detector is connected to a meter, a recorder, or a digital readout system preferably equipped with variable time constants and signal averaging.

2.6.3.2 Repetitive Wavelength Scanning in the Derivative Mode

One of the most important additions to FE techniques in recent years is the use of modulated systems which for simplification will from hereon be referred to as ac FE (as opposed to conventional FE systems providing dc signals). In the author's opinion this is a major breakthrough and is the reason for devoting this section to the subject. The system developed by Snelleman *et al.*[181] has been tested in the author's laboratory with great success and is the only one that will be described here.[182] The same system has been used by Lichte and Skogerboe[167] and produced excellent results when combined with a microwave-induced plasma.

The ac FE system used by the author consists of a quartz refractor plate placed immediately inside the exit slit of a Jarrell-Ash 0.5-m monochromator, which is electromagnetically driven to scan about a 0.5-Å wavelength interval at 172 Hz. A PAR lock-in amplifier is tuned to the same frequency and signals are observed in the first derivative mode. This system does not detect signals from continuum or broad bands such as that of CaO but only signals from sharp peaks such as atomic lines. Thus, such interferences are minimized, line to background is improved, and detection limits for many elements are 3- to 30-fold better than those reported by Christian and Feldman[183,184] or Pickett and Koirtyohann[185] using similar instrumentation in the conventional dc mode. Many of the limits given in Table 2.2 were obtained with this instrument. Outstanding FE detection limits are 0.7 ppb for Cr, 2 ppb for V, and 3 ppb for Mo.

2.6.4 Atomic Absorption (AA)

2.6.4.1 Instrumentation

Requirements for AA are somewhat different than for atomic emission. Although identical spectrometers and nebulizer–burner systems can be used for both FE and AA, only ground-state atoms are needed for AA. Nonflame reservoirs may also be used for AA to provide better sensitivity (Section 2.6.4.2). Since AA is based on adsorbance of radiation by ground-state atoms, a source of radiation is required. Although a continuum light source can be used, a monochromatic light source such as a hollow cathode lamp provides better sensitivity and is normally used. Dual-light-beam systems, although they provide less light throughput, may offer some advantages for AA in compensating for lamp fluctuations. Such arrangements can also utilize a continuum light source such as a xenon discharge or hydrogen lamp to provide automatic correction for molecular absorbance interferences which readily occur for AA. Flexible dual-channel instruments can also provide certain advantages. Combinations of two channels for simultaneous AA and/or FE determinations, compensation for molecular absorbance, and the use of internal standard methods is possible. Although multichannel AA–FE spectrometers are more expensive and more difficult to design than ESD instruments, the future should see more use of them. Such a spectrometer[186–189] should provide a wider scope of applications and improved analyses compared to spark techniques using ESD, especially if combinations with FE and possibly plasma torches are used, such as the instrument developed by Mavrodineanu and Hughes.[189]

2.6.4.2 Nonflame and Boat in Flame Methods

In more recent years a significant advance in AA has been the development of nonflame methods to provide better detection limits (up to 50-fold better compared to conventional flame AA). Much research has also been devoted to increasing the atomization (the production of free atoms) in flames as compared to spraying liquids into flames. The techniques involved include the boat, cup, and rod in flame approaches, graphite furnace, electrically heated carbon rods, and metal filaments. Syty[190] thoroughly reviewed these techniques, citing 360 references. Other reviews include those by Amos,[191] Kirkbright,[192] and Robinson and Slevin.[193]

In using these approaches one should be aware that there exist a greater susceptibility to interferences and less precision as compared to flames, which may severely limit their applications. Cruz and van Loon[194] made a critical study of graphite furnace applications in complex heavy-matrix sample solutions and found limitations for inorganic mixtures. Regan and Warren,[195] however, found that adding water-soluble organic compound (ascorbic acid) reduced these matrix effects markedly. Corrections for background absorption should also be made.[196] Alger *et al.*[197] studied matrix errors using a carbon rod, and Maessen and Posma[198] studied some fundamental aspects, including the separation of analyte and matrix peaks. Robinson and Wolcott[199] claimed advantages for a hollow-T furnace, including less interferences. Montaser and Crouch[200] claim similar advantages for a graphite braid atomizer. Lagesson[201] devised a simple, economic electrically heated tantalum boat system for which interferences were claimed to be lower. Ward *et al.*[202] found that a nitrous oxide–acetylene flame with a molybdenum boat eliminated interferences. Earlier studies of interferences are reviewed in Ref. 180.

New developments include continuous sample introduction into graphite furnaces for both AA[203] and AF[204] and the use of pyrolytic graphite tubes[205] or coatings.[206] Talmi[207] improved the original rf furnace used by Morrison and Talmi[208] and used both absorption and emission modes of operation. McCullogh and Vickers[209] developed an electrically heated tantalum filament atomizer which connected to a conventional nebulizer–burner. In addition to those in Ref. 190, more recent references discussing injection of solids into a flame include an iron-screw-rod technique[210] and the use of a cavity in a rod for molecular emission cavity analysis (MECA) for As, B, P, Sb, Si, and metal halides.[211–214] Skogerboe *et al.*[215] used a chloride generator–furnace attached to a burner or plasma torch to obtain improved results, eliminating solid matrix particles.

2.6.5 Atomic Fluorescence (AF)

Atomic fluorescence is a newer technique that has not seen much use as yet. The technique was developed by Winefordner and Elser, who also have presented a good review.[216] Bailey[217] and Ellis and Demers[218] have also reviewed the use of AF. According to these reviews, Alkemade first suggested the use of AF as an analytical technique in 1962.

The instrumentation is that used for AA. The technique is based on AF radiation of samples sprayed into flames, which is caused by irridation with an intense light source. Hollow cathode lamps used for AA are generally not bright enough, whereas microwave-excited electrodeless discharge lamps appear most suitable. Instrumentation using nondispersive systems can also be used with good results and have been reviewed by Larkins and Willis[219] and Elser and Winefordner.[220] Multichannel instruments have been reviewed by Norris and West,[221] who propose wavelength scanning for sets of four elements as being easier.

Requirements of burners are slightly different because radiational quenching processes should be minimized (Ref. 185, p. 29A). Flames using nonhydrocarbons, such as hydrogen diffusion flames are quite effective.

There appears to be considerable variation in detection limits reported for AF. Barnett and Kahn[222] compared AF to AA and claimed the technique to be less sensitive and subject to more interferences. Winefordner and Elser,[216] however, report excellent detection limits (those given in Table 2.2), which incidentally appear to be 20-fold better than previously reported in many cases and up to 50-fold better.

2.6.6 Burner-Nebulizer Systems

The types of burners used for AA and FE can be divided into two general classes: (1) total consumption-turbulent flow, and (2) partial consumption-laminar flow burners. The latter were designed with long, narrow flame slots to increase the absorbing path length, thereby increasing sensitivity for AA. Total consumption burners were designed for FE use prior to the development of AA.

2.6.6.1 History

With the advent of commercial flame spectrometers after 1948, integral burners were developed to spray liquid samples directly into the flame. The Beckman, Ditric Corp. or HETCO, Weishelbaum-Varney, and Zeiss burners are of this type. The main disadvantage of the total consumption burner is its irritating noise, especially when spraying organic solvents. (They also give less sensitivity for AA compared to long slot burners.) They were tolerated for most FE determinations until the late 1960s, when laminar flow, nitrous oxide-acetylene burners were developed, and Pickett and Koirtyohann[185,223] showed that their configuration was suited for FE and that they gave superior FE detection limits for many elements. The author found this flame to provide even better detection limits using ac FE (Section 2.6.3.2). This appears to be the best flame developed to date for most FE applications, but improvements to produce better stability (less fluctuations) should be possible.

2.6.6.2 Atomic Absorption and Partial Consumption—Laminar Flow Burners

After the development of AA in 1955, much research was done to provide better burners for AA. To provide increased sensitivity, burner heads with long, narrow slots were required. The burner head served as a mixing chamber for the flame gases, thus providing a premixed, laminar flow, quiet flame. To introduce sample into this flame a nebulizing chamber is attached to the burner head and only the finer spray from the sample (about 10%) reaches the burner.

The best earlier burners (used today for nonrefractory elements) had 10-cm-long slots and used acetylene-air flames. Unfortunately, these burners could not use oxygen or flashbacks (explosions) would occur, and, therefore, these burners were poor for refractory elements. In 1966, Amos and Willis[224] developed a burner for using nitrous oxide-acetylene flames. This provided a hotter flame with special reducing properties and was a major advance for determining refractory elements by AA and FE. A shielded, nitrous oxide-acetylene flame had been used as early as 1961[225] but was not developed for analytical applications. Shortly after this, Fiorino *et al.*[226] developed a versatile long-path slot burner for nitrous oxide–acetylene and oxyacetylene flames. With their special design, results were about the same with both flames and the burner was capable of sustaining a nitrous oxide–acetylene flame 7.6 cm long. Despite the good features of this burner, it does not appear to have been commercially exploited.

Commercial nitrous oxide–acetylene burners have a narrow 5-cm slot in order to avoid flashback. To minimize carbon buildup on the slot (a disadvantage for such flames), the best burners are made of titanium with grooved or ridged slots.[227] They also are equipped with an additional port on the nebulizer for auxiliary oxidant, which can help reduce unwanted luminosity when burning organic solvents (Section 2.6.6.4).

There has been much additional research on burner systems, some of which is summarized below. Beginning in 1967, Kirkbright, West, and coworkers published a series of papers on the use of separated flames, claiming some signal-to-noise advantages[228–235] and reviewed applications of separated flames.[236] Amos *et al.*[237] also studied shielded (separated) flames. Burners with heated spray chambers have also been developed by Hell *et al.*,[238] Venghiattis,[239] and Ury *et al.*[240] that are claimed to give up to 30-fold better AA response. Aldous *et al.*[241] discussed applications of capillary burners, Wang[242] developed a simple, multiport burner designed to handle high solids as easy as a Boling head,[243] Denton and Malmstadt[244] discussed ultrasonic nebulizer improvements and developed a burner using the low gas flow rates needed for such systems.

2.6.6.3 Flame Emission and Total Consumption Burners

Total consumption burners, compared to partial consumption burners used for AA, have the advantages of (1) greater freedom from contamination by memory effects from previous samples, (2) less fluctuations, (3) better detection limits for certain applications, (4) smaller sample consumption, (5) use of oxygen-supported flames, and (6) versatility for use with organic solvents (Section 2.6.6.4).

Contrary to many statements given in the literature (seldom backed by data) that these turbulent flow flames contribute greater fluctuations, there is evidence that they contribute less fluctuations. Perrin and Ferguson[245] compared different burners and flames for the AA determination of six elements in animal feed samples. They found that a Zeiss total consumption burner using oxygen–acetylene gave the best pooled precision for sample determinations. Using pure standards, they obtained coefficient of variations from 0.45 to 1.80 for six elements with this burner. See Section 2.9.5.1.2 for better AA results for Pb in gasoline via the air–hydrogen flame and a total consumption burner.

In the author's laboratory, comparisons were made using an oxygen–hydrogen total consumption burner (burner A) and a 5-cm-slot laminar flow burner using nitrous oxide–acetylene (burner B). For nickel at 3525 Å using ac FE, a 30-fold greater amplification was required to give blank fluctuations for burner A equivalent to those obtained with burner B. For Ni in an organic solvent, burner B provided about a 20-fold greater intensity, and detection limits were about the same. For antimony at 2598 Å, burner A gave a 3-fold greater intensity and a 5-fold better detection limit (0.2 ppm). Flame A also gave a better detection limit for the AA determination of Cd. This indicates that one type of flame will not provide the best result for all elements and also that the optimum burner for FE has yet to be developed.

2.6.6.4 Organic Solvent Nebulization

Many of the samples encountered in the P.I. may be organic, varying from light hydrocarbons such as gasoline to tarlike samples. Direct nebulization of solvent solutions and the problems involved thus become of extreme importance. Although spraying organic solutions into flames can increase signals for both AA and FE, it can also lead to serious errors. For FE a background radiation may occur, but usually this can be corrected. For AA a positive error caused by molecular absorption can occur and special correction methods must be used (corrections through the use of continuum correctors or nonabsorbing lines).

Concentration by extraction into organic solvents is a valuable technique for AA and FE (Section 2.9.2).

Because of the importance of solvent response in these applications some of the important variables involved will be reviewed here. The magnitude of errors and usefulness of a specific solvent will vary widely according to the burner system used and whether AA or FE is used.

2.6.6.4a Total Consumption Burners. Almost any type of solvent (including gasoline) can be sprayed into a total consumption burner without causing excessive luminosity in the outer flame brush such as occurs for AA. Changes that occur when spraying hydrocarbon solvents into flames are most pronounced for an oxyhydrogen flame. The use of organic solvents for FE in this flame was reviewed by Buell[246,247] and appears to provide overexcitation.

2.6.6.4b Reduction of Matrix Errors through the Use of Hydrocarbons. In addition to enhancing emission the use of hydrocarbon solvents also eliminates many matrix interferences. Some matrix interferences occur in solvents containing oxygen in their molecules and the matrix errors become larger as mixed solvent–water systems approach 100% water. As solvent systems approach 100% hydrocarbons, interelement (condensed phase) interferences disappear. The effect is discussed further and data are presented in Section 2.7.3.1.

2.6.6.4c Atomic Absorption and Laminar Flow Burners. The use of organic solvents to increase AA sensitivity is reviewed and discussed by Allan[248] and by Chakrabarti and Singhal.[249] The latter authors also studied enhancing effects and the use of various chelating agents, and observed the largest enhancing effects for refractory metals such as Al using a total consumption burner.

The response of laminar flow slot burners to organic solvents is quite different from that of total consumption burners. One of the major disadvantages for slot burners using air–acetylene is that many solvents (especially lighter hydrocarbons) produce a whitish, bright luminosity throughout the flame. This causes excessive molecular absorbance and/or scattered light interference for AA.

The absorption for normal flames (with no whitish luminosity) becomes progressively larger below 2700 Å. Abnormal luminosity, whether caused by organic solvents or increasing the fuel of the flame, merely increases this absorption. Whitish luminosity has far less influence on FE than on AA. On the other hand, emitting C and N species for wavelengths greater than 3500 Å that influence FE do not cause AA interference. Characteristics of flames for AA and FE are summarized in Table 2.1A. The nitrous oxide-acetylene (N–A) flame has peculiar properties with fuel adjustment. Absorption below 2500 Å and the OH-band emission (2800–3300 Å) increase greatly upon reducing fuel to form a very lean flame (no pink secondary cone is formed, even with isopropanol). A rich N–A flame (high pink cone, 40 mm), however, has the least absorbance of any flame below 2200 Å. The C and N emitting species above 3500 Å are maximum in the rich N–A flame and interfere some with refractory elements for FE. There are no problems caused by using organic solvents that cannot be rectified by reducing fuel flow. Interferences caused by solvents (differences in absorption or emission) are caused primarily by changes in the secondary, pink cone height. The differences are similar whether caused by organic solvents or changes in fuel flow. The best flame for determining nonrefractory elements in most organic solvents is nitrous oxide-hydrogen (N–H) in a 10-cm single-slot burner. Almost any solvent can be used in the N–H flame with little or no change.

Some further observations concerning organic solvents useful for AA and FE as applied to petroleum samples may be worthwhile. Much of this author's experience prior to AA concerned FE and a technical grade of cleaner's naphtha or AMSCO lactol spirits (boiling range 218–222°F). This solvent had good burning characteristics, gave large enhancements in total consumption burners, and was a good solvent for petroleum samples. When used in the laminar flow–slot burners developed for AA, this type of solvent was found to be unsuitable because it contributed excessive luninosity in air–acetylene flames. For use in slot burners the best solvents are cyclohexanone (Cx) (as recommended by Pye-Unicam) and special naphtholite [(SN) boiling range 240–290°F] supplied by American Mineral Spirits Co. (AMSCO). The best compromise is to use a mixture of cyclohexanone and special naphtholite. This type of mixture is the only one, to the author's knowledge, in which shale oil and heavy crude oil residuum samples will be completely soluble and stable. Mixtures up to 32% special naphtholite (with cyclohexanone or other solvents or samples used in the mixture)

Table 2.1A Flame Characteristics

<table>
<tr><th rowspan="2">Flame</th><th colspan="2">Characteristics</th></tr>
<tr><th>AA</th><th>FE</th></tr>
<tr><td rowspan="2">Nitrous oxide–acetylene</td><td>Best for refractory elements and arsenic. Lean flames cause increased OH absorbance, and absorbance below 2400 Å becomes progressively larger.</td><td>Best for most elements. Lean flame causes OH[a] emission. Rich flame causes C- and N-band emission. Slight luminosity occurs for heavy petroleum.</td></tr>
<tr><td colspan="2">Organic solvents cause no real problems except some flame and pink-cone expansion which must be adjusted by reducing fuel flow. Gasoline and light solvents may create a very small pink cone with the leanest flame possible.</td></tr>
<tr><td>Air–acetylene</td><td>Best for nonrefractory elements. Organic solvents cause interference, which is especially bad for gasoline and light solvents. Reducing nebulizing rates reduces interference but also decreases sensitivity and may increase fluctuations.</td><td>Satisfactory for Na and K.</td></tr>
<tr><td>Nitrous oxide–hydrogen</td><td>Good for some nonrefractory elements. Absorbance increases below 2300 Å becoming large below 2100 Å. Excellent for all organic solvents. Best for nonrefractory elements in solvents with 10–cm single slot burner.</td><td>Poor except for B via BO_2 emission and alkalies. Excellent for organic solvents, especially for heavy petroleum samples. Good for Na in residuum oils.</td></tr>
</table>

[a]Lean flames may cause second-order overlap of OH emission, causing severe interference over 5000 Å with certain spectrometers (30% of signal with Perkin-Elmer 306 at 5000 Å and 50–60% at 5300 Å for lean flame background).

Table 2.1B Organic Solvent Characteristics

Solvent	Petroleum solubility[a]		Burning qualities[a]		
	Gasolines and lubricating oils	Crude oil residuum and shale oil	Air–A	N–A	N–H
Cyclohexanone	1	2	1	1	1
64% Cyclohexanone–32% special naphtholite	1	1	2	1	1
Special naphtholite	1	3 (some insolubles)	4	1	1
Xylene	1	3	4 (streaked flame)	1	1
Toluene	1	3	4	1	1
Gasoline and light hydrocarbons	2	4	4	2	1
Methyl isobutyl ketone	3	4	4	1	1
Ethylpropionate	3	4	4	1	1
Isopropanol	4	4	3	2	1
Ethanol	4	4	2	1	1

[a] 1, excellent; 2, good; 3, fair; 4, poor.

can be used in the air–acetylene flame. Thus, no luminosity occurs if a very lean flame with excess oxident and auxiliary oxidant is used, even at full nebulizing rate. The solvents ethylpropionate and methyl isobutyl ketone (MIBK) have good burning characteristics but are poor solvents for petroleum and metaloorganic standards. However, they are excellent for chelate-solvent extractions. Xylene and even toluene have good burning characteristics in N–A flames. With an auxiliary oxidant and a very lean N–A flame, even gasoline causes only a small expansion of the pink cone. Characteristics for various solvents burning in slot burners and as solvents for petroleum are summarized in Table 2.1B as excellent, good, fair, or poor.

2.6.7 Detection Limits for Various Techniques

The best detection limits obtained for various techniques are compiled in Table 2.2. Sensitivities (the concentration giving a signal of 1% absorption) are often used in reference to AA. The range of sensitivities, which include the best reported by Amos and Willis, Instrumentation Laboratories, Jarrell-Ash, and Perkin-Elmer using a nitrous oxide–acetylene flame, are also given in this table.[224,250–252]

Table 2.2 Detection Limits for Atomic Spectroscopy[a]

Element	AA sensitivity (mg/liter)	Detection limit (μg/liter) AA flame	AA nonflame	AF	FE	Plasma torch	Emission Arc	Emission Spark
Ag	0.03–0.08	2	0.04	0.1	2		0.9	2
Al	0.4–1	20	6	100	1.5	0.2	15	10
As	0.3–1	50(1)[b]	20	100	2,000	30	1,800	400
Au	0.1–0.5	10	2	5	500			200
B	12–50	1,500	600		50	10	10	1
Ba	0.2–0.4	8		7,000	0.3	0.02	1.3	100
Be	0.006–0.03	0.8	0.2	10	1,000	0.4	1.2	2
Bi	0.2–0.8	5	1	5	20,000		28	20
Br	–				3,100			
Ca	0.03–0.07	<0.5	0.06	20	0.01	0.02	0.1	200
Cd	0.005–0.04	2	0.02	0.001	2,000	0.4	68	100
Ce	–				10,000	2	130	200
Cl	2–5000				7,000			
Co	0.06–0.15	10	1	5	7	3	26	20
Cr	0.06–0.15	3	1	50	0.7	0.3	30	4
Cs	0.05–0.5	20	4		5			500
Cu	0.03–0.1	1	1	1	10	0.1	0.5	2
Dy	0.7–1.5	50			50			200
Er	0.85	20			40			100
Eu	0.3–1.8	20	20		0.5			20
Fe	0.05–0.15	5	0.6	8	5	0.3	32	10
Ga	1–2.5	50	0.4	10	60	0.6	17	100
Gd	10–38	1,200			4,600			100
Ge	2.5	200		100	400	4	24	180

Hf	15	2,000			18,900	10		500
Hg	2–15	250(0.2)[b]	20	0.2		1	1,200	50
Ho	1.4–2.2	40			20			200
I	8–25,000	200			5,000			
In	0.9	20		100	2		52	40
Ir	9–12	600			380			5,000
K	0.005–0.1	<2	0.2		0.5			100
La	30–100	2,000			6,000	0.4		50
Li	0.03–0.07	0.3	1		0.02	0.3	83	1
Lu	12–15	700			1,000			10
Mg	0.004–0.008	<0.1	0.01	1	5	0.05	0.7	4
Mn	0.01–0.08	2	0.1	6	1	0.06	8	1
Mo	0.3–1	20	8	500	3	0.2	17	50
N	–				14,000	30–700		
Na	0.004–0.04	<0.2	0.02		0.5	0.3	49	100
Nb	19–24	1,000			1,000	10	80	200
Nd	10–35	1,000			200			200
Ni	0.05–0.1	2–10	2	3	6	0.4	20	4
Os	0.5–1	80			1,700			
P	–				10–100	70		30
Pb	0.06–0.5	10	1	10	50	2	29	50
Pd	0.07–0.3	10	40	40	10	2		500
Pr	13–72	5,000			690			200
Pt	0.3–2.5	50	40		Poor			20
Pu	0.03–0.02	5	1		1		1,100	200
Re	12–20	500			30			1,000
Rh	0.1–0.5	4–20		3,000	4			700
Ru	2	70			20			2,000
S	–	10–30,000			100–5,000	20		
Sb	0.03–1	40	6	50	200	200	97	40
Sc	0.3–0.8	20			30			5
Se	0.2–2	50(1.3)[b]			Poor			

(Continued)

Table 2.2 (*Continued*)

Element	AA sensitivity (mg/liter)	Detection limit (μg/liter)						
							Emission	
		AA flame	AA nonflame	AF	FE	Plasma torch	Arc	Spark
Si	1–2.5	20	10	600	3,000		4.3	
Sm	8–21	2,000			230			200
Sn	0.4–3	10	10	50	100	30	15	83
Sr	0.07–0.02	2	1	30	0.03	0.02	5.5	2
Ta	11–30	1,000			4,000	70	970	40
Tb	8	600			30			
Te	0.3–2	50		5	2,000			
Th	–				10,000	3	120	200
Ti	1.1–3.5	40			200	0.2	9.7	10
Tl	0.1–0.8	30	0.6	8	20		320	100
Tm	1	10			20			50
U	120–200	30,000			5,000	30	140	1,000
V	0.7–1.5	40	20	70	2	0.2	5	4
W	5–35	3,000			500	1	73	50
Y	1.1–5	50			40	0.06		5
Yb	0.17–0.25	5			2	0.04		10
Zn	0.01–0.04	<1	0.02	0.02	Poor	0.6	220	40
Zr	15–46	1,000		4,000	3,000	0.4	13	1

[a] Detection limits for atomic-flame techniques are primarily for aqueous solutions. Organic solvents may provide detection limits which are better by 3-fold or more.
[b] Using reduction–aeration methods.

2.6.7.1 Flame AA

Detection limits for flame AA are taken from Ref. 253, with a few from Ref. 250. A word of caution should be added here concerning these detection limits. Some of them are quite optimistic and were obviously obtained by skilled operators using optimum conditions. Some of the limits are outstandingly better than reported earlier from the same laboratory.[252] A practical worker will be skillful if some of his limits are within 2- to 3-fold of the best reported.

2.6.7.2 Nonflame AA

Detection limits for nonflame AA methods are primarily those reported by Varian-Techtron using their new carbon rod.[254] The literature was not searched for better detection limits which may have been reported (see Ref. 255 on the use of a NIM carbon rod for detecting 0.5 ppb of Au compared to 5 ppb with the carbon rod cited). These nonflame methods are relatively new and developing rapidly; therefore, many future improvements can be anticipated.

2.6.7.3 Atomic Fluorescence

AF detection limits are taken primarily from a table compiled by Winefordner and Elser.[216] As mentioned in Section 2.6.5, the detection limits for AF reported in the literature have been quite varied and those given are often 20-fold and up to 50-fold better than previously reported.

2.6.7.4 Flame Emission

Many of the FE limits are taken from Buell[182] using ac FE (Section 2.6.3) which are 3-to 30-fold better than those reported by Christian and Feldman[183] or Pickett and Koirtyohann.[185] Some data are from Boumans and de Boer,[256] who also reported excellent FE detection limits using the nitrous oxide–acetylene flame (three values were used in Table 2.2).

2.6.7.5 Plasma Torch

The limits for the plasma torch are also primarily from[256] and additionally from Dickinson and Fassel[257] and Lichte and Skogerboe.[167] Some of the plasma torch limits are outstanding compared to the other techniques and certainly should indicate more future use of such excitation sources (see Section 2.6.2.3a).

2.6.7.6 Arc and Spark

The limits for ES arc and spark excitation are the best found in the literature and are taken from Kaiser,[258] Fred *et al.*,[259] and Goleb *et al.*[260] using the copper spark technique and Morris and Pink[261] using the graphite spark technique. Calculations assumed that 0.1-ml sample volumes were evaporated on the electrodes. Those values for the dc arc are from Kaiser using an instrument with a resolution of 500,000. Normally, with various lesser quality instruments, detection limits may be poorer by 10- to 100-fold for arcs and on the order of 10-fold poorer for spark excitation. For example, in the author's lab using spark excitation with one instrument gave a detection limit of 40 ppb of V, whereas another provided a limit of about 10 ppb.

2.6.7.7 Detection Limits for Nonmetallic Elements

The nonmetallic elements N and S and the halogens (Cl, Br, and I) are also included in Table 2.2. The literature was not thoroughly searched to obtain the data. The elements C, O, and F are omitted because of the complex chemistry involved but can also be determined by emission techniques. Emission techniques are useful for determining such elements, primarily through molecular emission. AA is not as applicable but can be used for molecular absorption. L'Vov and Khartsyzov, however, used a graphite furnace and the I atomic line at 2061.63 Å to detect 200 ppb.[262] Kirkbright *et al.*,[263] using a separated nitrous oxide-acetylene flame, report detection limits of 8000 to 25,000 ppb of I, depending on the compound and solvent used. Detection limits were somewhat better for volatile alkyl iodides. Morrison and Talmi[208] report a limit of about 5000 ppb of I using an rf furnace in the emission mode, and briefly review other methods for nonmetals.

Herrmann and Gutsche,[264] using a special burner with an indium metal grid, detect 0.014 μg of Cl (about 7000 ppb) via indium chloride at 3599 Å and 0.0062 μg of Br[265] (about 3100 ppb). Tomkins and Frank[266] use a special burner and obtain limits for AA of 2000–5000 ppb of Cl via a CuCl band.

The determination of P and S has received considerable attention recently. Taylor *et al.*[267] reported the best detection limit for S using a plasma torch in the emission mode, 20 ppb, and a limit of 700 ppb using the AA mode. Boumans and de Boer[256] detected 70 ppb of P with a plasma torch. Aldous *et al.* (Ref. 268, p. 421) reported the best limits (10 ppb of P and 100 ppb of S) for FE using a filter photometer and a nitrogen–hydrogen diffusion flame. They used an HPO band near 5262 Å and an S_2 band near 3837 Å. A filter photometer (broader wavelength bandpass) apparently gives better detection limits, although possibly greater interferences. Dagnall *et al.*[269] also reported a limit of 100 ppb of P. Syty and Dean,[270] using a Beckman small quartz monochromator and an oxygen–acetylene burner, only achieved a limit of 6000 ppb of P and 5000 ppb of S but cite the use of an interference filter as a means of improving limits. Kirkbright *et al.*[271] used a nitrous oxide–acetylene flame with inert gas shielding to determine S in oils by AA. They cite a sensitivity of about 2000 ppb of S. Morrison and Talmi[208] report an ESD limit of 1000 ppb of P and S using an rf furnace. For determing P directly in oil via the PO band at 3270.5 Å, a detection limit of 6000 ppb was obtained in the author's laboratory using the ac FE and a nitrous oxide–acetylene flame.

Using the CN band, 14,000 ppb of N has been detected in the inner cone of an O-H flame.[272] Taylor *et al.*[273] detect 30 ppb of NH_3 and 700 ppb of N_2 in argon gas using a plasma torch.

2.6.7.8 Special Techniques to Increase Sensitivity

In order to determine certain elements at low levels by AA, special techniques using gaseous evolution have been developed. Special concentration techniques and long absorption tubes can also be used to increase sensitivity.

2.6.7.8a Cold Vapor Method. One of these elements, Hg, has received a vast amount of attention as a result of environmental aspects and the Hg poison scare. There are numerous references using the cold vapor method for Hg which is capable of detecting 0.2 ng or 0.2 ppb in a 10-ml sample. The Hg in the sample is reduced to the metal and swept into a long path cell by air.

2.6.7.8b Gaseous Evolution. For the determinations of As and Se, reactions to produce arsine and hydrogen selenide gases are utilized. The gases are swept into an Ar–H flame to produce detection limits of about 1 ppb. Fernandez and Manning have reviewed the methods and equipment used.[274,275]

2.6.7.8c Special Concentration Procedures. Concentration procedures (Section 2.9.2) and indirect methods can also be used to achieve better detection limits. These will not be reviewed here but two special examples will be given. Reichel and Acs[276] use an AA measurement of Ag to detect 220 ppb of Cl via AgCl pptn. Hurford and Boltz[277] detected 100 ppb of P and Si via extraction of their heteropoly complexes with Mo and an AA measurement of Mo.

2.6.7.8d Long Absorption Tubes. One last technique might be mentioned that has been used to increase AA detection limits. Fuwa and Vallee[278] are credited with developing the long-absorption-tube technique. Tubes up to 100 cm long can be used for the most volatile elements while tubes longer than 40 cm provided no further sensitivity increases for other elements. Rubeska and Moldan[279] used the technique and Iida and Nagura[280] studied its use recently. The latter authors report sensitivities 5- to 20-fold better than the best using conventional flames, Table 2.2, for elements such as Co, Cu, Fe, Hg, Ni, Pb, and Sn. Since it is more difficult to use than some of the nonflame methods, it may not see much future use.

Table 2.3 Detection Limits Reported by Various Reviewers for Atomic Absorption and Flame Emission Spectroscopy

Reviewer (chronological order)	Comparison of number of elements best by: Atomic absorption	Flame emission	Equal	Total number of elements
Slavin, 1966	27	10	22	59
Fassel and Golightly, 1967	25	10	33	68
Koirtyohann, 1967	17	26	26	69
Buell, 1973	22	30	22	74

2.6.7.9 Comparison of Flame AA and FE Detection Limits

Comparison of AA and FE detection limits was a game played in the literature in past years. At the 1969 Pacific Conference on Chemistry and Spectroscopy, Buell reviewed several previous comparisons of AA and FE detection limits, showing vacillating results (summarized in Table 2.3). Again the author would like to present a current comparison as shown in Table 2.4. For 74 elements, 20 are best by flame AA, 30 are best by FE, and 24 are essentially equal (detection limits are within 3-fold of each other). See Section 2.9.3 for outstanding detection limits for oil analysis using ac FE.

Table 2.4 Comparison of Detection Limits for Flame Atomic Absorption and Emission

Elements best by AA[a]		Equal		Elements best by FE[a]	
As	Pt	Ag	Mn	Al	N
Au	Sb	Co	Na	B	Nd
Be	Se	Cu	Nb	Ba	P
Bi	Si	Dy	Ni	Br	Pr
Cd	Sn	Er	Pb	Ca	Rb
Gd	Ta	Fe	Pd	Ce	Re
Hf	Te	Ga	Rh	Cl	Ru
Hg	Ti	Ge	Sc	Cr	S
Mg	Zn	Ho	Tl	Cs	Sr
Os	Zr	Ir	Tm	Eu	Sm
		La	Y	I	Tb
		Lu	Yb	In	Th
				K	U
				Li	V
				Mo	W

[a]Elements for which detection limits are better by more than 3-fold.

Table 2.5 Elements for Which Plasma Torch Excitation Provides the Best Detection Limits

Best	Best by 40-fold or more
Al	Ce
Ba	Ge
Cr[a]	Hf
Cu	La
Fe	Th
Mn	Ti
Mo	W
N	Y
Nb	Yb
Ni	
Pd	
Zr	

[a]Close to FE detection limit.

Plasma torch excitation provides exceptional detection limits which are best for 21 elements listed in Table 2.5. The second column lists nine elements which are better by 40-fold or more than by any other technique. Nonflame methods also provide exceptional detection limits which are best for 11 elements, and decisively so for six elements.

2.6.8 Additional Readings on Instrumentation Techniques*

Books on Atomic Absorption

Billings, G. K., and E. E. Angino, *Atomic Absorption Spectrometry in Geology*, American Elsevier Publishing Company, Inc., New York (1967).

Christian, G. D., and F. J. Feldman, *Atomic Absorption Spectroscopy: Applications in Agriculture, Biology and Medicine*, Wiley–Interscience, New York (1970).

Dean, J. A., and T. C. Rains, eds., *Flame Emission and Atomic Absorption Spectroscopy*, Marcel Dekker, Inc., New York, Vol. 1 (1969), Vol. 2 (1972), Vol. 3 (1975).

Hoda, K., and T. Hasegawa, *Atomic Absorption Spectroscopic Analysis*, Genshi Kyuko Bunseki Kondanska, Tokyo (1972).

Kirkbright, G. F., and M. Sargent, *Atomic Absorption and Fluorescence Spectroscopy*, Academic Press, Inc., New York (1974).

Pinta, M., *Atomic Absorption Spectrometry*, Adam Hilger Ltd., London (1975).

Price, W. J., *Analytical Atomic Absorption Spectrometry*, Heyden & Son, Spectrum House, London (1972).

Ramirez-Munoz, J., *Atomic Absorption Spectroscopy*, American Elsevier Publishing Company, Inc., New York (1968).

*The more recent books listed in any of the three section subdivisions include material on all three topics.

Robinson, J. W., *Atomic Absorption Spectroscopy*, 2nd ed., Marcel Dekker, Inc., New York (1975).
Slavin, W., *Atomic Absorption Spectroscopy*, Wiley-Interscience, New York (1968).
Welz, B., *Atomic Absorption Spectroscopy*, Verlag Chemie GMBH (1972).

Books on Flame Emission

Burriel-Marti, F., and J. Ramirez-Munoz, *Flame Photometry*, American Elsevier Publishing Company, Inc., New York (1957).
Dean, J. A., *Flame Photometry*, McGraw-Hill Book Company, New York (1960).
Gaydon, A. G., *The Spectroscopy of Flames*, 2nd ed., Halsted Press, New York (1974).
Herrmann, R., and C. T. J. Alkemade, *Flame Photometry*, American Elsevier Publishing Company, Inc., New York (1957).
Mavrodineanu, R., and H. Boiteux, *Flame Spectroscopy*, John Wiley & Sons, Inc., New York (1965).
Mavrodineanu, R., ed., *Analytical Flame Spectroscopy*, Macmillan Publishing Co., Inc., New York (1970).
Parsons, M. L., and P. M. McElfrish, *Flame Spectroscopy: Atlas of Spectral Lines*, Plenum Press, New York (1971).
Slavin, M., *Emission Spectrochemical Analysis*, Vol. 36 of Chemical Analysis Series, Wiley-Interscience, New York (1971).

Books on Emission Spectroscopy

Arhrens, L. H., and S. R. Taylor, *Spectrochemical Analysis*, 2nd ed., Addison-Wesley Publishing Company, Inc., Reading, Mass. (1961).
Barnes, R. M., *Emission Spectroscopy*, Halsted Press, New York (1976).
Boumans, P. W. J. M., *Theory of Spectrochemical Excitation*, Plenum Press, New York (1966).
Clark, G. L., ed., *Encyclopedia of Spectroscopy*, Reinhold Publishing Corporation, New York (1960). Many articles on "Emission Spectroscopy-Light," pp. 99-330.
Grove, E. L., ed., *Analytical Emission Spectroscopy*, Marcel Dekker, Inc., New York, Vol. I, part 1 (1971); part 2 (1972).
Harrison, G. R., R. C. Lord, and J. R. Loofbrourow, *Practical Spectroscopy*, Prentice-Hall, Inc., Englewood Cliffs, N.J.
Harvey, C. E., *Spectrochemical Procedures*, ARL, Glendale, Calif. (1950).
Mika, J., and T. Torok, *Analytical Emission Spectroscopy* (English transl.), Crane, Russak, & Co., Inc., New York (1974).
Nachtrieb, N. H., *Principles and Practice of Spectrochemical Analysis*, McGraw-Hill Book Company, New York (1950).
Pinta, M., *Detection and Determination of Trace Elements*, Ann Arbor Science Publishers, Inc., London (1962); 3rd printing (1971), Chaps. XI-XVIII.
Sawyer, R. A., *Experimental Spectroscopy*, Prentice-Hall, Inc., Englewood Cliffs, N.J. (1944; 1951).
Schrenk, W. G., *Analytical Atomic Spectroscopy*, Plenum Press, New York (1975).
Scribner, B. E., and M. Margoshes, *Treatise on Analytical Chemistry*, Part 1, Vol. 6, Wiley-Interscience, New York (1965).
Slavin, M., *Emission Spectrochemical Analyses*, John Wiley & Sons, Inc., New York (1971).

2.7 INTERFERENCES FOR AA AND FE

Since interferences are so prevalent in both AA and FE, this separate section will be devoted to such. Interference systems are often very complex and varied as to type of interference; thus, systems to classify them and nomenclature have also varied. That given by Gilbert(281) in a discussion of FE interferences in which he classified interferences as spectral, physical, and chemical will, in general, be used here. Further subdivision into vapor and condensed phases will be referred to, but defining exact borderlines is complex and will not be attempted. In general terms a concomitant is any substance, including the solute, present in the sample other than the analyte (element sought).

2.7.1 Spectral Interferences

Spectral interferences can occur for AA as well as FE, despite repeated published claims that AA is free of spectral interferences (see discussion in Ref. 282). Spectral interferences for AA can be divided into two types: atomic line interferences and molecular interference.

2.7.1.1 Spectral Line Interferences

Interferences in AA from overlapping atomic lines occur more frequently than the earlier literature indicates. These have been reviewed by Pickett and Koirtyohann(283) and Lovett *et al.*(284). In Ref. 284, 13 cases, one of which was new, are cited, 49 additional spectral overlaps are predicted, and 30 examples of overlap are cited for neon in hollow cathode lamps. Applications for analysis using spectral overlap are reviewed by Frank *et al.*(285) for iron lamps, citing examples for 17 elements, and by Norris and West,(286) who cite 18 published examples of overlap and 11 new examples which they found gave sensitivities of 250 ppm or better. Panday and Ganguly(287) reported two new instances of spectral interference for Mg and Os. Thus partial interference may be more probable than is realized by most workers. Fassel *et al.*(282) discussed the "wings" and broadening of lines in this respect. Interference possibilities increase if a trace element is determined in a given inorganic compound or even at a 1000-fold excess of interferent.

For example, there is a Rh line at 3280.55 Å (with a transition above the ground state) and a line used for Ag analysis at 3280.68 Å. The difference is 0.13 Å, which is far greater than the width of the Ag line emitted by a hollow cathode lamp. Measurements in the author's laboratory showed that 1000 ppm of Rh (prepared from pure rhodium chloride) gave an absorption equivalent to a little over 0.3 ppm of Ag.

Spectral interferences for FE and methods of compensating for them are thoroughly covered by Buell.(288) As with AA, actual line interferences are infrequent, but molecular interferences are common (particularly those contributed by organic fragments in the flame background). Detection and corrections, however, are generally simple and routine. Continuum and molecular broad-band interference can essentially be eliminated using Lichte modulation or ac FE techniques.[289] Recently, Buell[290] reported exceptional detection limits using ac FE (Section 2.6). Lichte and Skogerboe[291] also reported exceptionally good detection limits using the technique with a plasma torch excitation. The technique as used by Buell can eliminate the interference of CaOH band emission for the FE determination of Ba even at Ca/Ba ratios of 100 : 1. However, even with ac techniques, as detection limits are approached, a vibration-rotation peak from flame background (Bk) may overlap most principal analyte lines used for FE.[290] Again, correction techniques are routine. With the nitrous oxide-acetylene flame, many C and N molecular fragments as well as OH may be involved.

Some of the interference from flame background (especially for the nitrous oxide-acetylene flame) and for excesses of extraneous (sample matrix) elements may be caused by stray light. Larson *et al.*[292] discussed the importance of stray light on measurements made at low levels and ways of minimizing such interference, with emphasis on the use of the plasma torch. Good optics and instrumentation are required for best results and modulation techniques (for example, ac FE) are cited as one excellent approach for minimizing stray light. Stray light may be of more importance than previously realized and in the past has received too little attention.

Measurements made in the author's laboratory using a spectrometer with excellent stray-light qualities (a Jarrell-Ash 0.5-m monochromator with ac FE) and a spectrometer with relatively poor qualities (a Perkin-Elmer Model 306) showed extreme differences between the two instruments (A and B, respectively, in Table

Table 2.6 Flame Emission Comparison of Stray Light in Two Spectrometers[a]

Measurement	Ppm, element	Wavelength (nm)	Spectrometer A	Spectrometer B		
			0.05 exit	0.1(2)	0.3(3)	1.0(4)
Distance in nm,	200 Mn	Mn 279.48	0.06	–	0.2	0.9
analyte wavelength	200 In	In 451.13	0.08	2	4	15
giving a 10% signal						
Interference cal-	2000 Ca	Rh 369.24	0.03	1.1	3.0	–
culated as ppm	200 Ca		–	0.15	0.34	1.2
analyte caused	2000 Ca	Cr 425.44	0.13	1.9	5.3	–
by Ca	200 Ca		–	0.42	1.1	2.6

[a] A, Jarrell-Ash 0.5-m spectrometer equipped with repetitive optical scanning in the first derivative mode. Estimated spectral band width = 0.08 nm. B, Perkin-Elmer Model 306 spectrometer (2, 3, and 4 designate knob settings; band width varies from 0.07 to 0.7 nm for UV grating, 0.14 to 1.4 for visible grating).

2.6). Interferences at various wavelengths caused by the nitrous oxide-acetylene flame background and excess Ca were up to 1000-fold greater with spectrometer B. Some typical results for such interferences and for line broadening are given in Table 2.6. Unfortunately, the use of many spectrometers with acceptable qualities for AA but with poor stray-light qualities for FE cause many workers to underestimate the value of FE and to slight its use. Instrument makers could rectify this situation to some extent by spectrometer improvements.

2.7.1.2 Spectral Molecular Interferences

A form of molecular spectral interference occurs for AA that is caused by molecular absorbance and/or scattered light. Because this type of interference appears to the unaware as an atomic absorption signal, it can be a particularly insidious and a serious source of error. Pickett and Koirtyohann[(283)] briefly reviewed interferences and call attention to the fact that "the failure to apply background absorption corrections (or to be unaware of the possibility of line inter-

Table 2.7 Molecular Absorbance Interference Log[a]

Element	References
Al	297
Au	294
Ba	301(CaOH), 302, 304
Bi	297, 302
Cd	293[b], 294, 302(NaCl[c])
Co	302, 303(FeO)
Cr	304
Cu	302(NaCl[c])
Fe	302
Hg	308, 310
Li	304
Mg	297
Mn	302(NaCl[c])
Na	299 and 300(organics), 300
Pb	302(NaCl[c])
Zn	293[b], 295, 302
Many elements, 2138–7665 Å	305(scattered light)
Many elements, 2025 Å–7665 Å	297, also 296(organics, PO, SO_2)

[a] All elements may be affected (see Ref. 299 and 305), but only some specific, selected examples are listed by element.
[b] Long path O-H.
[c] From Buell.[290]

ferences) in AA is, or was until recently, a rather serious shortcoming of AA as it was commonly practiced." High concentrations of elements such as Na can interfer with Cd,[293, 294] Zn, [293, 295] and other elements. Also, many organic fragments originating in the flame or from nebulization of organic material can cause molecular absorbance interferences.[293, 295–300] Inorganic elements can contribute molecular species, causing various molecular absorbance interferences, some of which are very difficult to differentiate from scattered light interference (equally as serious and also appearing as spurious atomic absorption). One example is interference caused by CaOH species on Ba,[301] which incidentally can also cause FE interference. In the latter case, ac FE can be used to eliminate the CaOH interference[289, 290] Alkali halides and additional inorganic molecules can cause AA interference and have been studied by Billings,[302] Flemming,[303] and Koirtyohann and Pickett.[304, 305] Provided that the interference is not prohibitive, such as is caused by nebulization of many hydrocarbons, corrections can be made using nonabsorbing lines,[293, 299, 300] continuum sources,[293, 304, 306, 307] or a unique technique using the Zeeman effect.[308–310] Marks *et al.*,[311] however, discuss correction methods and cite examples wherein some of these correction methods may fail. Thus, they must be used with caution and with the knowledge that potential errors exist.

A unique application of a nonabsorbing line has been used by Zeeman and Brink[312] for noble metal analyses. With a multichannel spectrometer they integrate, with time and ratio, all absorbance readings to a nonabsorbing Pt line. Since they used a dc system, they also encountered some OH emission interference emitted by the flame. A log of molecular absorbance interference references is given in Table 2.7.

2.7.2 Physical Interferences

Physical interferences are those effecting rates of spraying and desolvation in the flame. An entire chapter is devoted to the subject by Koirtyohann,[313] and the book by Mavrodineanu and Boiteux[314] also has an excellent section. Corrections for physical interferences can be accomplished by dilution, making standards similar to samples (at least partially) or by using internal standard techniques, especially with dual-channel instruments. Care must be used, however, not to assume the interferences as too simple. An example of a complex interference which is partially physical and partially spectral and chemical can be contributed by organic material. An organic solvent can change viscosity, thus increasing rates of spraying and also increasing rates of desolvation due to increased volatility and the production of finer droplets (a physical interference). The same organic solvent can also cause molecular spectral interference and overexcitation, causing enhancement (a chemical interference).

2.7.3 Chemical Interferences

Chemical interferences may occur in the condensed phase or vapor phase (a commonly used subdivision). Despite some published statements to the contrary, they influence AA and FE equally. West, Fassel, and Kniseley,[315, 316] in a recent study of the nitrous oxide-acetylene flame, show many strong interferences for Al, Mo, and V, which are about equal for AA and FE. Their study also revealed a new type of interference due to lateral diffusion in the flame, which has also been noted by Koirtyohann and Pickett[317] and by Marks and Welcher.[318] This type of interference could be judged as a combination of physical and chemical interference and appears difficult to reproduce quantitatively. In Ref. 315, 15 measurements of the effect of Ni on Mo produced results varying from +26 to +67% enhancement. This perhaps explains some of the differences and discrepancies found in the literature for various interference studies.

A typical vapor phase interference is ionization, which is more pronounced in higher-temperature flames, especially the nitrous oxide-acetylene flame.[319, 340] The addition of an easily ionized element such as K is generally used to correct for ionization interferences. Rubeska[321] devotes an entire chapter to vapor phase interferences.

Condensed phase interferences are discussed in a chapter by Rains[322] and are generally thought to be caused by the formation of high-boiling compounds in the flame. Alkemade[323] discussed several types of solute vaporization interferences. Rubeska and Moldan[324] investigated such interferences for Ca, Mg, and Sr determination, their mechanism, and elimination. These types of interferences on alkaline earth elements, and particularly Ca, have been studied more than any other system.[325, 326]

West *et al.*,[316] however, more recently provided some evidence that phosphate may act as a lateral diffusion interference on Ca rather than a condensed phase interference. All of these explanations are complex and may perhaps occur simultaneously. For example, the author finds some of the interferences studied in Ref. 315 to occur with a slot burner turned at an angle which would lead one to think that lateral diffusion was not entirely responsible.

Recent studies have shown that solution equilibria can influence interferences and that M—O and M—OH bonding is of importance.[327, 328] Many other studies concerning the mechanism of interferences included catalytic effects,[329] response surface via atomization,[330] and analyte reduction in the solid state.[331]

Many other chemical interferences have been recorded, including tabulated FE interferences in the books by Dean[332] and by Herrmann and Alkemade.[333] An updated tabulation of chemical interference references[334–393] is given in Table 2.8 and emphasizes recent studies and use of the nitrous oxide flame. Some of the early references[303, 312, 315–336] are also included and Ref. 343 is for an AF and FE study. These interferences, primarily recorded for AA, may apply

Table 2.8 Atomic Absorption Chemical Interference Log

Element	References	
	N_2O flame	Air
Ag		360[a]
Al	315–318, 327, 337, 343, 344, 351, 360, 382, 384	
As	(401–403)[b]	
Au	377	378[a]
Ba	317, 319, 337, 382, 385	325, 328, 382
Be	327, 337, 345, 346, 358, 360	334
Bi		[a]
Ca	317, 319, 326, 327, 337 383	325–328, 334, 383, 386
Cd	359 (O–H), 400[b]	334, 360
Co	303, 327	303, 327, 334, 342, 360
Cr	318, 327, 338, 349, 351, 354,[a] 365, 380, 381	327, 328, 334, 338, 354, 365, 380, 387
Cu	338, 354,[a] 366, 380	334, 338, 354, 360
Eu	351	366, 380
Fe	354,[a] 366	328, 334, 344, 353–355, 360, 366
Ga		[a]
Gd	351	
Hf	327, 328, 337, 339	
Hg	397 (cold vapor)	412
In		[a]
Ir		379
K		368
Li	317	[a]
Mg	326, 327, 337, 347	325–328, 347
Mn	338, 349, 380	363, 364, 389
Mo	315, 343, 349, 352, 357, 363, 390	363, 364, 390
Na		334, 335
Nb	327, 370	
Ni	318, 327, 338, 380	327, 338, 360, 373, 381, 391
Pb		350, 360, 388
Pd	378[a]	379[a]
Pt	312 (air–gas), 378	379, 392
Rb		369

(Continued)

Table 2.8 (*Continued*)

Element	References	
	N_2O flame	Air
Re	373 (O_2–C_2H_2), 374, 375	
Rh	312 (air–gas), 376–378	377–379
Sb		372
Se	(401, 402)[b]	
Si	344, 382	
Sn	(335, 399–401)[b]	
Sr	337, 382, 393	325, 328, 383
Ta	327, 337, 339	
Ti	318, 327, 328, 337, 339, 343, 348, 382	
V	315, 339, 343, 349, 351, 356, 361, 362, 382	
W	327, 349	
Y	327, 351	
Yb	337	
Zn	359 (O–H)	360, 388
Zr	328, 337, 339, 340, 343, 351	

[a]Tested for interferences and none found.
[b]Using hydrogen flames.

equally to FE (especially since the laminar flow, nitrous oxide-acetylene burner is also excellent for FE). Reference 334 concerns matrix errors caused by small percentages of organic material, and Ref. 375 explores organic solvent effects for determining Re by FE.

2.7.3.1 Hydrocarbon Systems Reduce Chemical Interferences

Another important aspect of eliminating interferences should be mentioned in the use of organic solvents. Condensed phase (chemical) interferences do not appear to occur normally in hydrocarbon solvent systems. They do occur to some extent in solvents containing oxygen in their molecular structure, particularly if some water is present. To obtain some measure of the degree of interferences, FE measurements were made using Ca and an oxyhydrogen total consumption burner in the author's laboratory. In this system, interelement effects severely depress Ca emission in aqueous solutions. The effect of P, and particularly of Al and Si, are pronounced. In pure hydrocarbon solutions an excess of these three elements produced no effect on Ca emission. To further measure de-

Table 2.9 Matrix Effect of 10 ppm of Al on Flame Emission Signals for 2 ppm of Ca in Various Solvents Using an Oxyhydrogen Flame

% H_2O	% Acetone	% Hydrocarbon	Ratio of signal to that with no interference
50	50		0.25
10	90		0.31
0.1	99.9		0.50
0.1	49.9	50	0.90
0.1	19.9	80	1.0

grees of interference, mixed solvent systems of a hydrocarbon (AMSCO lactol spirits), acetone, and water were used. Emission was measured in an oxyhydrogen flame for 2 ppm Ca in the presence of 10 ppm Al, added as the interferent. The magnitude of depressing effect varies from 75% for 50% aqueous solution to none for primarily a hydrocarbon solution, as shown in Table 2.9.

2.7.3.2 Compensating for Interferences—La as Releasing Agent

Other methods for eliminating interference are given in the reference list and texts and may include techniques based on buffers, addition and dilution techniques, separation-concentration procedures (Section 2.9.2), and so on. One technique that has been used much with excellent results is to add La as a releasing agent for eliminating interferences for alkaline earth determinations. Such use has become common practice in AA but was first used for FE applications by Yofe and Finkelstein(394) and studied later by Dinnin.(395) A note of caution should be added to check all techniques for specific applications, since none is foolproof. As an example, Shatkay(396) presents experimental and theoretical data showing that the standard addition technique can fail.

2.7.3.3 Interferences for the Hg Cold Vapor Method

Various interferences may also influence the cold vapor determination of Hg. Organic vapor (especially hydrocarbons) can cause a serious molecular absorbance interference. In addition, other interferences can occur, as discussed by Lindstedt.(397)

2.7.3.4 Interferences in Hydrogen Flames (As, Se, etc.)

Special consideration should be given to interferences using cool flames such as the argon-supported hydrogen flame (Ar–H). This type of flame appears susceptible to certain interferences discussed by Rubeska and Miksovsky[398] and Nakahara *et al.*[399] Kahn and Schallis[400] use this flame to obtain 5- and 10-fold better detection limits for As, Cd, Se, and Sn because of the lower flame background. For As and Se the percent absorption for the flames was 15 and 13 for Ar-H and 62 and 53, respectively, for air-acetylene. Kirkbright *et al.*,[401] however, cite the cool Ar-H as being very susceptible to interferences and found that a nitrogen-separated, air-acetylene flame gives less interferences for As and Se and improved Bk compared to a conventional air-acetylene flame. Despite this, out of 28 potential interferences they found 10 for As and 12 for Se (their detection limits were also much lower than reported in Ref. 396). Kirkbright and Ranson[402] found no interferences for As and Se with a nitrogen-separated nitrous oxide-acetylene flame. Smith and Frank[403] studied interferences for As in another higher temperature flame (oxygen-acetylene) and found some. Smith[404] discussed interferences for seven elements using hydride evolution techniques.

Emphasis has been deliberately given to the possibilities of interferences as a warning to check specific applications. Considering these possibilities leads one to think more seriously of using some of the separation-concentration methods (especially if high sensitivity is needed). Generation as a gaseous compound, solvent extraction, and coprecipitation are cited in Section 2.9.2 for applications to As and Se.

2.7.3.5 Interferences for Nonflame Methods

Nonflame AA, one of the newer techniques, although it provides exceptional increase in sensitivity is especially susceptible to chemical and molecular absorption interferences. Amos *et al.*[405] discuss interferences and cite some for Pb and Mg that are not encountered in flames. They show that surrounding a carbon rod with hydrogen flame (a reducing atmosphere) greatly reduces interferences. Alger *et al.*[406] made extensive matrix studies for nonflame AA, and Jackson *et al.*[407] studied the effects of 22 metals on the nonflame determination of V. Maruta and Takeuchi[408] studied interference mechanisms of acids on Cu, Cr, and Fe using a tantalum filament atomizer. Talmi and Morrison[409] studied induction furnace AA and emission applications and include many data on emission interferences and some on AA interferences. Newer techniques for background correction include Zeeman techniques using a dual furnace[410] and wavelength modulation for graphite furnace atomic emission.[411]

The Delves cup technique, although employing a flame, is similar to nonflame techniques using atomization after evaporative deposition of liquid samples. Clark *et al.*,[412] although they obtained a detection limit of 0.17 ng of Hg compared to 1 ng using conventional flame AA, found multiple Hg peaks versus time and severe interferences for the cup method.

2.8 ANTIPOLLUTION

The petroleum industry (P.I.) has been quite aware of environmental considerations and has spent millions of dollars on antipollution. The API has formed a committee for air and water conservation with five committees, four subcommittees, and many task groups. They represent part of the P.I. effort and will continue to devote much research to the subject as they have in the past. For example, the API prepared a background paper on lead in the environment that reviews possible hazards from lead in gasoline.[413]

New antipollution regulations[414] have resulted in increasingly large antipollution programs by industry, especially the P.I. Because of this, some background information and analytical methods will be briefly reviewed.

2.8.1 New Regulations

In the United States, awareness of pollution and the problems involved has led to a national effort toward antipollution. The Environmental Protection Agency (EPA) was formed and we now have many new and proposed antipollution regulations. Among new regulations is the California Ocean Discharge Plan. Some of the limits set for heavy metals in water (Table 2.10) are more strict than by the U.S. Public Health Service limits for drinking water. Table 2.10 also includes the latter limits and the abundances in natural waters, seawater, and the earth's crust.

2.8.2 Wastewater Disposal and Cleanup

One of the problems of concern is how to remove heavy metals from wastewaters. This has been the subject of several recent reviews.[415–417] All of this has led to an expanding business in consultants and equipment manufacturers for water antipollution[418] and has also led to many new books on environmental and pollution subjects. For example, CRC is publishing a three-volume series; Vol. I, *Air Pollution*,[419] Vol. II, *Solid Waste*, and Vol. III, *Water and Waste Water*. There are also books on water treatment,[420] odor, noise, and thermal pollution,[421–423] and Cd and Hg in the environment.[424–426] The increase in the scientific literature on environmental pollution studies, analytical methods, and so on, has also been tremendous. The *Water Pollution Control Federation Journal* reviews the literature each year, and their coverage has expanded 12% per year,[427] with the 1976 review including 285 references on just the inorganic contents of water.

2.8.3 Analytical Problems

The current environmental concern has led to increased research on improving and developing new analytical methods. The situation concerning analytical

Table 2.10 Concentration (in ppb) for Various Types of Water

	Limits for water			Abundance of elements		
Element	Drinking water, USPHS	Discharge to ocean, Calif.	Irrigation	Natural fresh water	Seawater	Earth's crust
Al	—		1000	200-5000[a]	5-1000	81,300
Ag	50	20		0.01-0.07	0.1-2	0.07
As	10,50[b]	10	1000	1-30	1-20	1.8
B	1000		750	1-10,000	4600	10
Ba	1000		500	4-35	10-100	425
Be	—			0.1	0.001-0.03	2.8
Cd	10	20	5-50	Low	0.02-0.2	0.2
Co	—		200	0.03-10	0.04-0.7	25
Cr	50[b]	5	5-20	0.5-40	0.04-2	100
Cu	1000	200	200	0.2-30	1-20	55
Fe	300			40-1500	10-1000	50,000
Hg	—	1		0.3-3	70-200	20
Li	—		5000	0.01-0.1	0.03-0.15	0.08
Mn	50		2000	0.3-300	0.5-10	950
Mo	—		5-50	0.05-4	1-20	1.5
Ni	—	100	500	0.02-10	0.2-7	75
Pb	50	100	5000	0.3-3	0.1-10 Deep water 0.03	13
Se	10		50	?	0.1-6	0.05
V	—		10,000	1	0.2-7	135
Zn	5000	300	5000	1-2000	1-20	70

[a]Approximate.
[b]Grounds for rejection.

methods was discussed in an editorial by Laitinen,[428] which can be summed up in his conclusion: "Standard methods undeniably have an essential role to play in environmental science, as in other applications of analytical chemistry. Every effort should be made, however, to facilitate the introduction of alternate methods and to encourage the development of new approaches."

Water and land pollution by heavy metals is currently of great concern. Antipollution in the P.I. has been reviewed by Wilson,[429] including a survey of 12 refineries. The EPA has sponsored a comprehensive review on the inorganic contents of oils as potential pollutants in fossil fuels (Section 2.9.3, Ref. 831, on the analysis of crude oils). In an issue of *Chemical and Engineering News*, "Trace Metals Unknown, Unseen Pollution Threat" was presented.[430] The article listed sources of possible heavy metal pollutants and concluded with the need for more knowledge and basic research. Cheremisinoff and Habib[431] reviewed the occurrence, toxicity, and analytical determinations of Cd, Cr, Pb, and Hg in water pollution. The National Science Foundation sponsored a report on baseline studies of heavy metals, halogenated hydrocarbons, and petroleum hydrocarbon pollutants in the marine environment and research recommendations.[432] A book on the atomic spectrometric analysis of heavy metal pollutants in water has been authored by Burrell.[433] A book by Gallay *et al.*[434] *Environmental Pollutants: Selected Analytical Methods*" includes AA methods for metals in air and water. The U.S. Geological Survey (USGS) has completed a nationwide survey of metals in streams and reported that "only a few" of the 720 samples contained heavy metal concentrations in excess of (PHS) drinking water standards.[435] In addition to the PHS standards, we now have a 1974 Safe Drinking Water Act to be regulated by the EPA.[436]

It is indicated by such references as those above that good analytical techniques are mandatory for maintaining antipollution programs and that analyses of trace heavy metals in waters is of major importance. For this reason the following discussions will emphasize analytical techniques for water. More detailed discussions are given in Section 2.9.1. Since Cd, Cr, Hg, and Pb have received the most attention, comments will be mostly limited to them. Each element will be discussed individually.

2.8.3.1 Chromium

Because of the new limit of 5 ppb for disposal of wastewaters under the California Ocean Discharge Plan, the element Cr currently is receiving considerable attention. Because Cr compounds have been used as anticorrosion additives in cooling tower waters, making the transition to water with almost nil Cr may be difficult. Another difficulty involves waters taken for industrial purposes from natural sources that may have Cr in greater quantities than 5 ppb. As examples Durum and Haffty[437] analyzed 15 major rivers of America in 1961 and found a range of 2-30 ppb of Cr and an average of 8 ppb. The average content of the earth's crust is 100,000 ppb of Cr.[438]

The most popular method for the determination of Cr in water is AA, but FE is equally sensitive and ac FE is more sensitive.[439] In such cases one would think that FE should be used more often, but this does not appear to be the case. Fassel *et al.*[440] and Johnson *et al.*[441] discuss this "unjustified neglect" of FE and show applications where results by FE are equal or superior to AA results. Good precision is provided by ac FE for determining Cr directly in water at the 5-ppb level. Using ac FE for direct measurements of Cr in seawater, however, is complicated by the high brine content. Although it is still easy to detect 5 ppb of Cr directly, a small negative background is encountered and the high solid content tends to clog burners and leads to a contaminating sodium memory. To avoid this and to provide better precision at 5 ppb and lower levels, a precipitation method based on aluminum hydroxide carrier has been used[442, 443] and provides good results. A synthetic sample prepared by adding 5 ppb of Cr to a clean seawater sample was analyzed by the precipitation ac FE method and gave an average recovery of 91%, with an average deviation from the mean of ±3.3% relative.

2.8.3.2 Mercury

Because of the mercury in-fish problem and instances of reported pollution and poisoning, the determination of Hg has received much attention. Extensive reports on Hg hazards in the environment are given in Refs. 444-447, which also include many Hg values from natural sources. Additional references report the Hg content of coal,[448, 449] urban soils (Ref. 450–also includes other metals), urban woody plants (Ref. 451–also includes lead) and surface water.[452] Isolated cases in which crude oils are located near Hg ore deposits, may cause concentrations to exceed 1 ppm,[453] but in general crude oils contain less than 0.2 ppm.[454] Oil in the Cymeric Field in California appears to contain Hg in the range 1-21 ppm,[455] as docs the nearby, shallow North Belridge oil.[454]

The most sensitive methods for determining Hg are generally based on special AA techniques. A cold vapor technique capable of detecting fractional nanogram amounts is much used. Manning[456] reviewed nonflame methods for determining mercury. Mercury in crude oil has been determined by Hinkle[453] and by Bailey *et al.*[455] Huffman *et al.*[457] used a Ag screen collection method combined with cold vapor evolution to eliminate interference from organic material and increases sensitivity.

An example of the volume of literature on Hg deteterminations was shown in our recent library search, which culminated in a bibliography of 134 references. One should be aware that determining Hg (or any tracc metal) at such low levels is touchy. One problem is that of sampling and stability of samples. Coyne and Collins[458] studied the stability of Hg solutions stored in polyethylene and recommend fixing with HNO_3 at pH 1. In addition, one should be aware of the interference problems in the use of the cold vapor method.[459] Organic material can cause severe interferences, so methods that compensate for this molecular absorbance interference are best. Such methods include the use of continuum lamps to correct for the interferences,[460-462] incorporation of differentiation by absorbing traps,[463] the Ag screen method cited above,[457] and even the use of the Zeeman effect.[464]

There may be some advantage in using emission methods with plasma torches[465-467] in such cases, since they are not influenced by molecular absorption by organic material. A unique, new instrument has just been announced based on changes in the resistance of gold foil caused by Hg.[468] If proved to be trouble-free, this method could replace some of those described above.

2.8.3.3 Cadmium

Atomic absorption is generally used for determining Cd, because of its excellent sensitivity. The California limit of 20 ppb in water can be met by direct analysis but is complicated by interference from Na in high-brine samples.[460,469] A correction can be applied if the Na content is known. Flameless (C rod, etc.) AA techniques offer a considerable increase in sensitivity, which is probably great enough so that molecular absorbance interferences (which are larger for this technique) may not be a limiting factor.

2.8.3.4 Lead

Lead has received the most attention in the P.I. because of its use as a gasoline additive. For this reason, in addition to water samples, various other petroleum-derived samples are also monitored for Pb content. The method used most commonly is AA; however, good detection limits are provided by ac FE.[439] Techniques using the Delves sampling cup or nonflame methods provide better detection limits and are especially suitable for water analysis. A nonflame, AA method using a carbon rod technique was reported for determining Pb in petroleum products by Bratzel and Chakrabarti.[470] The distribution of Pb in soil with emphasis along highways was determined by Seeley *et al.*[471] using ES and AA methods, which gave equivalent results. Lead poisoning has been briefly reviewed by Varian,[472] pointing out that a unique source of contamination in coffee is from the pottery used in cups. The carbon rod technique is cited as a very sensitive method for determining Pb. The Pb contents and isotope differences in coal and gasoline have also been reported by Chow and Earl.[473]

2.8.3.5 Accuracy of Analytical Methods for Water Analysis

In addition to the words of caution given in Section 2.1.1.4, it should be reemphasized that care must be used to obtain correct results for trace elements in water. At a recent symposium on the identification and measurement of environmental pollutants, Chau,[474] in a review of determining trace metals in water, urges the reader against "overexaggerated performance and the freedom of interference for AA methods." He also cites a common practice in solvent extractions which can lead to errors. Carpenter[475] reviewed analytical problems applied to oceanography and intercalibration analysis for Zn in seawater reported by Brewer and Spencer.[476] The results varied from 0.8 to 44 ppb for 15 laboratories. These examples illustrate the problems involved and the extreme care needed for accurate determinations at trace levels.

2.8.4 Metal Analysis as a Means of Identifying Crude Oil Spills

There has been much concern about oil spills and their threat to the environment. As a consequence, analytical techniques have been applied in a number of schemes to identify oil pollution sources. A useful parameter in such schemes is the magnitude of Ni and V content and the Ni/V ratio. Since these elements are predominant and indigenous to crude oils, analyses for Ni and V are quite helpful. In 1968 Brunnock *et al.*[477] studied the analysis of oils as beach pollutants and presented a scheme for detecting their source. They include Ni and V data for 27 crudes from various sources which were obtained by x-ray methods. Some oil companies use ES methods for such purposes. Other schemes for such applications have been reviewed by Adlard[478] and proposed by Stuart and Branch[479] and Thruston and Knight.[480] The latter two references use AA for determining the metal content.

2.8.5 Heavy Metal Analysis Applied to Air Pollution

New regulations are developing and antipollution programs are in progress to keep the air as clean as possible. Industrial and automobile exhausts are major sources of pollution. It is quite interesting to note that there is also natural air pollution, which is greater than man-made pollution for carbon monoxide,[481] and that sulfur pollution is released into the air by bacterial action in the muds of lakes.[482]

Analytical Chemistry reviews air pollution literature in the odd years; see the reviews for 1975 and 1977 for additional references.[483] Although the major pollutants are nonmetals, there is also interest in air particulates and heavy metals analysis. Hwang[484] reviewed trace metal analysis in atmospheric particulates and the use of AA methods. He provides tables summarizing air pollution sources, pollution data for air over various cities, and tables of sensitivities for conventional and nonflame AA techniques. He discussed the problems involved in the preparation of samples and cites the outstanding sensitivity of nonflame methods for application to air pollution analysis. In addition to the previously cited[419] book on air pollution, Hwang cites still another.[485]

In addition to the references cited by Hwang, others that can be added are one by Woodriff and Lech[486] applying a nonflame, AA method for Pb in air, and those by Loftin *et al.*[487] and by Christian and Robinson[488] for the AA determinations of Pb, Cd, and Hg in air. More recently, Scott *et al.*[489] developed an optical emission method for the EPA to analyze 24 trace elements in air particulates. Additional studies of metals in the atmosphere have been made, for example those abstracted in Ref. 490. Some of these studies involve potential pollutants from the burning of fossil fuels. Therefore, a consortium approach to establish metal contents of petroleum products has been made by a group of oil companies (Section 2.9.3; Ref. 836).

2.9 METHODS OF ANALYSIS GROUPED BY PRODUCT TYPE

The heart of the P.I. is crude oil, and the largest number of references for analyzing any one type of petroleum product concerns metals in oils (a major heading in the *Analytical Chemistry* annual reviews on petroleum). For this reason organization of the products in this section will place crude oil and lubricating oil second and third.

Water and concentration methods will be placed first for three reasons: (1) tremendous volumes of water are used in P.I. operations and produced along with oil from oil wells, (2) antipollution control of wastewater disposal has become a large and critical subject (Section 2.8), and (3) any other product which can result ultimately in an aqueous solution (oils ashed and dissolved in dilute acid, etc.) may be analyzed by one of the methods used for water analysis.

2.9.1 Water and Oil-Field Brines

Of any one type of sample, none is more universal than water. Although the analysis of water samples is thoroughly covered in the literature, additional coverage will be presented here for several reasons. Water is of extreme importance to the P.I., since millions of gallons are used in refineries and produced as brines from oil wells. Since many other types of samples may end up as aqueous solutions during preparation for analysis, all analytical techniques used for water analysis may be applicable to such samples. Development of new techniques and stricter antipollution regulations have led to a boom in water analysis and associated studies (Section 2.8).

Although fairly detailed coverage is given here, it is not intended to be all-encompassing; more complete coverage can be obtained through the selected literature references given.

2.9.1.1 Water Produced and Used

Formation brines are produced in large volumes along with crude oil and natural gases. Collins[491] reviewed the involvement of brines in drilling and producing oil wells and the problems connected with pollution. Brines produced may vary from none or little for some new oil wells to 95% for older wells. The sodium chloride content of such brines may vary from 4000 to 200,000 ppm and the trace element content may also vary widely. Coburn and Bowlin[492] presented a thorough review of water use by the petroleum industry. Their data show that secondary recovery operations for oil fields use less than 0.2% of the total water used in the United States, and that over 90% of this water is saline (thus not depleting freshwater supplies). California is the largest user of water—67,775 million gallons per day (mgd) in 1960, and the largest use is for irrigation, at 18,500 mgd. The comparable figure for the P.I. is 54.6 mgd, of which 53.2 mgd was brine. Of all the brines produced from oil wells, Collins[491] says: "Currently, some produced brines are being discharged into approved surface ponds, whereas most brines are returned underground for disposal or to repressure secondary oil or gas recovery wells." Smith and Olson[493] reviewed this disposal by subsurface injection and presented analytical data for 44 California wells.

The elemental composition of oil-field brines is of interest for many reasons, including pollution involvements and geological significance in production and exploration. Additives added to brines in oil-field formations may include the following: acid to make formations more porous, thus producing more oil; corrosion inhibitors, for which their effectiveness is followed by iron analysis in the produced water; and scale inhibitors, to keep alkaline earths from forming deposits which reduce production by plugging oil wells (their effective use may be followed by Ba and Ca increases). In the study of oil-field formations, tracers may be added to waters and injected back into the formation.

In addition to field operations, industrial water may also be used in refinery cooling towers. At times seawater may be pumped from the ocean and ultimately returned from such cooling operations. In all these operations it is important to comply with antipollution laws for the disposal of waters; therefore, the heavy metal content is monitored to ensure compliance.

2.9.1.2 Reviews and Books

There are many reviews and books referring to methods of water analysis. Only a selected cross section of these is given here. References to combined soil, plant, and water analysis are given in Section 2.9.9.1.

Foremost of the comprehensive reviews are those that appear in *Analytical Chemistry* in odd years[494] and the annual reviews by the Research Committee of the Water Pollution Control Federation.[495] The more than 600 references[494] cover the topics of short reviews, trace cations and anions, dissolved gases, detergents, pesticides, herbicides and fungicides, and organics. The extensive annual reviews (600–700 pages) by the WPCF[495] deal with trace inorganics, major inorganics, organics, water characteristics, continuous monitoring, automated analysis and sampling procedures, and such topics as wastewater treatment, chemical and physical methods, industrial wastewaters such as those from petroleum processing, and the water pollution effects of heavy metals; thermal, surface, and ground waters; and stream purification. The number of references have increased about 12% per year; in the section on inorganics in water alone, the 1976 review included 285 references.

In addition to the Water Pollution Control Federation (WPCF), other groups providing extensive data on water analysis are as follows: the Analytical Reference Service (ARS), composed of 302 laboratories devoted to contaminants in the environment; the American Public Health Association (APHA), which includes federal and state offices; the American Water Works Association (AWWA); the U.S. Bureau of Mines (USBM); and the U.S. Geological Survey (USGS). The Environmental Protection Agency (EPA) has also become involved in this area and also publishes a manual of methods.[496] The book *Standard Methods for the Examination of Water and Wastewater* is published jointly by the APHA, the AWWA, and WPCF.[497]

The ARS conducts extensive cooperative studies and provides recent data on an evaluation of AA methods.[498] Some of their cooperative results are evaluated in a book publishing the results of a symposium on *Trace Inorganics in Water*[499] and in the chapter by McFarren and Lishka.[500] In many cases they preferred chemical to AA or other instrumental methods. Various official public health departments have published studies such as that in California reported by Greenberg *et al*. on the use of reference samples in their many water laboratories.[501,502] In one study,[501] out of 92 laboratories perfect scores were reported by 38 laboratories, and 54 had one or more unacceptable results.

The USGS has published extensive data on water analysis, including manuals of methods. A recent manual by Brown *et al*.[503] contains mainly spectrophotometric (colorimetric) and AA methods and no emission methods, although FE is mentioned under a discussion of instrumental methods. This is rather a peculiar turnabout, compared to the extensive use of ES and some FE methods by the USGS in many of their water surveys.[504–508] *Water-Supply Papers 1473* and *1454* by Hem[509] and Rainwater and Thatcher[510] are extensive works devoted to water characteristics and analysis. The latter,[510] printed in 1960, contained mainly colorimetric methods but contain some FE methods and a section discussing their use. At about this time ES methods and some FE methods were developed by the USGS through the works of Durum and Haffty[504,505] and Silvey and Brennan[506,507] and appeared to be used considerably through the 1960s. Livingstone[511] reviewed the chemical composition (including minor trace metals) of worldwide rivers and lakes, which provided extensive data. Durum and Haffty[504] presented extensive data for principal rivers in North America. More recently, a USGS reconnaissance of U.S. surface waters emphasizing metropolitan water supplies has been completed which involved analyses of 720 water samples.[512] Some additional conception of the extensive involvement of the USGS in their job, stated in the preface in Ref. 510 as "responsibility for measuring and evaluating water moving through that portion of the hydrologic cycle between the time the water from the atmosphere reaches the earth's surface and the time it is returned to the atmosphere or enters the ocean," can be obtained from the following facts. The number of sampling sites involved in our National Water Resources Data System includes 28,000 groundwater observation wells, 18,000 stream flow stations, and 5700 general sites.[513] The USBM publications also provide extensive data on water, including some on oil-field brines (discussed later).

An evaluation of instrumental methods for water analysis is given by Hume[514]; Kopp and Kroner[515] reviewed ES methods for the analysis of natural waters. Burrell[516] has prepared a book on atomic spectrometric analysis of heavy metal pollutants in water.

The oceans are a special subject for which studies of elemental composition are numerous. The chemistry of trace elements in seawater and other brine samples can be quite tricky and complex. Høgdahl[517] prepared a compilation, *The Trace Elements in the Ocean: A Bibliographic Compilation*. Other books containing extensive data on seawater (including analytical methods) are by Barnes,[518] Martin,[519] and Riley and Skirrow.[520] The state of our knowledge on the abundance of trace metals in the oceans was reviewed by Richards.[521] He emphasizes our lack of adequate data. Nehring *et al.*[522] reviewed analytical procedures used in oceanography, including FE. Other reviews of trace metals in oceans and analytical methods used are given by Hitchcock and Starr,[523] Newton and Atkins,[524] and Schutz and Turekian.[525] Riley and Taylor,[526] in their application of AA analysis to seawater, summarize some of the more recent references. Books on the analytical methods in oceanography[527] and marine chemistry[528] have been published. The National Science Foundation sponsored a report on baseline studies of heavy metals, halogenated hydrocarbons, and petroleum hydrocarbon pollutants in the marine environment with research recommendations.[529]

Books on water analysis include one by Khalil[530] on instrumental methods, another by Traversy[531] for the Canadian Water Quality Division, and a four-volume work on water pollution by Ciaccio.[532] A book on analytical chemistry in industry[533] includes chapters on "Potable and Sanitary Water" by Kroner and Ballinger and "Analysis of Industrial Wastewaters" by Mancy and Weber. It is interesting to note the contrasting opinion of these two works; the former includes more references to ES than any other instrumental method, while the latter favors AA, slights FE, and claims that the ES is little used.

Skougstad[534] has discussed worldwide water-quality standards and expressed concern that agreement be reached on an international level for the standardization of methods and techniques. Laitinen[535] cited standard methods as essential but stressed development of new approaches.

Extensive bibliographies have been presented for FE and AA by Mavrodineanu[536] and by Perkin-Elmer[537] that segregate references to waters in a key index.

2.9.1.3 Variations in Elemental Composition of Waters

Water sources may be divided into three arbitrary categories: (1) surface waters (rivers, lakes, and shallow wells) relatively low in total dissolved solids (TDS); (2) subsurface brines, such as those associated with oil fields; and (3) seawater.

Variations in the composition of worldwide surface waters was discussed in detail by Livingstone.[511] He summarized the mean composition for rivers of the continents in Table 81, p. G41,[511] for four anions and six cations that comprise the major constituents. The average mean of these are listed individually below and will add to 130 ppm:

Constituent	HCO_3	SO_4	Cl	NO_3	Ca	Mg	Na	K	Fe	SiO_2
Mean ppm	58.4	11.2	7.8	1	15	14.1	6.3	2.3	0.67	13.1

He also reviewed variations in minor constituents, few of which exceed 1 ppm, with many under 0.1 ppm. Ranges of some minor elements of interest to antipollution are given in Table 2.11. Generally, the total dissolved solids (TDS) content of rivers increases from source to the mouth of the river and may vary regionally according to the geology of the formations traversed by the rivers. Roberson[(538)] discusses selected rivers of western United States and some notable differences in Ca, NaCl, and HCO_3/SO_4 ratios. He also provides a map showing a distinct line of demarcation for changes in the latter ratios.

2.9.1.3a Geological Studies of Oil-Field Brines. Although most rivers and lakes are relatively low in total dissolved solids (TDS) content, some inland water bodies and hot springs have a high brine content or unusual mineral content. Although these occurrences are more the exception for surface waters, a higher brine content is typical of subsurface waters associated with oil fields. Rittenhouse *et al.*[(539)] made an extensive study of minor elements in oil-field waters, involving 823 samples from the United States and Canada. Gullikson *et al.*[(540)] discussed geochemical relations of oil-field brines and compared them to seawater, as was done by Rittenhouse *et al.*[(539)] Oil-field brines are considerably different than seawater in several ways. They may vary from a NaCl content of about 4000–200,000 ppm, whereas seawater has an average TDS of 34,400 ppm

Table 2.11 Range of Minor Elements in River Water[a]

Element	μg/liter	Element	μg/liter
Ag	0.1– (1–200)	Mo	0.5–7
		Ni	4–80
Al	30–1500		
		Pb	1–50
B	2–50		
	(50–80,000)	Rb	1–8
Ba	10–150	Sr	10–800
Co	0.1–6	Ti	2–100
Cr	2–30	Si	100–100,000
Fe	100–1500–10,000+	V	0.1–6
Li	0.1–40	Zn	(est.) 100–50,000
Mn	3–200		

[a]From Durum and Haffty[(504)] and Livingstone.[(511)]

(mostly NaCl). Seawater is unique in having a high ratio of Mg/Ca (1272:400 ppm), whereas oil-field brines have a much lower ratio, and in fact can become concentrated in Ca. Bailey *et al.*,[541] in a study of the occurrence of Hg in crude oil from the Cymeric, California field, shows an unusually sudden increase of Ca brine content versus depth from about 370 ppm to over 3000 ppm. The corresponding Mg content dropped from 115 to 49 ppm. High Ca content also occurs for certain hydrothermal wells, such as those near the southern end of the Salton Sea in California.

Studies of elemental variations in oil-field brines have been utilized and are important in geological studies of oil fields. In addition to Ref. 539-541, other studies have been reported by Bredehoeft *et al.*,[542] Clayton *et al.*,[543] Graf *et al.*,[544] and Meents *et al.*[545] Reference 545 has been applied to a distribution and study in relation to ancient basins where SO_4 content in particular increased basinward. Bonham,[546] on the other hand, found that Ni and V concentations were higher in crude oil as the ancient shorelines were approached. Correlations of such studies may be of considerable significance.

In addition to the major elements, minor elemental variations have also been studied and may be of significance.[539] Bailey *et al.*[541] reported unusually high, heavy metal contents (0.4–40 ppm) for the shallower of the three wells analyzed in his study. In general, oil-field brines may contain higher concentrations of the minor trace elements than seawater or surface waters. As one example, Silvey[508] reported that the occurrence of Ge was unique because it was found only in oil-field brine and spring water. Dong[547] developed a special concentration–spectrographic method for determining Ge to eliminate interferences from heavy metals in certain water samples. The relative standard deviation was 8.0%. Another example, where interferences were studied in detail, were the unusually high contents of 0.3 ppm of Zr and 0.03 ppm of Sn in Saratoga mineral waters as studied by Strock and Drexler.[548] These elements were enriched on the order of 25-fold relative to Fe than the earth's silicate crust, while the elements Co, Ni, Mn, and especially Ti and V, were relatively depleted.

2.9.1.3b Seawater. The average composition of major elements and ranges for minor trace elements found in seawater are given in Table 2.12. A number of references pertaining to seawater are cited in Section 2.9.1.2, including those by Martin,[528] Goldberg,[529] and Richards,[521] who reviewed the state of our knowledge of trace elements in seawater. He emphasized the fact that at that time, contrary to the belief of some oceanographers, the concentrations of trace metals in the sea are not very well known. Considerable additional work has been performed since his report. The analytical methods used for such applications have received much study but may still provide questionable accuracy and certainly cannot be used to establish abundance of trace elements in the sea unless a carefully planned extensive sampling is made. Carpenter,[549] has written a chapter on problems in applications of analytical chemistry to oceanography. Among other problems he discussed the lack of accuracy and poor agreement

Table 2.12 Approximate Abundance of Elements in Seawater[a]

Major elements		Minor elements		
			μg/liter	
Element	mg/liter	Element	Average	Range
Cl	19,000	Ag	0.3	0.1–2
Na	10,600	Al	10	5–1000
Mg	1,300	As	3	1–20
SO_4	2,600	Ba	30	10–100
Ca	400	Be	0.05	0.001–0.1
K	380	Cd	0.11	0.02–0.2
Br	65	Cr	0.05	0.04–2
HCO_3	140	Co	0.5	0.04–0.7
Sr	8	Cu	3	1–20
B	4.8	Fe	10	5–1000
Si	3	Hg	0.3	0.03–0.15
F	1.3	Mn	2	0.5–10
Li	0.2	Mo	10	1–20
Rb	0.12	Ni	0.5	0.2–7
PO_4	0.2	Pb	3 (deep sea)	0.1–10
I	0.05	Se	4	0.1–6
		Sn	3	1–10
		V	2	0.2–7
		Zn	10	1–20

[a] From Carpenter[549] and Krauskopf.[550]

among different laboratories. He shows a table of unpublished data compiled by Brewer and Spencer of Woods Hole for a cooperative study of Zn in two seawater samples. For 15 laboratories using different methods, four results were discarded and the computed mean value for sample 1 was 4.92 μg/liter, with a standard deviation of 4.60 μg/liter; for the other sample the mean was 5.6 μg/liter and the standard deviation was 5.3 μg/liter. This clearly indicates poor accuracy for such determinations and the doubt which must be placed on some literature data.

Riley and Taylor,[526] in addition to reviewing recent analyses references, used a chelating ion exchange concentration and AA for analyzing 101 samples for Cd, Cu, Fe, Mn, Mo, Ni, V, and Zn. In the area studied, the tropical northeast Atlantic Ocean, they showed random variations for Mo, V, and Zn and to some extent for Cu, whereas the remaining elements were distributed uniformly. Considerable data were also obtained for six elements by Shutz and Turekian,[525] which show variations for Ag, Co, and Ni in worldwide oceans but little for Au, Sb, and Se. For additional elemental analyses of seawater using AA, see Ref. 589.

Although the Cl, Na, and especially Mg contents of seawater are unusually high, the trace metal content is quite low, usually less than 20 ppb, and for many elements less than 1 ppb. Krauskopf[550] reviewed the concentrations for 13 trace elements in seawater and the reasons for their low concentrations. He estimated the amounts that could have been present based on the amounts supplied through all geologic time as contrasted to the current concentrations. An extreme example is 120 ppm Cr supplied to the oceans versus the current abundance of 0.04-1 ppb. Much of this apparent depletion is explained by natural processes in the ocean, including absorption on solid materials and particularly plankton. Enrichment factors for trace metals, especially V, in such substances appear to be very large. Such relations to plankton (the largest biomass bulk on the earth) were discussed in Section 2.3 as a possible source of V in crude oil. Since seawater is relatively high in Mg content and the Mg/Ca ratio, it is intriguing to speculate on the Mg content of this large plankton mass as a possible source. It could be most significant, since according to Mason,[551] the total biosphere mass (which is mostly phytoplankton from the ocean) produced in all geological time is estimated as 6×10^{25} g and the total ocean mass as 1400×10^{21} g. The chlorophyll content of plankton can be estimated to be about 3%,[552] and thus the Mg content of plankton must be at least 0.1%. On this basis the total planktonic mass could have contributed all of the 0.13% Mg in seawater, and more.

2.9.1.4 Water Analysis—Emission Methods

Spectrographic methods have great utility for application to water analysis. They are especially useful for qualitative and semiquantitative surveys of minor trace contents. A sample can be evaporated to dryness or a concentration procedure may be used, after which an arc spectrum using a cup electrode can be photographically recorded for all detectable elements that are present.

In America, ES methods were used as early as 1933 by Braideck and Emery[553] for analyzing minor constituents in U.S. water supplies. They have been used extensively by the USGS[504-508] and others.[515] Uman[554] has prepared a critical review of emission methods used for water-quality control and served as chairman of an ASTM task group that studied spectrographic analysis of water.

Direct solution excitation methods are generally sensitive enough for only major constituents in water. However, such techniques can be applied to water samples concentrated by evaporation[515] or to the end products of concentration procedures which have been redissolved in acidic solutions. Because of their very low amounts, concentration procedures, as reviewed in Section 2.9.2, are required for determining minor elements in water.

Regardless of whether concentration or direct procedures are used, the excitation techniques and electrode arrangements used are the same. The techniques used will be reviewed briefly here. They may be divided into two general types: solution methods and methods employing solid residues.

2.9.1.4a Solution Excitation and Analysis Techniques. Young[555] reviewed ES solution methods any of which could be applied to water analysis, and Hitchcock and Starr[523] reviewed ES methods as applied to the analysis of seawater, including solution methods. Various techniques used are itemized as follows:

1. *Rotating disc electrode (Rotrode).* The most popular solution method appears to be that using the Rotrode, as tested by Hitchcock and Starr[523] and utilized by Kopp and Kroner[515] for the ESD determination of 19 elements in natural water. Using spark excitation, they achieved relative standard deviations varying from 3.3 to 6.1%. They used evaporation to a volume of 5 ml for samples varying from 100 to 2000 ml to concentrate analytes. The Rotrode method has been used in the P.I. since the late 1940s and has seen much use for the direct analysis of lubricating oils. Pagliassotti and Porsche[556] in their many uses of the technique since 1951 have developed the technique to a high degree, achieving some outstanding precision, as demonstrated in applications to a solution technique for alloy analysis. They achieved coefficients of variations ranging from 1.5 to 2.7% for Cr, Cr, Mn, Mo, Ni, and V, and 4.1% for Si. The technique has been used with good results in the author's laboratory, especially for lubricating oil analysis, but the vacuum cup electrode is preferred for aqueous solutions.

2. *Vacuum cup electrode.* In the author's laboratory a modification of the vacuum cup electrode as originally developed by Zink[557] is preferred for analyses of aqueous or acid solutions. The modification[558] to the use of a flat tip increases the rate of sample consumption about 2-fold, thus improving line-to-background ratios, and consumes less sample. Precision appears better, emission intensity versus time stabilizes faster (compared to the Rotrode), the technique is easier to use (requiring no complicated attachments), and the electrodes may be reclaimed and reused to economize on cost. The technique is especially useful for the ESD determinations of elements such as B and Si which are difficult to determine by other methods and also provides excellent sensitivity for Ba and Sr. Determinations are influenced by Na and Ca, major components of water. Therefore, a buffer composed of 1.5% NaCl and 0.2% Ca is used to level matrix effects. Calibration at ppm levels for which precision is still good and determination limits are as follows:

Element	B	Ba	Sr	Si
Calibration levels	4	2	0.4	6
Determination limits	0.2	0.1	0.02	0.2

Using phosphoric acid buffer with the technique stabilizes Si emission, providing the best precision of any method tested for determining Si, although sensitivity is reduced somewhat. Precision data given in Table 2.13 show that the average precision was improved 8-fold, from 5.6 to 0.65%.

Table 2.13 Improved Precision Using Phosphoric Acid Buffer for the Spectrometric Determination of Silicon

With phosphoric acid		Without phosphoric acid	
Net response[a]	Deviation from mean (%)	Net response[a]	Deviation from mean (%)
5230	1.95	7360	0.14
5110	0.39	7530	2.45
5000	0.58	8390	14.20
5170	0.78	7000	4.76
5130	0.0	7030	4.35
5120	0.20	6780	7.74
Av. 5130	0.65	7350	5.60
Av. background = 2230		Av. background = 600	
Lines/background = 2.3		Lines/background = 12.2	

[a]30 mg/liter of Si used in each case.

3. *Porous cup electrode.* In 1949, Feldman[559] reviewed ES solution methods in use at that time and developed a thin-walled hollow upper electrode which allowed sample solution to seep into the electrical discharge. He preferred this technique for analyzing catalyst via a solution method. Other workers, including Gunn[560] preferred the technique to using the Rotrode. Hitchcock and Starr[523] found the technique poor for seawater analysis because of the need to control the electrode wall thickness and seepage rate. Modern closely machined electrodes of special low-porosity material have improved this situation, but there may still be some limitations because of the high TDS content.

4. *Cu-spark or graphite-spark methods* (flat-top electrodes). The Cu-spark method as used by Fred *et al.*[561] and Goleb *et al.*,[562] or the graphite-spark as used by Morris and Pink,[563] can be applied to aqueous solutions to obtain increased and, under certain conditions, outstanding detection limits (see Table 2.2, Section 2.6). A small amount of water (0.1 ml) or aqueous concentrate is evaporated to dryness on a flat-top electrode, thus concentrating the analyte prior to excitation. Hitchcock and Starr[523] tested the technique for application to seawater analysis. Durum and Haffty,[504] Ko,[564] and Skougstad and Horr[565] applied the technique to water analysis. A special version of the technique has been applied to the ESD determination of metals in oil by Hoggan *et al.*[566]

The method does not appear to provide enough sensitivity for determining many of the minor elements in natural water or seawater, and precision is somewhat less than that provided by other solution methods.

5. *Plasma arc and plasma torch.* Collins and Pearson[567,568] applied plasma arc excitation to the ES determinations of B, Ba, Fe, Mn, and Sr in oil-field brines with sensitivities of about 0.4, 2.5, 0.8, 0.4, and 0.08 mg/liter,

respectively. Precision data at low levels were not given, but they appeared to vary from 5.5 to 10.1% at a level of 100 mg/liter. As experienced in the author's laboratory, the type of plasma arc used by Collins has poor stability upon the introduction of sample solution and is tricky to operate. The improved plasma arc discussed in Section 2.6 is probably better.

The plasma torch (also discussed in Section 2.6; see Table 2.5 for detection limits) as used by Boumans and de Boer,[569] Dickinson and Fassel,[570] Lichte and Skogerboe,[571–573] and others[574–579] provides the most promise for direct analyses of minor trace elements in water. Some of the detection limits obtained with plasma torch excitation are outstandingly superior. References 572, 573, 578, and 579 have even applied the technique using special procedures to As and Hg determinations. For As[572] using arsine generation, determinations were restricted by about a 5-ng. As reagent blank, but precision was about ±10% at this level. At levels of 0.063 and 0.125 mg/liter, the relative standard deviations were 2.3 and 2.9%. For Hg,[573] using a reducing aeration technique, determinations were made at a level of 0.01 ppb using a 10-ml sample with a precision in the range ±10–12%.

2.9.1.4b Excitation and Analysis Techniques Using Solid Residues. Various techniques and electrode types have been used for water analysis on samples evaporated to dryness or solid concentrates. The techniques used can be placed in three arbitrary classes: (1) solid, evaporated sample mixed with graphite and pressed into an electrode shape (briquette method); (2) dc arc excitation of solid, evaporated residue in a cup electrode; and (3) concentration procedures using pptn techniques. These techniques will be reviewed next.

1. *Briquette method.* Hitchcock and Starr[523] tested the use of $\frac{1}{4}$-inch by $\frac{1}{8}$-inch briquettes for the analysis of seawater. They used 1 : 1 mixtures of sample residue and graphite and a dc arc excitation for determining Al and Ni. They experienced difficulty preparing dilute standards for other elements but did not use appropriate techniques for their preparation. We have found successive dilutions of oxides mixed with graphite in plastic vials plus a plastic ball with a vibratory mixer to be quite adequate. They[523] also apparently slighted spark excitation, which can increase precision and also line-to-background ratio for determining certain elements by using solid samples incorporated into briquettes.

Helz and Scribner[580] used the technique as early as 1947 for application to Portland cement analysis. The briquette, $\frac{1}{2}$- by $\frac{1}{8}$-inch thick, improved arc precision and was particularly promising using an overdampened-condenser discharge. Key and Hoggan[581] used the technique in a unique application for analyzing catalyst by preparing a Rotrode briquette with $LiCO_3$ added as buffer. They cited a maximum deviation of 10% and a considerably better average. The technique has been used in the author's laboratory with good precision provided by a high-voltage spark excitation. Briquettes weighing about 0.4 g, about $\frac{1}{4}$- by $\frac{3}{8}$-inch thick, were prepared by diluting various nonconducting, solid samples 5-fold with graphite. A hand press and $\frac{1}{4}$-inch-diameter mold were adequate for preparing briquettes, which are then inserted into a brass electrode holder for excitation. Spark excitation can thus be applied to non-conducting samples to obtain better precision, for which arc excitation would otherwise be required.

2. *Arc excitation.* Durum and Haffty[504,505] have applied simple dc arc excitation to water residues mixed with graphite. Concentrations were as low as 0.001 ppm in water, and values for Fe, Mn, and Sr deviated 7–10% from other methods. Some matrix influence was apparently encountered, but a suitable buffering technique might minimize such errors. Hitchcock and Starr[523] also tested the technique for application to seawater but encountered some difficulties. LeRoy and Lincoln[582] applied this technique to the spectrographic analysis of 36 elements in industrial effluents and claimed good results as compared to that for AA.

Techniques using ES-arc methods generally provide lower precision than is obtained by spark excitation. Deviations in the range 10–20% are not uncommon, and exceptional technique and good instrumentation must be used to approach a precision of 5%.

3. *Concentration–precipitation methods.* Concentration methods, to be reviewed in Section 2.9.2, include the use of mixed organic precipitants. Heggan and Strock[583] applied the technique to various types of samples, including oil-field brines. The technique is especially useful for waters high in brine content, since the analyte is concentrated more than a simple evaporation could accomplish and is separated from the bulk of the major sample elements. In Ref. 583 the sample precipitate is mixed with 2 parts of carbon containing 12% In_2O_3 buffer. A modification of the Heggan and Strock method has been applied to considerable water analyses by the USGS.[504,506,507] Their modification consisted of ashing the precipitate and adding a greater amount of indium arcing buffer. Silvey[507] found that this improved detention limits and diminished average deviations for the low concentrations of seven elements analyzed from ±64% to ±19%. Later Silvey and Brennan[506] applied the technique to 320 determinations of 17 minor elements in waters with an average precision of ±15%. At concentrations in the range 0.0025–0.25 mg in the electrode they obtained a standard deviation of 3.3%.

2.9.1.5 Water Analysis—Atomic Absorption Methods

The technique of AA can provide rapid results with good precision for the analysis of many elements in water. Applications of the technique for the analysis of water and seawater are given in a book by Burrell,[516] a symposium on trace inorganics in water,[584] and another on atomic absorption spectroscopy.[585] In the latter, Brewer and co-workers reviewed some methods applied to seawater and successfully applied a solvent extraction procedure to Cu, Ni, and Zn but had difficulty with Co, Fe, and Pb. Riley and Taylor[526,586] reviewed analysis of seawater and applied AA methods using their concentration procedure with chelating ion exchange resins. Cd, Cu, Fe, Mn, Ni, and Zn were determined by AA, but Mo and V were determined by colorimetric methods. Burrell[587] reviewed AA methods as applied to marine research, including water analysis and Kuwata *et al.*[588] determined Cd and Cu in river and seawater via an extraction with sodium diethyldithiocarbonate. Windom and Smith[589] used AA to determine the distribution of Cd, Co, Ni, and Zn in the continental-shelf seawaters off the southeastern United States. Burd[590] reviewed AA techniques and applications to water analysis. Farnsworth[591] and Platte and Marcy[592] applied AA to waters and water-formed deposits. Paus[593] presented a brief review of water analysis using solvent extraction, ion exchange, the sampling boat, and graphite furnace techniques. AA methods are included in a 1970 USGS manual by Brown *et al.*[503] as the only methods other than chemical-colorimetric methods. Parker[594] authored an AA manual for water analysis published by Varian-Techtron.

Galle[595] applied AA to the determination of trace elements in oil-field brine. He used a chelating ion exchange resin to separate elements from the high brine matrix, citing the technique as providing superior stability for concentrates compared to solvent extraction. Mansell and Emmell[596] also applied AA to the determination of trace metals in brines. They use an oxine extraction for Co, Cu, Mn, and Ni and APDC for Cr, Mn, Mo, and V, but found that heating was required for the latter extraction, particularly for Cr.

The application of AA to pollution problems has been of special value. A recent international symposium on identification and measurement of environmental pollutants includes papers on water analysis.[597] Included is a review on trace metals in water by Chau, which includes concentration procedures and analytical methods. Solvent extraction followed by AA is cited, but Chau cautions against common errors and reliance on "overexaggerated performance" for AA. At the same symposium, Petersen and Manning[597] reviewed applications of the graphite furnace for water analysis. Surles *et al.*[598] made a comparative AA study of trace elements in lake water. Simultaneous determinations of seven elements in potable water were made by Aldous *et al.*[599] using solvent extraction and a Vidicon AA spectrometer. Detection limits were 10-fold less than for conventional AA but were adequate except for Cr and Pb.

2.9.1.5a Nonflame Techniques. This flameless technique has excellent possibilities for the analysis of trace and ultratrace metals in water due to the 10- to 100-fold better detection limits that it provides compared to conventional AA flame methods. Although these methods provide outstanding sensitivity, caution must be observed in using them. The small volumes used (0.5–100 μl), the very short time of atomization, possibilities of lower precision, and chemical interferences and interferences caused by scattered light and/or molecular absorption make for a tricky operation of nonflame devices. Various nonflame devices were reviewed in Section 2.6.

In addition to Petersen and Manning,[597] Paus,[593] Parker,[594] Edmunds *et al.*[600] Fernandez and Manning,[601] Donnelly *et al.*,[602] and Shkolnik and Shrader[603] have discussed applications of nonflame methods to water analyses.

Pickford and Rossi[604] developed an automated sampling system using a Perkin-Elmer furnace and a Beckman spectrometer with a detection system for five channels. They made a detailed study of precision for which the automated system produced outstanding results. Using a multielement lamp and 100-μl sample volumes they obtained relative standard deviations from 0.6 to 2.4% for 50–100 ppb of Co, Cr, Cu, Fe, and Ni and 10 ppb of Mn. They discussed precision obtained by other workers, which varied from 3% to 12%. Using manual sample injection for Mn, with care they obtained a precision of 2.2%, but found that with less care values as poor as 11% were obtained and that an average value of 4% was obtained under normal conditions. The precision for Mn with the automated sampling method was 1.7%. This is good evidence that injecting the small volumes used in nonflame methods, especially volumes under 10 μl, is a critical factor influencing precision.

2.9.1.5b Aeration-Reduction Methods for As, Hg, Sb, and Se. Flame AA methods for As, Hg, Sb, and Se have poor sensitivity and are not adequate for antipollution applications. As a consequence, chemical methods have been devised which introduce these elements into the flame or an absorption cell as a means of increasing sensitivity.

Since the world became aware of the hazards of Hg, many publications have appeared for Hg determinations using the cold vapor method. Method references and background information are discussed in Section 2.8. Only a few selected references will be given here. See Ure[605] for a review with 442 references on nonflame and AF for Hg analyses.

The cold vapor or aeration technique for producing Hg vapor after its reduction was first used by Kimura and Miller[606] to concentrate Hg for a colorimetric determination. Manning[607] reviewed methods, including the Hatch and Ott[608] method, which has been very popular and much used. The method can detect fractions of a nanogram of Hg but is subject to certain interferences, such as hydrocarbons and an excess of easily reduced metals such as Cu.[609] Caution must also be used in employing some of the various procedures designed for analyzing nonionic, organic forms of Hg. Many of these procedures have used compounds such as methyl mercuric acetate or chloride to establish accuracy, when actually these compounds are ionic in nature. Generally speaking, analysis of water samples is straightforward, with few interference problems.

Aeration methods for As, Sb, and Se are based upon producing arsine, stibine, and hydrogen selenide, respectively, which are introduced into argon-supported hydrogen flames. These techniques were reviewed and described by Hwang *et al.*[610]

2.9.1.6 Atomic Fluorescence Methods

This is a relatively new technique and applications have been limited. Technicon Industrial Systems has designed a multichannel AF spectrometer which has been applied to the simultaneous determination of trace elements in water and lubricating oils by Demers *et al.*[611] For Ag, Cd, Co, Cr, Cu, Fe, Mn, Ni, and Zn, direct detection limits were found equal to or a factor of 2 lower than federal water quality limits except for Co and Ni. Results were reported to be comparable to AA results. However, a very high sample volume was necessary to justify its purchase, so this unit has been discontinued.

2.9.1.7 Flame Emission Methods

With the availability of commercial instrumentation for FE after World War II, various applications began to appear in America, including water analysis. The technique was especially valuable as a replacement for the more tedious chemical determinations of K and Na. In 1950 and 1951, determinations of alkali and alkaline earth elements in water by FE were studied by West *et al.*[612] and by Scott *et al.*[613] Skougstad and Horr[565,614–616] and Durum and Haffty[504] applied FE to an extensive study of Sr in natural waters. A unique application is that of Agg *et al.*[617] using Li as a tracer to follow flow rates of water or sewage. There are many other published applications for determining the alkali and alkaline earth elements in water including over 140 references listed by Mavrodineanu[536] and FE should be accepted today as one of the best techniques for analyzing these elements. Folsom *et al.*[618] applied FE to the precise measurements of Cs in ocean water.

Marsh[619] extended FE to the determination of Fe as well as Ca, Mg, and Na. At the level of about 1 ppm of Fe, the average standard deviation for four samples was 0.015 ppm. With the development of modern, improved instrumentation and burners, FE can be applied to the determination of many additional elements (Section 2.9.1.6b). Prager and Seitz[620] used FE for analyzing P in air and water, and Prager[621] applied FE to pollution monitoring for the nonmetallic elements Cl, N, P, and S at concentrations of 0.07 ppm or less via molecular emission. Belcher *et al.*[622] applied molecular emission cavity analysis (MECA) to sulfur anions, and Ghonaim[623] reviewed its use for B, As, and Sb analyses. Milano *et al.*[624] applied a Vidicon scanning spectrometer to the simultaneous determination of Ca, K, Li, and Na, with standard deviations ranging from 0.01 to 0.25 ppm. As in Ref. 599 using a Vidicon AA, they also found some restrictions but potential value for certain applications.

2.9.1.7a Analysis of Oil-Field Brines and Seawater. Considerable study has been undertaken by the USBM of applications of FE to the analysis of oil-field brines. The importance of elemental analysis of oil-field brines to geological studies and the application of FE were reviewed by Gullikson, Caraway, and

Gates.[540,625] Analyses of certain brines and seawater were compared to establish that the brines did not come from the ocean. Gates and Caraway also applied FE to the analysis of oil-well scales.[626] In a series of USBM reports and other publications, [627-633] Collins applied FE to the analysis of oil-field brines. The elements determined included Ba, Cs, K, Li, Mn, Rb, and Sr. Mihram and Catto[634] devised a quick identification of formation brines by comparing flame spectrograms. A unique application to brine analysis is a 24-hour continuous flow for monitoring Ca devised by Bissett *et al.*[635]

The problems for the FE analysis of seawater and oil-field brines are similar. In both cases high TDS content can cause burner clogging and depressing effects, and high Na content can cause unwanted background. Background corrections, however, are easy to make for Na interference. Frequent rinsing or dilution of samples can minimize difficulty from high TDS content. Another approach using the standard addition technique has been used by Chow[636] for determining Sr in seawater. This compensates for TDS or matrix interferences, which depress emission. Chow and Thompson also determined Ca[637] and K[638] in seawater and marine organisms. Hitchcock and Starr[523] used FE spectrography for determining Ca, K, Mg, Na, and Sr in seawater. In comparison to ES spark methods, they found FE especially suited to K analysis and remarkably free from spectral or background interferences using the 4044 to 4047-A doublet. Joensson[639] and Riley and Tongudai[640] determined Li in seawater, and Honma[641] determined chloride in seawater via molecular emission.

2.9.1.7b New Developments That Extend Applications. Modern instrumentation and techniques to improve FE detection limits have recently extended the scope of this technique for analyzing concentrations below 1 ppm. A 1969 review by Pickett and Koirtyohann[642] showed that the use of a 5-cm laminar flow burner using nitrous oxide-acetylene improved FE detection limits and made them highly competitive with AA. Flame emission should be used more often, especially when it may be better than AA. See, for example, only the exclusive use of AA in addition to chemical colorimetric methods in a recent EPA manual.[496] Fassel *et al.*[643] discussed this "unjustified neglect" of FE as applied to solution methods for alloy analysis, citing many references and the fact that FE can provide equal and even superior results to AA. They applied FE using a 5-cm, nitrous-oxide acetylene flame to the sequential, multielement analysis of Al, Co, Cr, Cu, Mn, Ni, and V in steels with relative standard deviations from 0.1 to 2% for the first six elements and 4% for V.

By combining the use of nitrous oxide-acetylene flames with the ac FE technique discussed in Section 2.6, potentials for FE are even greater. Buell[644] discusses the advantages of the ac technique and applications to water analysis. Out of 23 elements of current interest to antipollution, Ca, Mg, and Na are not problems, both AA and FE are inadequate for direct determinations of B at low levels, AA is superior for seven elements, and FE is adequate for the remaining 12 elements, even at the most stringent limits, as imposed by the new California

Table 2.14 Water Analysis by ac Flame Emission versus the Most Stringent Limits Required by Various Regulations

Element	Disposal limit (ppb)	Current ac FE detection limit (ppb)
Al	1000[a]	2
Ag	20	3
Ba	1000	0.3
Co	200[a]	7
Cr	5	0.7
Cu	200	10
Fe	300	6
Li	5000[a]	0.5
Mn	50	3
Mo	5-50[a]	3
Ni	100	6
V	10,000[a]	2

[a]Limits for irrigation.

Ocean Discharge Plan. A list of these elements and ac FE detection limits and pollution limits are given in Table 2.14. Precision of the ac FE method is excellent; 0.5–2% and detection limits for many elements, including Al, Cr, Mo, and V, are better than those provided by AA, and equivalent for many others. The technique also eliminates spectral interferences encountered using conventional dc Fe.

Of the elements discussed in Ref. 644, Cr is of great interest to the P.I., and the new antipollution limits are beyond the detection limits of some analytical methods. A colorimetric method is used as a standard method in most references but severe interferences can be encountered in certain P.I. effluents. Direct AA or FE methods are faster and eliminate these interferences, but AA detection limits are pushed by the new California limit of 5 ppb Cr. Buell[(644)] cites a conservative detection limit of 0.7 ppb Cr using ac FE and a precision within 5% for continuous measurements at a level of 20 ppb.

By combining the technique with a precipitation separation method, a rapid method for determining Cr in seawater has been developed (see Section 2.8.3.1). Using 250-ml samples of seawater from an ocean bay, the average obtained for four determinations was 0.46 ppb of Cr, with an average deviation from the mean of 4.8%.

2.9.2 Concentration and Separation Methods

When detection limits for analyzing water samples or other aqueous solutions are inadequate, one must concentrate the desired analyte to provide adequate sensitivity. Concentration procedures are often also separation procedures, in which case they may be used to eliminate interferences. For emission and AA methods, the degree of separation very often need not be high to eliminate interference. For these reasons concentration-separation procedures can become a very important part of analytical methods. Desirable features of concentration methods are as follows: (a) procedures should be rapid and efficient; (b) they should provide high concentration factors and a suitable form for analysis; and (c) procedures should provide quantitative recovery (in one step if possible) and the degree of desired separation if interferences are present.

The concentration methods reviewed here will include evaporation, solvent extraction, ion exchange, and precipitation.

The coverage in this section is not intended to be an exhaustive review of the subject. The references selected are, however, intended to present a reasonably thorough coverage. A general discussion of concentration-separation procedures is given in a chapter by Mizuiki[645] and a discussion by Rottschafer *et al.*[646] as applied to neutron activation techniques.

2.9.2.1 Evaporation

The simplest concentration method for aqueous samples and some organic samples is evaporation, although it may be somewhat slow because of the elapsed time required. It is quite effective for samples low in total solids but it is poor for brines with a high total dissolved solids (TDS) content. Qualitative spectrographic analysis on the TDS has been a routine approach to water analysis of trace elements for many years. It has the advantage of the simultaneous determination of many elements.

For the quantitative analysis of natural waters, concentration to a given volume followed by AA has been used by Farnsworth[647] for determining seven elements and by Kopp and Kroner[648] for the simultaneous ESD determination of 19 elements using spark excitation; the latter achieved precisions of 3-6%. In a review of methods applied to analyzing high-purity chemicals, Joy *et al.*[649] cite evaporation as a technique for the spectrographic analysis of acids. LeRoy and Lincoln[650] evaporated industrial effluents at 120°C for a simultaneous spectrographic determination of 36 elements and claimed good results compared to AA. Both Kroner[648] and Joy *et al.*[649] reviewed evaporation and other methods of concentration.

The copper electrode-spark method of Fred, Nachtrieb, and Tompkins,[651] also used by Goleb *et al.*,[652] and the graphite electrode electrode-spark method as used by Morris and Pink[653] are essentially evaporation-concentration methods. Sample solution is dried on the top of a flat or slightly dished electrode and spark excitation is used to achieve detection limits in the ppb range (see the last column in Table 2.2, Section 2.6).

2.9.2.2 Solvent Extraction

Solvent extraction is well suited to AA and FE methods. Usually, a chelating agent is added and the analyte is extracted into an organic solvent. This provides a 3-fold advantage as follows:

1. The analyte is concentrated in the organic phase, thereby increasing practical detection limits.
2. The organic solvent used for the extraction enhances response, giving an additional sensitivity increase and also reduces interelement interferences by eliminating water (Section 2.7).
3. Potential interferences are eliminated, since the analyte is generally separated away from alkaline earth elements and possibly other elements.

2.9.2.2a Literature and General Systems. There are many solvent extraction systems, which complicates selecting the best system. Useful general books on this topic include those by De *et al.*,[654] Morrison and Frieser,[655] and Stary.[656] A series of articles by Frieser arranged in a useful periodic-table treatment include the use of dithizone,[657] 8-quinolinol or oxine,[658] cupferron,[659] and acetylacetone.[660] Katz[661] reviewed solvent extractions and cites over 5000 procedures given in other reviews.

Since spectral interferences caused by other elements are few for AA and FE, solvent extraction systems that extract as many elements as possible (excluding alkaline and alkaline earth elements) should be the best. Such systems use mixtures of chelating agents such as that by Sachdev and West[662] for AA applications (it should be equally useful for FE applications). They extract 10 metals of environmental significance using a mixture of diphenylthiocarbazone (dithizone), oxine, and acetylacetone in ethyl propionate. Kirichenko *et al.*[663] used a mixture of diethyldithiocarbamate (DEDC) and oxine in 4:1 chloroform isoamylalcohol at pH 4.3–4.9 for AA applications.

2.9.2.2b Applications for Emission Spectroscopy. There has been little use of solvent extractions (compared to precipitation methods) for ES applications, but they have been used to some extent (cf. 8 and 14 in Ref. 649).

2.9.2.2c Applications for Flame Emission. Solvent extractions have been used extensively by Dean[664] and coworkers for FE. Many other workers have also applied the technique to FE, including Stander,[665] who extracted V and used a V band at 5550 Å with a Beckman Model DU. Since those early days of FE, technique and instrumentation have developed markedly. With a modern monochromator and an ac FE technique, Buell[666] has detected 1 ppb of V in organic solvent using the 4379.2-Å atomic line. Using this technique combined with solvent extraction can lead to a detection limit of less than 0.01 ppb. Since V is of extreme importance in the petroleum industry, another chelate extraction, with high V specificity, used by Jeffery and Kerr[667] for a colorimetric determination is cited. By using a final FE measurement, the yellow interference mentioned for Ti can be eliminated. Oxine has also been a standard chelate used for V extractions[654,655] and has been applied to a FE determination of V by Goto and Sudo[668] and also to FE determination of Ca, Co, Cu, Ga, In, Mn, and Ni (See Section 2.9.2.2e for additional V extractions).

Nickel is also of importance to the P.I. Dimethylglyoxime is the chelating agent most often used for chorimetric determinations,[654-656] and can also be used for solvent extractions and separations from Co. Other systems, including the mixed chelate system,[662] can also be applied to Ni if the degree of interference separation is adequate.

2.9.2.2d Applications for Atomic Absorption. Since 1961 solvent extractions for AA have grown until now it seems to be the favored method used in conjunction with solvent extractions. Much of the early use of the technique for AA stressed the use of ammonium pyrrolidine dithiocarbamate (APDC). Mulford[669] reviewed solvent extraction techniques for atomic absorption. He studied APDC using methyl isobutyl ketone as extractant, indicating additional elements that might be extracted and other elements for which extraction is only partial. Brooks *et al.*[670] used APDC for extracting trace elements from brines. Mulford also used cupferron to extract Cu, Ti, and V and cites oxine as reacting with even more metals than APDC, yet cites only one and two references for their use while 16 references are given for APDC. This appears to be rather a curious exclusion of other extraction systems for AA, giving the use of APDC approval by popularity rather than scientific selection. Here, again, early AA users ignored previous extraction systems used for FE and other applications. This exclusive popularity for APDC does not appear to be currently universal. In the Analytical Chemistry 1971 review of water, only 1 out of 6 solvent extraction methods cited for AA used APDC. Orren[671] and Mansell and Emmel,[672] in addition to APDC, used oxine as an extractant for AA applications, as did Fishman[673] for the AA determination of Al in waters and Yanagisawa *et al.*[674] for Ca. The ultimate technique would appear to be the use of mixtures, as mentioned in Section 2.9.2.2.1.

One unique application for a nonflame, AA application is the determination of Ag in snow via dithizone extraction by Woodruff *et al.*[675] as an aid in evaluating the success of cloud-seeding programs with AgI.

2.9.2.2e AA Applications for Vanadium and Additional Extractions. Since V is of such great importance to the petroleum industry, additional references are cited in this section as applied to AA. The same systems can be applied to ac FE or nonflame AA to achieve better detection limits (Section 2.6).

Chau and Lum-Shue-Chon[676] studied 13 chelates for the solvent extraction–AA determination of V in lake waters. They preferred dichlorooxine as the chelate and butyl acetate as the solvent (better than MIBK). Both Crump-Wiesner and Feltz[677] and Pearton *et al.*[678] used cupferron to extract V into

MIBK and butyl acetate, respectively. Satyanarayana *et al.*[679] extract V^{5+} into 1:1 acetylacetone–butanol.

Jacubiec and Boltz[680] developed a unique, indirect AA determination of V via a heteropoly complex with a P/Mo/V ratio of 1:11:1 with the determination of Mo. They extracted the complex into a mixture of 1-pentanol ether but back-extracted the Mo into an aqueous buffer for the final measurement. Johnson *et al.*[681] used the same procedure, but developed a technique to eliminate interferences.

2.9.2.2f Extractions of Heteropoly Compounds. In addition to V, the elements As, P, and Si (also Mo) can be extracted as heteropoly complexes. Because of the poor AA sensitivity for these elements, such extractions should be invaluable. Kirkbright *et al.*[682] developed an indirect, sequential method for the AA determination of P and Si. As with the V complex in Ref. 683, a high ratio of Mo to P and Si (12:1) is formed in the complex, which is extracted into an organic solvent. The Mo is then determined by AA, thus providing increased sensitivity via the favorable ratio in the complex and the solvent extraction.

Any of the similar complexes, well known for colorimetric determinations, can be applied for such applications. These have been used by Ramakrishna *et al.*[683] for As, P, and Si by AA and by Danchik and Boltz[684] for As determinations.

2.9.2.2g Other Special Extractions. Another element with poor flame sensitivity is B, which has been analyzed via solvent extraction by Pau *et al.*[685] and Holak[686] with 2-ethylhexane-1,3-diolinto in $HCCl_3$ or MIBK and by Agazzi[687] as tetrabutylammoniumborontetrafluoride ion in MIBK.

Mercury (of special environmental interest) has been extracted from brines with quaternary amines by Moore[688] and as HgI_2 into CCl_4 by Morris and Whitlock.[689]

2.9.2.3 Solid Extractions or Cocrystallization

An improved technique for extractions which provides very high enrichment factors was reviewed and studied by Flaschka *et al.*[690] It consists of using molten organic compounds as solvents, after which removal of the extract is facilitated after solidification of the solvent. This technique provides high enrichment factors without some of the problems encountered when attempting similar enrichments using conventional liquid organic extractions. This technique, although it has seen little use, may offer some advantage in certain cases, especially if combined with micro techniques for the final measurements. Perhaps sampling boats or nonflame methods could be utilized in conjunction with the technique to provide super detection limits.

Cocrystallization with an insoluble organic material can be considered to be a similar technique and has been used by Riley and Toppin[691] for trace elements in seawater.

2.9.2.4 Ion Exchange

Ion exchange is another technique for concentration that can provide certain advantages. One notable advantage is that stable solutions can be provided by chelation–extraction procedures. Galle[692] evaluated this advantage and describes the use of chelating ion exchange resin for determining Ca, Co, Cu, Fe, Mg, Mn, Ni, Pb, Sr, and Zn in oil-field brines using AA. Other selected references pertaining to the use of chelating ion exchange resins are those of Biechler,[693] Freudiger and Kenner,[694] and Riley and Taylor.[695,696]

Other advantages of ion exchange are the possibility of very large concentration factors and separations of elements in the presence of large concentrations of elements, such as Na, Ca, Al, and Fe. The use of anion exchange to separate chloro complexes in such a situation was reviewed by Le Riche.[697] He separated Ag, Mo, Pb, Sn, and Zn, followed by dry ashing of the resin plus sample prior to ES analysis of rocks and soils. This technique is also excellent for other applications. In the author's lab it has proved especially valuable for the AA analysis of gold in brines and seawater. Gold in such solutions is very strongly held on the resin, and dry ashing was employed to break the bond. Treatment of the ash with aqua regia, evaporation to dryness, and dissolution in HCl was used to prepare sample solutions. Brabson and Wilhide[698] determined Ca in wet-process phosphoric acid by FE. The many interferences for Ca are well known; in this case they are eliminated by the use of cation exchange to separate the Ca from phosphate. Sr was used to minimize Al interference on Ca and as the internal standard. Hinson[699] also eliminated interferences by using ion exchange for an AA determination of Ca.

An approach similar to that of Ref. 698 would be to remove phosphate with anion exchange resin. Since P and Si have poor sensitivity by AA or FE, an anion exchange resin such as Amberlite IRA-410 (Cl form) can be used prior to their determinations to concentrate phosphate and silicate.

Ion-exchange-resin loaded papers are also commercially available and may have some advantages in certain cases[700] for concentrating analyte. A general text on this topic is that by Samuelson[701] on ion exchange separation in analytical chemistry. Another type of ion exchange, uses of liquid ion exchangers in inorganic analysis, was thoroughly reviewed by Green.[702]

2.9.2.5 Precipitation Methods

Precipitation methods can be useful in achieving very high enrichment factors and also in separating the analyte away from alkali, alkaline earth, and other elements which might cause interference. Both inorganic and organic reagents can be used, although the latter using mixed reagents appears more universal. Silvey[703] reviewed the history of some different precipitation techniques and the use of combined organic precipitants as applied to the ES analysis of waters. Precipitation techniques can be very fast and easy to use under certain circumstances (see the discussion later on the determination of Cr).

More recently, a unique precipitation for trace metals from highly saline water for AA analysis has been reported by Buono *et al.*[704] They achieved precipitations using poly-5-vinyl-8-hydroxyquinoline which were quantitative and faster than those provided by conventional chelating agents such as 8-hydroxyquinoline. Another unique separation was reported by Vanderborght and Van Grieken[705] using multielement chelation with 8-hydroxyquinoline, with subsequent adsorption on activated carbon. Enrichment factors up to 10,000 were cited for activation analysis and x-ray fluorescence.

In 1960–1961 the *Chemist-Analyst* published a series of articles prepared by Hoffman and by Cheng presenting useful precipitation reactions in a periodic-table form. The series covers the use of hydrogen sulfide,[706] ammonia,[707] cupferron,[708] and 8-quinolinol or oxine.[709]

2.9.2.5a Mixed Organic Precipitants. Oxine as a precipitant (reviewed by Hoffman) has seen considerable use, and early users included Scott and Mitchell.[710] Mitchell and Scott later employed a mixture,[711-713] adding tannic acid and thioanalide in addition to oxine to achieve quantitative precipitation of more elements. They applied the technique to ES determinations of as little as 2 μg of elements in agricultural studies. Farquar *et al.*[714] reviewed the technique, use it for target values of about 6 ppb of impurities in KCl, and extend the application to 39 elements determined by ES. Heggan and Strock[715] applied the technique to ES analyses of various samples, including oil-field brines. The technique has been used considerably by the USGS for water analysis by ES, including Silvey,[703] Durum and Haffty,[716] and Silvey and Brennan.[717] It has also been applied by Husler and Cruft[718] to the AA determination of 19 elements in calcium sulfate minerals, and also for ESD methods.[719] They used Cd as a collector and a pH of 5.9 in an attempt to include Mn. Small amounts of Ca were carried down and interferred with the AA determination of V and Ti unless Ca was added to the standards. This use of mixed organic precipitants appears to be an excellent means of preconcentration for analyzing waters and other samples.

2.9.2.5b Carrier or Coprecipitation. The technique of adding a slight excess of certain elements as an aid in achieving complete recovery by coprecipitation should be used whenever possible. The technique of coprecipitation is well known and, for example, has been used by Marshall and West[720] for the AA determination of elements in Al salts. They coprecipitated Ca and Mg on iron(III) hydroxide and Fe and Ni on hydrated manganese(IV) oxide and in each case extracted the analytes from the precipitate with oxine in MIBK. Ko and Anderson[721] used lanthanum hydroxide as a carrier precipitate for the ES semiquantitative determination of 31 elements (including P and Si) and the quantitative determination of five elements. Severne and Brooks[722] applied AA to the determination of Se by coprecipitation on As. Krishnamurty and

Reddy[723] obtained AA enrichment factors up to 400 for Pb with APCD added to Co.

A simple hydroxide precipitation followed by AA or FE analysis of the dissolved solid can provide a very simple, fast procedure. In a method developed in the author's lab for the determination of Cr in waters (including brines and seawaters), the technique of Strohal *et al.*[724] and Ishibashi and Shigematsu[725] using aluminum hydroxide as carrier precipitation is utilized (see Section 2.8.3.1). Such techniques are discussed in the book by Sandell.[726] Fukai[727] precipitated total Cr and Cr^{3+} with Al or Fe hydroxides to determine Cr and the distribution of its valance states in seawater using a colorimetric method. The procedure developed in the author's lab was applied to the analysis of Cr in seawater using ac FE. In a sample from an ocean bay, 0.46 ppb was found, with an average deviation from the mean of 4.8%.

2.9.3 Crude Oils

The importance of a crude oil supply to the P.I. has been discussed in Section 2.3 as well as the origin of crude oils (crudes) and their metals. Because of this importance, in addition to analyses of metals in crudes, this section will discuss the importance and nature of the metals and porphyrins in crudes and methods of analysis for metal porphyrin compounds. All the items discussed may be of some importance in developing analytical methods for determining metals in crudes. For example, Garwood *et al.*[728] applied the knowledge obtained from the literature to remove metals and nitrogen from petroleum stocks using an approach similar to those used in demetalation studies of porphyrin compounds (Ph) as discussed in Section 2.9.3.6a. They treated crudes and coker stocks with hydriodic acid in an autoclave treatment at higher temperatures, and although they found V more resistant, they were able to extract 97 to 99% of Fe, Ni, and V with an 89% recovery of oil. Most demetalation methods discussed in Section 2.9.3.6a do not remove metals quantitatively, although the chemistry may be similar but is utilized in perhaps milder treatments. Another example is the study of Barney and Haight,[729] who compared four extraction procedures to three ashing procedures. In the determination of five metals in oil, extractions with HBr–acetic acid and H_2SO_4-HCl gave results equivalent to a wet-ashing procedure.

2.9.3.1 Importance of Crude Oil Metal Content

The trace metal content of crude oils (generally less than 0.2%), although small in comparison to the organic content, is of considerable importance to the P.I. The unique preponderance of Ni and V and occurrence of their porphyrin compounds in crudes is of great interest to geochemical studies that might aid exploration for crude oil and its production. This has resulted in many studies relating to the origin and migration of crudes that involve the associated metals.[730-732]

In addition to geochemical studies, metal content is important to refining operations and to finished products, since it could be detrimental (Ref. 1 and Sections 2.9.6 and 2.9.7). Bieber *et al.*[733] in their introduction review the subject and its impetus on research by the P.I. Certain metals, especially Ni and V, cause problems in both refining procedures and final products. Some of the metals are carried into the feedstocks (gas oils) fed to the catalytic cracking units for producing gasoline. In these units metal deposits may build upon catalyst beds poisoning them and thereby reducing distillate yields (Section 2.9.6). After World War II, increased gasoline demands led to special procedures to use heavier oil fractions and correspondingly greater problems in refining procedures. As a result, much research has been conducted by the major oil companies which has revealed that not only the metal quantity but the type of metal compound has a bearing on refining procedures. Larson and Beuther[734] showed that hydrodesulfurization of crudes removes V faster than Ni, until at 99% metals removal the ratio of V/Ni becomes 0.1, as compared to 3.8 in the starting material. They also studied the chemical nature of metal complexes as coordinated with S removal and discussed the distribution of Ni and V in micelle formations in oil. They cite the greater surface-active powers of V as causing it to concentrate at the outer micelle edges.

Metals concentrate in the higher-boiling, residual oil fractions and in petroleum cokes. Metal residues in such products cause corrosion of boiler tubes and firebricks (Section 2.9.7).

The nature of metal compounds and their volatility is also important to analytical chemists and the development of analytical methods. Analysis of metal content (especially Ni/V ratios) has also been of interest as an aid in identifying sources of oil spills (Section 2.8).

2.9.3.2 Metallic Elements Found in Crude Oils

Many elements have been identified in crudes, including 36 elements detected in a survey of 120 U.S. crudes by Hyden.[735] Other surveys of metals in crudes are given in Section 2.9.3.7. Of all the elements, Ni and V are of the greatest importance and significance. In most crudes their concentration is higher than that of other metals and they also occur as porphyrin (Ph) compounds. They are the only Ph compounds that have been identified, and all Ph compounds appear to be complexed with Ni or V, but all of the Ni and V is not complexed as Ph. The V concentration in certain crudes can be relatively high, greater than 100 ppm, and reaches the exceptional value of 1100 ppm in Boscan crude. Crudes can be considered high in Ni content for concentrations over 50 ppm and range to over 100 ppm. Concentrations can also be very low in some crudes, <0.1 ppm. Many of the other rarer metals in crudes also occur at <0.1 ppm. The V/Ni ratio can vary widely through the range 0.1-10 and occasionally can reach 0.02-0.05. An exceptional ratio of 220 is recorded by Hyden[735] for a Half Moon crude oil (sample D96083).

Although Ni and V are usually predominant, other metals that have been recorded as equivalent or higher in concentration include Ca, Cu, Fe, Mg, Na, and Zn (Ref. 735 and other references in Section 2.9.3.7). In such cases the concentrations may vary from 1 to 100 ppm and may reach 200 ppm. The predominance of other elements, especially Cu, seems to be more frequent when Ni and V are under 2 ppm. These occurrences are usually infrequent but seem to be recorded more often for Cu, which in these cases may range from about 0.5 to 10 ppm. An unusual case of 60 ppm of Zn is recorded in Ref. 735, p. 11). In the studies of metal compound types, no one seems to have selected any of these unusual crudes. Other unusual oils include a Baxterville crude with 100% of the V as Ph, and a Red Wash crude with no Ph. Extensive reviews of metals in oils are given by Dunning[736] and Sugihara,[737] who present a concise review of metals in oils, emphasizing Ph studies.

2.9.3.2a Correlation of Ni and V with Nitrogen and Sulfur Content. It has been noted in the literature that high V content occurs in crudes with high sulfur content. Nelson,[738] in a study of demetallation of residuum oils to aid refinery processes, plotted S content versus V + Ni content, which showed a definite correlation. Sugihara *et al.*[739] also cited this correlation and also the possibility that higher N and Ni contents may correlate.

The S content of crudes varies from about 0.1 to 6%, and the largest V content for any crude is 1200 ppm V for Boscan crude, with 5.4% S. Additional values are given in Table 2.15 for the largest V contents recorded. For the 11 additional crudes selected, the V contents range from 370 to 170 ppm. The S content is greater than 3% for 8 of the crudes and ranges from 1.5 to 2.6% for the remainder. Selecting values below 0.3% as low S content, data for many crudes were examined. The V contents were all low, ranging from 0.3 to 3 ppm. Exceptions can occur since low V content is recorded for crudes with S in the range of 1.5-2%.

The N content of crudes varies from about 0.01 to 1% and over 0.4% could be considered high. For 20 of the highest Ni content crudes the % N ranges from 0.48 to 0.88%, with the values for Ni ranging from 50 to 120 ppm. Eight of these crudes are among those with high S content listed first in Table 2.15; note the lower Ni and N content for the remainder. The additional crudes, with low Ni content, are listed last in Table 2.15. Note the lower V content, 17-94 ppm. correlating with S in the range 1.0-2.0%. The Ni content of crudes low in N, less than 0.15% is invariably low. Some examples of low-content California crudes is also given in Table 2.15.

To further test these correlations, especially for Ni, plots of S versus V

Table 2.15 V, Ni, S, and N Contents of Various Crudes

Crude oil	ppm V	%S	%N	ppm Ni
Venezuela				
Boscan	1200	5.4	0.8	105
Bachaquero	370	2.6	0.4	48
Laguinillas	295	2.2	0.3	38
Mara	190	3.1	0.26	17
Tijuana	190	1.5	0.2	22
California[a]				
Santa Maria Valley	220	5.1	0.6	110
Cat Canyon, North	209	5.4	0.54	102
Guadalupe	200	5.7	0.52	102
Lompoc	200	5.2	0.48	90
Gato Ridge	190	3.6	0.68	80
Cosmallia	180	3.8	0.73	92
Castaic Junction	170	2.5	0.65	70
California crudes high in Ni content[a]				
North Belridge	28	1.1	0.8	104
Huntington Beach	30	1.3	0.88	104
McKittrick	85	1.5	0.86	77
San Joaquin Valley, heavy	65	1.2	0.73	68
Lost Hills	17	1.0	0.57	61
Wilmington	40	1.5	0.7	60
San Ardo	94	2.0	0.88	53
Beverly Hills	58	1.6	0.57	53
McKittrick	54	1.5	0.75	52
Oakridge	53	1.0	0.66	52
Ventura	55	1.2	0.73	50
Stewart Station	27	1.3	0.58	50
California crudes low in elemental content[b]				
Dominguez-Rosecrans	18	0.68	0.28	24
Coles Levee, North	10	0.54	0.33	20
Railroad Gap	16	0.65	0.34	16
Coalinga Nose	10	0.44	0.25	6
Riverdale	0.8	0.32	0.22	5
McKittrick, North East	2.5	0.31	0.20	2.5

[a]Most of the data for California crudes are those provided by the Union Oil Co. files for 1970–1976.

similar to those in Ref. 738 and plots of N versus Ni were prepared. Data were plotted for 47 California crudes with the following compositions:

ppm Ni	% N	% S	ppm V
2–110	0.09–0.88	0.2–5.7	0.8–230

The correlations were good, with perhaps some distortion toward high Ni values in crudes with S over 3%. Plots of Ni versus N in the range 1–2.5% showed marked improvement in correlation. Plots of Ni versus S were also made, and there was a correlation, but the scatter of points was wider. A plot of V versus N, on the other hand, showed little correlation with a wide scatter of points and in fact showed essentially no correlation in the ranges of 0.44-0.88% N and 5-230 ppm V for 30 of the 47 crudes.

In conclusion, it appears that larger Ni contents occur in crudes with larger N, although there is some correlation with the S content. For V there is a good correlation with S content but little if any correlation with N content.

2.9.3.3 Nature of Metal Compounds in Crude Oil

There has been considerable debate concerning the nature of metals in crude oil. They may occur as organic compounds or may be inorganic compounds suspended as solids or aqueous (brine) solutions. Part of the difficulty in establishing which forms occur may be due to the fact that crude oil produced from wells may contain entrained solids and brine which are often very difficult to separate from the oil emulsions.

Simple filtering and centrifuging may not be too effective; therefore, conclusions made by various workers must be weighed carefully against their preliminary treatments of the crude oil. In producing oils, special treatments are used in field operations for separating water from crudes.[740]

2.9.3.3a Organic versus Inorganic Metal Compounds. An extensive study by Filby and co-workers[741-745] indicated that many elements occur as organic compounds, including Ca and Fe. Although these compounds could not be water washed away from oil components, one must consider the possibility that inorganic compounds might be trapped in "holes" in large asphaltene molecules.[734,746] Filby found only limited removal by water washing, centrifuging, and filtration techniques. This is in direct opposition to a statement (no supporting data) by Erdman and Harju,[747] who claimed that most metals can be removed by the techniques above and that Ni and V are the only metalloorganic compounds. There appears to be more data for than against the presence of many metalloorganic compounds in crudes. Skinner[740] identified a variety of elements present in solvent-extracted fractions of a Santa Maria Valley, California, crude oil. Dodd *et al.*,[748] in a study of metals absorbed at crude oil–water interphases, showed that many of them are soluble in organic solvents. Ball *et al.*[749] also showed semiquantitative analyses for many elements in the pentane-soluble portion of a Wilmington crude oil, and Moore and Dunning[750] identified more than 50 metals in pyridine extracts of oil shale. This appears to be quite ample evidence that a variety of metalloorganic compounds occur in oil despite failures to identify specific compounds other than Ni and V porphyrins.

2.9.3.4 Nature of Metalloorganic Compounds

Little is known about the nature of metal compounds except for Ni and V porphyrins. The presence of porphyrins, Ph, is well documented as discussed next in Section 2.9.3.5. The presence of nonporphyrin metal compounds as a class has been established, but specific compound types have not been identified. Howe and Williams[751] separated Ni and V compounds into three classes: (1) Ph, (2) stable non-Ph, and (3) non-Ph extracted by glacial acetic-hydrobromic acids. The technique of electron spin resonance (ESR) has been used by Saraceno *et al.*[752] to establish that the V compounds in petroleum exist mainly in the +4 oxidation state. Others who used ESR to study the compounds in crudes were Dickson and Petrakis[753] and Yen and coworkers.[754] Yen and Filby reviewed chemical aspects and the nature of metals in crudes in Ref. 754. Dicksen and Petrakis[753] briefly reviewed the use of ESR and used the technique to establish the presence of V compounds that were not Ph. They concluded that the non-Ph compounds corresponded to one of three possible environments: $VO(N_2OS)$, $VO(S_2O_2)$, or $VO(N_3O)$. The good correlation of V with the S content (Section 2.9.3.2) and lack of correlation with N adds support that much of the V compounds should contain S, such as in the first two compounds.

Most of the metal compounds in crudes are associated with the asphaltine or higher-molecular-weight fractions of crude.[731,739,740,755–757] Whereas the non-Ph compounds predominate in the heaviest fractions, their ratio to porphyrins becomes smaller for lower-molecular-weight fractions. For one crude oil high in Ni content, Filby[741] found 100% as porphyrins in the oil (maltene) fraction, 64% in the resins, and 49% in the asphaltenes (66% of the total Ni was porphyrin). Some of these compounds are volatile and distill with higher-boiling fractions.[731,733,741] Eldib *et al.*[758] and Ray *et al.*[759] have established that some of the metal compounds are colloidal in nature and can be concentrated by ultracentrifuging. Since these compounds precipitate with the asphaltenes, they are believed to be colloidal micelles associated with large asphalt and asphaltene molecules. There is considerable debate about the size of such molecules, which are believed to be no longer than a molecular weight of 2000 by Dean and Whitehead,[731] but which by newer gel permeation or other techniques are believed to be as large as 22,000[741] or even 40,000.[757] Altgelt, in the latter reference, explains some of the differences reported as being caused by the wide range in size of asphaltene molecules, including a distinct portion in the low-molecular-weight range of 700–4000.

2.9.3.5 Metal Porphyrin Compounds

The unique occurrence of porphyrin (Ph) compounds identified only as Ni and V compounds is of considerable importance in the study of crude oils and their origin. To date, these are the only compounds known, despite considerable research to identify other porphyrins and other metalloorganic compounds.

Metal Ph compounds in crude oil were discovered by Triebs[760] in 1934. In 1952 Skinner[740] identified vanadium porphyrin compounds in a California crude oil and studied their distribution (also that of other metals) in various fractions of the oil. The presence of such compounds in crude oil is highly significant and has received much attention. Hodgson *et al.*[732,761] reviewed the literature and the role of porphyrins in the geochemistry of petroleum. Additional references were given in Section 2.3.

Only a portion of the Ni and V contents of crude oils occur as Ph. There is some debate as to the accuracy of Ph determinations (see Section 2.9.3.6b), but evidence of excess Ni and V not complexed by Ph appears quite conclusive. Dunning *et al.*[762] report the percentage of total Ni plus V complexed as porphyrins for 11 crudes to vary from 3 to 45%. Sugihara and Bean,[763] however, found 41% metal Ph for Boscan crude compared to 19% found by Dunning *et al.*[762] Assuming that direct methods[760] give higher results, this might place the range from 10 to 70%. Dean and Whitehead[731] reported corresponding 5–34% for eight crudes, and Constantinides and Arich[730] reported a range of 10–44% for nine crudes. Other reports show higher results, however, such as 66% of the Ni in a crude as porphyrin by Filby,[741] 100% of the V as Ph in a Baxterville crude by Erdman and Harju,[747] and 100% for V in a crude by Skinner.[740] The latter results have been evaluated (Ref. 730, p. 73) as being too high as a result of the use of an erroneous extinction coefficient. Despite inaccuracies in quantitative comparisons (Section 2.9.3.6b) it is still quite apparent that excess metals not complexed by Ph are present in most crudes. It is also interesting to note that crudes are apparently not saturated by metals.[747]

2.9.3.6 Separation and Measurement of Porphyrin Compounds

In the study of metal porphyrins and other metal compounds in oil, various approaches to quantitative procedures have been made. Because of the highly complex nature of crude oil (especially the heavier fractions) and the low metal content, combinations of techniques are used. The techniques used start with separation and concentration procedures. Because they are important to studies on the nature of metal compounds and may be of importance in the analysis of metal content and the development of methods, they will be reviewed here. Before proceeding with specific techniques, one study using a large variety of techniques applied to both porphyrin and nonporphyrin V compounds will be reviewed.

Wolsky and Chapman[764] subjected a Boscan crude oil, asphaltene concentrate rich in vanadium (0.45%), to a variety of treatments. These included acid and base extractions, chemical reduction, dialysis, fractional precipitation, electron bombardment, thermal degradation, ultrafiltration, ultrasonics, and hydrocracking. Their results were negative for concentrating nonporphyrin vanadium compounds. They did find that vanadium porphyrins preferentially passed through dialysis membranes and that ultrafiltration concentrated them in the smallest particle-size class (less than 5 nm, 15-20% of the asphaltenes). They also identified vanadium trisulfide and trioxide in hydrocracked fractions.

2.9.3.6a Demetalation. The first measurements of Ph content in crudes were based on combined demetalation-extraction procedures, followed by measurements of molecular absorption spectra. Groennings[765] developed the demetalation procedure of Triebs,[760] using HBr in glacial acetic acid, into a quantitative determination. Sugihara and Garvey[766] obtained greater yields by substituting formic acid for acetic acid. Various other modifications have been used to improve the Groennings method.[730,755,767,768] Other procedures are those of Kinherler[769] using 90% HF, Dean and Girdler[770] using sulfuric acid, and Erdman[771] using alkyl sulfonic acids. Hodgson, *et al.*,[772] in a study of metal porphyrins in rocks and shale, used reactions with increasing concentrations of methanesulfonic acid and measured the amount of released porphyrin from fluorescence spectra. Very little sample was required, and sensitivity for detecting porphyrins was excellent. Baker,[773] in addition to liquid chromatography, also used a demetalation technique with methanesulfonic acid extraction for a mass spectrometric measurement of petroporphyrins. Sugihara *et al.* investigated an oxidative demetalization process.[774]

2.9.3.6b Direct Porphyrin Measurements. Sugihara and Bean[763] found that dematalation procedures gave low results and probably produced artifacts; thus, they proposed a direct determination. They utilized the soret area of the absorption spectrum and obtained greater Ph values than those of Dunning *et al.*[762] using demetalation. Dean and Whitehead[731] used a similar method, and Erdman and Harju[747] used a direct measurement based on different absorption bands. Constantinides and Arich[730] compared a direct method as in Ref. 763 and a modified Groenings method[765] and obtained 16-32% lower values (with one exception) with the latter method. They stated that discrepancies between the two methods is of little importance in view of the inaccuracies of the methods. As an example they cite the use of an erroneous extinction coefficient used by Skinner[740] to obtain 100% of the V in Santa Maria Valley, California, crude as Ph. Although results obtained by different workers using direct methods vary considerably, they are still higher than those obtained by demetalation methods. Results given in various references for four crudes are listed in Table 2.16 and show considerable variation.

Because the sensitivity for direct methods is less (compared to using demetalized extracts), concentration methods become very inportant and will be discussed next.

2.9.3.6c Precipitation in Asphaltenes. Asphaltenes are a designation in the P.I. which has become accepted as that fraction of crude oil which is insoluble in

Table 2.16 Comparison of Metal Porphyrin Determinations

Nickel and/or vanadium porphyrin content of crude oils expressed as percent of the total Ni and/or V content				
Boscan	N. Belridge	Mara	Wilmington, Calif.	Reference
41	—	—	calc. 70	763
—	44	13	33	730
25	—	12	—	731
31	53	27	—	747
19	45	—	22	762

n-pentane. Asphaltenes contain much higher concentrations of metals than the parent crude and are easy to prepare. As a consequence, many workers have used the asphaltene fraction as a starting point for studies of metals in crudes. Asphaltenes are not a definite class of compounds and do not contain all the total metals in crude oils. Also, techniques for the preparation of asphaltenes have not been uniform with regard to pentane/oil ratio, temperature, and so on. Moore and Dunning(755, 756) and Witherspoon and Winniford (Ref. 775, p. 263) have reviewed the subject. Howe and Williams(751) stated that reproducible weights of asphaltenes (±10%) can be prepared by refluxing 900 ml of pentane with 100 g of oil for 3 h at 43°C. Using this procedure for three oils, they obtained fractions representing 21, 9.6, and 17% of the crude (values for asphaltines in the literature range from 0.2 to 20% for different crudes). The Ni and V contents of these fractions represented about 20% for one oil and 70% for two oils. Although some studies such as this treat the remaining oil fraction (sometimes referred to as the maltenes), many studies have been only for asphaltenes containing about 60% of the metals in the original crude.

An additional reason for such procedures stems from the difficulty of analyzing low metal concentrations in oil. For this reason many studies have been conducted on crudes selected for their higher metal concentrations. Boscan crude oil has been popular for this reason, owing to a content of about 1100 ppm of V and 100 ppm of Ni.

Another fraction, termed resins, has also been accepted in the P.I. and represents the fraction of crude oil which is soluble in pentane but not propane. Some workers find treating samples with propane somewhat difficult(741) and therefore prefer the technique of Hodgson *et al.*(776) using methanol, which is claimed to produce similar fractions. Resins range from about 1 to 30% in crude oils and contain higher concentrations of metals than the parent crude oil but less than found in asphaltenes.

2.9.3.6d Solvent Extraction. In contrast to the preparation of asphaltenes (extraction of the bulk of crude oil away from the desired metal compounds with pentane), various solvent-extraction procedures have been used to extract metal compounds from oil fractions and asphaltenes.[739, 740] Constantinides and Arich,[730] in comparing different methods of metal compound separation, cite solvent extractions as best. Pyridine has been used considerably[731, 733, 777] and was shown by Beach and Shewmaker[777] to extract all the volatile metal-porphyrin compounds for Bachaquero crude, whereas Biebar *et al.*[733] show little similar correlation for a series of oils. Aniline[778] toluidine,[779] acetonitrile,[780] methanesulfonic acid,[781] and alcohols[740, 782] have also been used. Skinner[740] sequentially used a variety of solvents for extracting an asphaltene. Countercurrent extraction was tested by Dean and Whitehead,[731] who used dimethylforanamide and decoline, which was suggested as a promising technique by Constantinides and Arich.[730]

2.9.3.6e Chromatography. Liquid chromatography using a variety of different solid absorbents is a valuable technique commonly used to concentrate Ph fractions. Alumina and silica gel columns have been used most,[739, 783–785] although cellulose powder[786] and thin-layer chromatography[787] have also been used. Howe[783] developed an "improved chromatographic analysis" for demetalated concentrates showing separations for etiophyllo- and rhodo-type porphyrins. He used an integral absorption measurement for the Soret band and demonstrated that sequential fractionations on silica and then alumina are required to obtain good separations. Similar chromatographic techniques have also been used with good results by Sugihara and co-workers[739] in conjunction with solvent extraction and gel permeation chromotography.

In a study of Ph structures, Corwin and Baker[786] used the technique to avoid artifacts that might be formed by demetalation methods (Section 2.9.3.6.2). They use separations first on alumina and magnesol cellulose, then silica gel, and establish Ph in Wilmington crude as a range of compounds of the phyllo series predominating at a molecular weight of about 740.

2.9.3.6f Gel Permeation Chromatography. This technique is based on separation by molecular size and appears to have excellent applications for studying the distribution of metal compounds in crude. The technique has been reviewed by Altgelt[757] and applied to three asphaltenes (Section 2.9.3.4). He found molecular weights ranging up to 40,000 (higher than a previously reported maximum of 6000) and claimed there was ample evidence that the numbers represented the true molecular weight and not particle weights of aggregates. The technique has been used at Union Oil to study the molecular weight distribution of metals in crude oils.[788] The technique has also been studied by the API and USBM and was the subject of a symposium of which two papers were applied to crude oils.[789, 790] This technique has also been applied to Ph compounds by Blumer and Snyder[791] and Pohlmann and Rosscup.[792] Coleman *et al.*[793] applied it to the heavy fractions of a Wasson crude and Cogswell *et al.*[794] to a study of acids and phenols in crude oil fractions. Sugihara *et al.* applied gel permeation along with other techinques to a study and preseparation of Boscan Ph, thereby avoiding an initial demetalation step.[739, 785]

Filby[741] also used gel permeation as well as liquid chromatography in a study of 10 crude oils. He used activation analysis for the determination of 16 elements and the direct Soret band technique for measuring prophyrin content.

2.9.3.7 Emission Methods for Determining Metals in Crude Oils

Because of the low metal content of some crudes and their wide range of viscosities, emission methods for analyzing crudes generally utilize ashing techniques to concentrate the metals. The ash may be excited in a cup electrode using dc arc ES, or it may be dissolved in acid solution followed by spark excitation using ESD.

One method used in the author's laboratory is an ESD procedure using vacuum cup electrodes and a high-voltage spark to excite the acid solutions of sample ash. A buffer composed of Na and Mg is added to minimize matrix effects and Cu, Fe, Ni, and V are simultaneously determined. With careful calibration and technique, precisions obtained are in the range 2–6%. Results compare well to AA and FE methods.

The method was used recently in a Western Oil and Gas Association (WOGA) cooperative study for analyzing four petroleum samples. Out of six laboratories, two used ESD and ashing (one laboratory also used x-ray fluorescence), two others used x-ray fluorescence, two used chemical methods, and one used AA. The results were surprisingly good and for Ni in the range 15–167 ppm the percent deviation ranged from 3.4 to 9.3, while for 77–127 ppm of V the percent standard deviation ranged from 2.7 to 6.4 for three samples and was 30% for one sample containing 7 ppm of V.

2.9.3.7a History and Reviews of Metals in Crude Oil. Spectrographic methods were not used much for oil analysis in the United States until after the late 1940s despite early investigations which reported the presence of trace metals in crudes. Spectrographic analyses were reported by Shirey[795] in 1931 for the analysis of seven American crudes and by Thomas[796] in 1938, who summarized known metal analyses of crudes and cited ES as a likely technique. A summary of analyses for metals in the ash of crudes was presented as early as 1924 by Thomas.[797] The early history of metals in crudes dating from 1883 is reviewed by Southwick,[798] Whitehead and Breger[799] (who comments on Southwick's study), Hyden,[735] Witherspoon and Nagashima,[800] and Hodgson and Baker.[801] The latter[801] present a table summarizing ranges of Ni, V, and V/Ni ratio via age for known references from 1931 to 1956.

Reviews on general use and specific techniques of ES in petroleum labs are discussed in Section 2.9.4 as applied to lubricating oils. Much of this can be applied to crude oil analyses; detailed discussion of specific excitation techniques is presented in Section 2.9.4

2.9.3.7b Methods of Analysis and Metal in Crude Surveys. One of the earliest applications of ES for reduced crude oil analysis was discussed at a 1949 API symposium by Murray and Plagge.[802] They ashed samples in the presence of silica as an ashing aid and used dc are excitation. One of the most comprehensive surveys of metals in crudes was given by Hyden,[735] who presented data for 36 elements in 120 crudes, and 16 samples of refinery residues collected in the western half of the United States in a USGS *Bulletin.* The analytical methods used are discussed by Horr *et al.*[735] in the first part of the *Bulletin.* They gave quantitative results for 15 elements in three crudes and one asphalt by both a dry ashing and the wet oxidation technique of Milner *et al.*[803] Ash samples were mixed with graphite and silica and arced to complete sample burnup. They concluded that both methods gave equivalent results and elected to use the faster-dry ashing method. Semiquantitative determinations were also used. The USGS supplied semiquantitative determinations for 24 elements detected in 24 U.S. crudes and quantitative determinations for Cu, Ni, and U, and V in a study by Ball *et al.*[749]

Considerable data have also been obtained for Cu, Ni, and V by Bonham[804] for 66 U. S. crudes mostly from Oklahoma. He ashed the samples and added Co as the internal standard (I.S.) directly in graphite electrodes, excited the samples with a dc arc in a helium-oxygen atmosphere to reduce background intensity, and observed the cathode layer area[805] for enhanced intensity of spectral lines. In applying the analyses to distributional studies in the Seminole, Oklahoma, area, he found that metal concentrations were greatest near basin shorelines and decreased basinward. He found no significant differences according to geologic age, and the V/Ni ratio ranged mostly from 0.02 to 0.5.

Witherspoon and co-workers obtained data for 45 Illinois crudes[800] and 25 Illinois and Indiana crudes.[806] Witherspoon and Nagashima[800] ashed the samples, dissolved the ash in acid solution, and analyzed the resulting solution using a high-voltage-spark excitation with a rotating disc electrode (Rotrode) according to the method of Nagashima and Machin,[807] which has also been used for analyzing Cu, Ni, and V in crudes. This technique, using Co as an internal standard (I.S.), gave a precision within ±10% for Cu, Ni, and V in the range 0.2–5 ppm. Ratios of V/Ni were given and were quite consistent, varying from 0.2 to 0.07, and did not appear to vary with age of the crude as much as the concentration appeared to become smaller with age. They also reviewed some of the history of metal analyses in crudes. Mast *et al.*[806] used a wet-ashing procedure similar to that of Hansen and Hodkins,[808] precipitated Ni and V with mixed organic reagents according to the Mitchell and Scott method[809] as modified by Shimp *et al.*,[810] and used an arc excitation of the ashed precipitate. By this means they could use a smaller sample, 3 g, due to a 100-fold concentration factor and obtain coefficients of variation for Ni and V of ±9 and ±5%, respectively, for a range of 0.2–15 ppm.

Erickson *et al.*[811] analyzed 29 crudes, 22 asphalts, and 27 petroliferous rocks for metal content. They provide qualitative and semiquantitative data using a dc arc and chemical data for uranium. Jones and Hardy[812] presented quantitative data obtained by chemical methods and semiquantitative ES data for 29 crudes. Colombo and Sironi[813] presented data for 32 crudes and 38 asphalts from Italy and Sicily for Cr, Mn, Mo, Ni, and V using the wet ash–dc arc method of Gamble and Jones.[814] Later, Colombo *et al.*[815] analyzed four of these crudes and four asphalts by activation analysis for 11 elements and compared the results for Ni and V to ES and colorimetric analyses. The two latter methods agreed reasonably well, but results using activation analysis were considerably higher in some cases.

One of the earlier studies (1952) by Karchmer and Gunn[816] presented very little data on crude oils but discussed ES, FE, and colorimetric methods used in the P.I. They used an ashing method developed by Gunn and Powers.[817]

Bieber *et al.*[733] and Dunning *et al.*[762] use the wet-ash method of Hansen and Hodkins[808] and x-ray fluorescence for higher concentrations in their study of metals and Ph in 7 and 11 crudes, respectively.

Bailey *et al.*[818] studied the Hg and Cu, Ni, Fe, and V contents of crudes in the Cymeric field, California, and also the metal contents of associated waters. They reported Hg results which they obtained from other laboratories that varied from 2 to 3 ppm, except for high results on one sample by one of the labs. Other Hg determinations will be discussed in Section 2.9.3.8 under AA.

2.9.3.8 Atomic Absorption Methods

Compared to other methods, there has been little use of AA or FE for crude oil analysis. AA has been used in lubricating oil analysis (discussed next in Section 2.9.4), but perhaps the problems involved with direct methods for crudes have been a deterrent. Ashing procedures circumvent some of these problems, but of course this requires more time.

The problems involved with direct methods concern errors associated with sample matrix, viscosity, background problems, and possible variations in the response of different types of metalloorganic compounds. In the API Research Project 60 annual report[819] it was stated that direct AA determinations of Ni and V in crudes were erratic unless samples were first ashed (the report lists 10 selected crudes being studied by the API, but no results for metals were published). Similar erratic results have been noted in the author's laboratory. Nearly 2-fold discrepancies were observed for one particular oil, comparing direct and ashing methods for AA and FE. This led to the development of a direct method that was suitable for AA and FE (see Section 2.9.3.9).

Sebor *et al.*[820] reviewed the problems involved and from data obtained from their investigations claimed that the type of V bonding in standards and sample matrix can cause significant errors using direct methods. One of the compounds they tested as a standard was vanadyl tetraphenylporphyrin, which they prepared. No data are given to establish values for standards. Similar data obtained in the author's laboratory produced somewhat different results that could not be completely explained. Possibly matrix (perhaps ionization) effects were encountered. Nitrogen and sulfur compounds present in some of their standards in certain forms in the author's laboratory were shown to enhance V response. Smith *et al.*[821] investigated trace metal determinations in crudes and stated that "instrumental techniques did not always yield the quality of analytical results that might be expected."

Perry and Keyworth[822] applied AA to direct determinations of Ni in crudes and vacuum tower bottoms. They showed that one of the problems involved is the variable molecular absorption caused by organic material and pointed out that even the small amount of material added as an NBS Ni standard caused a difference. To compensate for this, they used a nonabsorbing line at 2316 Å to make a correction. They also reviewed the use of nonabsorbing lines as a general technique. To compensate for viscosity and matrix effects, they used a standard addition technique. They cited a standard deviation of about 1.5 ppm Ni for their direct method in the range 5–50 ppm. In comparison to an AA method using acid digestion of samples, agreement was good for 6 samples, but the results for 2 samples are high by the direct method. The results obtained were 43 and 23 ppm of Ni, compared to 33 and 15 ppm obtained by the alternative AA method and 32 and 15 ppm obtained by the colorimetric method. (Viscosity and matrix influences will be discussed in more detail in Section 2.9.3.9.) Manjarrez and Pereda[823] applied AA to the direct analysis of crudes and claimed good results. Vavrecka *et al.*[824] used porphyrin compounds as standards for the AA direct analyses of Ni and V in asphaltine and obtained lower results as compared to the use of "conventional oxygen-containing standards." Ibrahim and Sabbah[825] studied the influence of viscosity on determinations of Fe in crude oils and cited the possibility of large errors.

Alder and West[826] applied a nonflame, carbon filament method to the analysis of Ni in crudes and residual fuel oils. Although most of the results looked good compared to a spark emission method, the results for one sample were high by 30%. Also, the precision for one sample was ±14%, indicating the potential for considerable error using this method. Sampling could be a problem considering the small amounts used in such methods, and matrix errors are generally larger than those of flame methods. Bratzel and Chakrabarti[827] made a detailed study using a carbon rod method for determining Pb in petroleum products. For the two crudes analyzed, the standard deviation was ±0.02 ppm for the range 0.17–0.31 ppm. Omang[828] applied a graphite furnace method to Ni and V determinations directly in crudes and pollution samples and obtained good results compared to an ashing procedure. The carbon rod method was also applied to the determination of various elements by Araktingi *et al.*[829] and Hall *et al.*[830] (see Ref. 836 for additional use of nonflame methods).

Because of the potential pollutants in fossil fuels,[831] considerable effort has been devoted to the analysis of certain trace elements. The determination of Hg in crudes by AA was investigated by Hinkle,[832] who, using the technique of Hinkle and Learned,[833] collected the Hg by the aeration of the cold vapor onto a silver screen combined with a Shoniger flask digestion of the crudes. Abu-Elgneit[834] investigated the direct analysis of Cd, reporting results in the range 0.012–0.026 for four crudes. Gleim *et al.*[835] determined Mo in three crudes and a tar sand.

A consortium approach has been used by five petroleum laboratories, co-operating to develop methods and analyze for trace elements of environmental importance. Various papers were presented at the 1975 national A.C.S. Meeting and published in a book by the society.[836] This consortium developed methods for determining As, Be, Cr, Hg, Mn, Mo, Ni, Pb, Sb, Se, and V in the ppb range in petroleum samples. Nonflame methods with a graphite furnace were used, and accuracy was calculated to be on the order of ±20%. Wet-ashing procedures were used to prepare samples in cases where difficulty were encountered with direct measurements. For example, low-Ni results for a direct method were attributed to distillation loss of volatile porphyrin compounds during the ashing step. The consortium effort also resulted in additional published papers.[837–841] Reference 837 presents a study wherein the graphite furnace is pretreated with carbide forming elements such as La and Zr to provide large signal enhancements.

A unique application of AA is that for the determination of S in crudes and fuel oils by Kirkbright *et al.*[842] Errors compared to standard methods varied from −10.3 to −13.9 for four samples and from −9.7 to +2.0 for the other eight samples. A shielded nitrous oxide-acetylene flame was used to obtain transparency for the 1087-Å line of S and the sensitivity for different compounds varied from 0.6 to 2.7 ppm/1% absorption. An equivalent application should be possible using FE via molecular emission measurements with perhaps better sensitivity.

2.9.3.9 Flame Emission Methods

As with AA methods, FE methods also have been used only moderately for crude oil analysis. With the development of better instrumentation and use of nitrous oxide-acetylene, FE should be a useful technique for analyzing many elements in crudes. Sarkovic[843] has applied FE to the analyses of Cu, Ba, Mo, Ni, and V in Vojvodina crudes.

Prior to this time FE was used primarily for determining alkali and alkaline earth elements in crudes. Actually, there has been less interest in the Na content of crudes than in residual fuel oils. In his book, Milner[844] discussed FE applications for determining Na in petroleum products. He cites Schuknecht and Schinkel[845] for use of an Al–Cs buffer combination that allows Na and K to be determined in the presence of a several-hundredfold excess of alkaline earths and other elements. He also cites an ASTM standard method for Na applied primarily to residual fuel oils via an ashing procedure.[846]

Popescu *et al.*[847] applied FE to the direct analysis of Na in crude oil cuts, and Augsten[848, 849] applied FE to the determination of alkali and alkaline earth elements. Augsten used ultrasonic atomization of oil samples diluted in 1:1 alcohol to give improved sensitivity. Karchmer and Gunn[816] also used FE for determining Na but gave few data for crude oils.

Moore and Dunning[755] applied FE to the determination of Ni in a Tatums, Oklahoma, crude oil and its various fractions. Heemstra and Foster[850] determined V in crudes using plasma jet excitation.

In the author's laboratory, ac FE (Section 2.6) has been used for determining various metals in crudes and residual crude oil fractions. The elements Al, Ca, Cu, Cr, Fe, Mo, Na, Ni, and V have been determined, at times directly, but more often with a preliminary ashing of the sample. This special technique (ac FE) is more sensitive and more precise than ESD methods but slower if more than two elements are to be determined in a given sample.

The better detection limits provided by the technique, however, allow direct determinations to be made at lower levels, which provides for faster analyses as compared to the ashing-ESD methods.

Problems involved with direct methods were discussed in Section 2.9.3.8. A direct method developed in the author's laboratory using ac FE (Section 2.6.3.2) is accurate, fast, and provides determinations down to about 1 ppm for Cu, Fe, Ni, and V in heavy crude and residuum samples.[851] Detection limits down to 0.02 ppm of Ni and 0.01 ppm of V can be obtained for lighter oils (light crude oils, fuel oils, etc.). The method can also be used with AA, but the detection limits are correspondingly lower. Detection limits in solvent solutions for AA using a Perkin-Elmer Model 306 spectrometer and ac FE with a prototype Jarrell-Ash spectrometer were as follows: for AA, 5 ppb of Cu, 16 ppb of Ni, 12 ppb of Fe, and 15 ppb of V; and for ac FE, 1.3 ppb of Cu, 1.8 ppb of Ni, 1.6 ppb of Fe, and 0.7 ppb of V. Using the Perkin-Elmer spectrometer in the FE mode improved the detection limits to 5 ppb of Ni and 10 ppb of V, with 5-cm nitrous oxide-acetylene burners. Statistical detection limits were calculated for at least 2 runs each on 2 different days to provide a realistic, average value. A 10-s integration period was used for AA and a 10-s time constant on the lock-in-amplifier with a 3-s averaging on a digital readout were used for ac FE. The use of the digital readout and organic solvents improved detection limits compared to those previously reported (Sections 2.6.7 through 2.6.7.4).

Problems encountered in the development of the method involved:

1. Matrix problems caused by both the organic and inorganic nature of the petroleum samples.

2. Background problems caused by the organic nature of samples.

3. Viscosity problems influencing signals, especially background at the lower levels.

Viscosity problems have been discussed by Ibrahim and Sabbah,[825] using AA for Fe in crude oils; by Mashireva *et al.*,[852] who devised correction curves for B in oils by FE methods; and by Korovin *et al.*,[853] who developed a special nitrous oxide–acetylene burner for oils. For an AA determination of wear metals in lubrication oils, Burrows *et al.*[854] used a constant ratio of 20% oil in sample and standard solutions. Plotting absorption versus percent oil, however, they found peak signals at 5% oil. However, Lush[855] reported that at least a 17-fold dilution should be used for crude oils. In the author's laboratory over a 300-fold dilution was required for a solid, crude oil asphaltine to eliminate viscosity, plus organic, physical matrix effects.

Matrix effects and variations in response for different types of organic compounds can occur for direct methods. This has been reviewed and discussed by Šebor *et al.*[820] and Vavrečka *et al.*[824] Vavrečka *et al.* reported that accuracy was improved by using porphyrin compounds as standards.[824] A similar approach was used in the author's laboratory except that naturally occuring compounds were used. They were obtained as an asphaltine concentrate separated from a heavy crude oil high in Ni and V content. The use of this solid, asphaltine concentrate as a standard, in addition to compensating for matrix effects, provides a stable, homogeneous solid which is easy to handle and weigh as compared to heavy oils. Additional problems in developing the method were solved (1) by using large dilutions and a special solvent mixture, 80% cyclohexanone–20% special naphtholite, to minimize viscosity and background differences, and (2) by critically adjusting all burner and optical parameters. The solvent mixture above completely dissolves difficult samples such as crude oil asphaltine and residuum samples and shale oil. The solutions appear clean, whereas aromatic solvents will often produce dirty-looking solutions that are not stable. In the cyclohexanone mixture, the asphaltine standard, diluted to a level of 2 ppm V, was stable for over 20 months.

Sample composition and viscosity variations influence the height of the pink cone in a nitrous oxide–acetylene flame and thus the magnitude of flame background. Critical adjustments of parameters are required to optimize this flame background while nebulizing organic solvents. For example, using ac FE for Ni at 3525 Å, a pink cone about 5 mm high should be used. Competing background species, CH and OH, occur at this wavelength for which a very lean flame gives maximum OH background; a rich flame gives maximum CH background. With a 5-mm-high pink cone a minimum occurs for these backgrounds, so precision and detection limits are improved. This direct method has been used for more than 3 years and has provided not only faster, more economical analyses for many samples but also better accuracy and precision than for the ashing methods.

2.9.3.10 Metal Surveys Using Other Techinques of Analysis

Nelson[738, 856–861] has compiled considerable data, including Ni and V contents as well as S and some N contents for worldwide crudes. In connection with a survey of potential pollutants in fossil fuels, Ref. 831 compiled all values found in the literature for metals and N and S contents of petroleum.

Using colorimetric methods, Hodgson and Baker[862] determined Ni and V in 192 western Canadian crudes. This followed an earlier study in which Fe, Ni, and V were determined in 77 crudes.[863] Depth of the crude produced, API gravity, and S content were tabulated and the V/Ni ratios were calculated in a study versus age. Baker and Hodgson[864] also analyzed 37 western Canadian crudes for Mg as well as Ni and V. This study was made in an effort to correlate Mg present in crude with chlorophyll as a derivative for Ph and Mg, but all Mg contents were very low. Hitchon *et al.*[745] analyzed 88 Canadian crudes for 22 trace elements but only summarized the results.

Table 2.17a Range of Analyses by Different Laboratories for Various Crude Oils

	ppm of Ni		ppm of V		
Crude	Range	Av.	Range	Av.	References
Agha Jari	8–10	9	30–36	33	731, 779, 865
Bachaquero	42–53	48	320–430	370	731, 733, 762, 780, 857, 865, 867
Boscan	66–150 (100–110)	106 (105)	900–1400 (1134–1280)	1190 (1200)	731, 747, 762, 780, 857, 865
Gach Saran	33–35	34	104–114	108	730, 731, 865
Kuwait	5.6–9.6	7.0	20–31	26	731, 779, 803, 865, 867
Lagunillas	35–41	38	265–317	295	733, 762
Mara (La Luna)	13–19	17	164–220	190	730, 731, 747, 780, 857, 865
N. Belridge, Calif.	83–120	104	23–32	28	730, 747, 762
Ragusa	66, 68	67	10–12	11	730, 780
Santa Maria, Calif.	97–130	111	180–223	220	740, 762, 803, 866
Tijuana	16–24	22	170–200	190	730, 731, 762, 780, 865, 867
Wafra	7–13	10	52–37	44	780, 865
West Texas	3.8–5	4.5	8–23	14	733, 762, 803
Wilmington, Calif.	46–61	57	41–46	44	730, 762, 866

Table 2.17b Number of Crude Oils Analyzed by Selected References

Reference	Number of crudes	Reference	Number of crudes
730	9	779	10
731	8	780	10
747	4	783	4
751	3	732[a]	33

[a]Contains table summarizing ranges of Ni and V contents for 33 crudes taken from Refs. 17 and 18 in Ref. 732.

Baker[865] provided Ni and V analyses for 43 crudes with 29 from South America and 14 from the Middle East. With two exceptions the ratio of V/Ni was generally high, varying from about 2 to 14. He used colorimetric methods and stated that for crudes dry ashing was less troublesome and gave results equivalent to those using wet ashing. Milner *et al.*[803] also analyzed 15 crudes using colorimetric methods.

Brown[866] studied California crude oils showing great variety in properties and provided Ni, V, and Na values for 17 crudes with the ratios of V/Ni generally being less than 1 and Ni contents ranging from 11 to 106 ppm (methods of analysis were not given).

Brunnock *et al.*,[867] in presenting a scheme for the analysis of oil sources of beach pollutants, present x-ray fluorescence analyses for Ni and V in 27 crudes and also S contents. Adlard[868] also discusses methods for identifying beach and sea pollutants, summarized various methods of analysis, commented on their use, and cited ES and neutron activation analysis as likely techniques. One of the references they cited was Bryan *et al.*[869] who analyzed 22 elements in 20 crude oils and 20 fuel oils by neutron activation analysis. In his thesis on a study of crude oils Filby[741] also used this technique for the analysis of 10 crudes for the 15 elements Ni, V, Co, Hg, Fe, Zn, Mn, As, Au, Sb, Se, Sc, Cu, Na, and Ca. Shah *et al.*[742,743] developed an activation analysis method for 23 elements in crude oils. Data for 10 crudes were given, but some of the Ni values appeared suspiciously high. Neutron activation analysis was also used for determining As, Br, Cu, Na, Ni, and Zn in 10 crudes,[870] Cd in 9 crudes and 2 asphaltines,[871] Hg and Se in 46 Illinois crudes,[872] V in 53 Illinois crudes,[873] and Ni and V in a study of Iraq oil fields.[874] In the crudes analyzed, As ranged from 5 to 140 ppb, Cd from <0.5 to 29 ppb, Hg <0.03 ppb in Illinois crudes and Se < 0.04 ppb in 38 Illinois crudes and an average of 0.14 ppb for 8 Illinois crudes.

Other references, discussed earlier in Section 2.9.3, mainly provide Ni and V data using chemical methods and are listed below along with the number of crude oils analyzed. Some of the data from these references and others in this section have been used to compile results showing ranges of Ni and V values that were obtained for 14 crudes. Average values and references used are also given in Table 2.17.

2.9.3.11 References from Foreign Countries

The references reviewed have been primarily those written in English but some comments on foreign references should be made.

Arich and Constantinides[875] critically reviewed methods for the determination of Ni and V in crudes. Lakatos[876] reviewed elemental determination in crudes using spectrochemical methods. Biernat and Solecki[877] used a Rotrode made of copper for the ES analysis of crudes, lubricating oils, and brines. Farhan and Pazendeh[878] determined 10 elements in crudes and petroleum products using arc spectrography. Cornu[879] analyzed crudes using spark excitation for concentrations in the range of 0.1 to 100 ppm and an arc on evaporated samples for lower concentrations. Faulhaber and Liebetrau[880] studied the ratios of N and S to V for 8 oils. Hohmann *et al.*[881] found no relation between Ni and V to N in crudes. In a study of ashes from Polish crudes, Gregorowicz and Orzechowski[882] found Be in one case. Manjarrez and Pereda[823] in Mexico used AA in a direct analysis method for crudes and other petroleum fractions. Serbanescu[883-885] analyzed Romanian crudes for Co, Fe, Ni, and V using colorimetric and polarographic methods. Extensive work has been done by the Russians. Only a few of the many studies will be cited. A compilation by Aleksandrov *et al.*[886] covers applications of various methods, including spectral methods for petroleum products. Gulyaeva and Lositskaya[887] analyzed 81 crudes produced in the Soviet Union for Ge content which varied from 0.015 to 0.15 g/ton of crude. Poplavko *et al.*[888] determined Re in USSR crudes, solid bitumens, oil shales, and coal, finding < 0.01 g/ton in heavy oils. Extensive data for Russian and mid-Asian crudes are given in studies by Demenkova and coworkers[889-891] and by Katchenkov.[892, 893]

2.9.4 Wear Metals and Additives in Used and New Lubricating Oils

Of all methods used for the analysis of inorganic elements in petroleum products, the determination of wear metals in used oils has received the most attention. The analysis of additives in new oils added as P and S compounds containing cations such as Ca, B, Ba, Mg, K, and Zn is also important. Any technique applied to the analysis of wear metals in used oils could be applicable to the analysis of additives in oil. They might also be applicable to crude oils (Section 2.9.3).

The determination of wear metals and contaminating metals in used lubricating oil is of extreme importance in preventative maintenance programs for engines. Such programs can be applied to any type of engine that uses lubricating oil, and extensive maintenance programs are used routinely in many laboratories connected with varied industries. Such programs can predict engine failures from trace metal increases in oil due to various causes, such as excessive wear, bearing failure, and radiator leaks. Table 2.18 lists trouble areas and corresponding metals to which their increase in oil may be related. Various references in the next section discuss specific trouble symptoms, including those reviewing maintenance programs.

2.9.4.1 History and Applications of Emission Methods

McBrian[894,895] is generally credited with the discovery in the early 1940s that analysis of lubricating oil could be used to diagnose impending trouble in railroad diesel engine operations. Similar programs were employed by Cassidy[896] in 1950, and Burt[897] in 1950, who describes maintenance operations for

Table 2.18 Metals Analyzed in Lubricating Oils and Areas Associated with Possible Trouble

Element	Aircraft	RR diesel engines
Ag	Ag-plotted spline	Wrist-pin bearings
Al	Lubricating pump	Bearings
B		Water leaks
Cd		Bearings
Cr	Cr-plated parts	Water leaks
Cu	Bearings	Bearings
Fe	Excess wear → various parts	← excess wear
Mg	Gearbox housings	
Na	Saltwater leak	
Ni	7 ppm	Alloy constituent
Pb	Bearings	Bearings
Si	Filter failure → airborne dirt	← filter failure (8 ppm)
Sn	15 ppm	Bearings
Zn		Absence[a] (<10 ppm)
Oil additives		Depletion

[a]The Zn additives used in some oils corrode Ag bearings.

the Southern Pacific Railroad and the use of ES for analyzing additives and wear metals. They used a disc electrode (Rotrode) rotating in the oil. In 1951, Pagliosotti and Porsche[898,899] described applications of the Rotrode to oil analysis in more detail. Other ES techniques were applied to the analysis of oils by Calkins and White[900] in 1946, Russell[901] in 1948, and by Gassman and O'Neill,[902,903] Murray and Plagge,[904] and Hughes *et al.*[905] in 1949 (see Section 2.4 for use by the latter since 1946). From 1950 to 1952 applications increased, and by 1953 many laboratories were using wear metal analysis programs for preventive maintenance. Jackson,[906] in his comparison of ES and AA methods, reviewed the development of such maintenance programs and cited two symposia on the subject held in 1952 and 1953[907,908] and a notable monograph on the subject.[909] Southern R.R.[910] again described the use of their programs, citing analyses of 1800 oil samples per month. Rozsa and Zeeb[911] reviewed methods and applications and described a new, rotating platform electrode technique (Platrode). Since then, such programs have expanded tremendously and include many applications by other industries for any type of engine using lubricating oil. By 1954, ESD methods with modern instruments were being used.[912] For example, the simultaneous determination of 16 elements for a Chicago and North Western R.R. program was described by E. T. Myers.[913] Analyses for such programs became an important business creating two commercial laboratories using ESD.[915] Reviews in Refs. 914 and 915 presented an excellent story of such programs used by railroads, cement and sugar industries, the U.S. Navy and Army Air Force, and commercial airlines.

Not only do such programs save money but also save lives, owing to the avoidance of impending airplane failures.

Extensive data obtained by the Air Force program (SOAP) comparing emission and AA methods for lubricating oil analyses were reported by Kittinger and Ellis.[916] The NBS[917] briefly described the effectiveness of the SOAP program as annually saving $13 million of $50 million and air crew lives. The NBS provided a series of metalloorganic compounds for use as calibration standards, citing a considerable savings to American industry in the preparation of adequate standards (see Section 2.9.4.3.2 for comments on the difficulties in the use of these standards for routine analyses). Beerbower[918] also reported on this program for helicopter applications. Jolliff[919] reviewed application of ESD programs to ships of the Navy Pacific Fleet, and Barr and Larson[920] reviewed computerized programs for the Navy Oil Analysis Programs (NOAP). Maintenance programs have also been used by the trucking industry,[921] by the petroleum industry for pipeline operations,[922] and for similar applications such as maintaining pumping stations by the American Gas Associations,[923] and jet engine power plants.[924]

Such programs employing emission methods have received considerable attention from the ASTM, including a 1956 symposium on lubricating oil resulting in several papers.[925-927] ASTM summaries of methods and cooperative programs are given in Refs. 928-930.

Extensive studies of lubricating oil analysis have also been made in Russia. Kyuregyan and coworkers have prepared two books on the subject[931,932] and other papers.[933,934] Aleksandrov *et al.*[935] compiled a volume on methods applied to petroleum products, including ES and FE methods. Il'ina[936] reviewed the 1946-1960 literature on ES methods for analyzing metals in lubricating and fuel oils.

Books on inorganic analysis of petroleum have been written in English by McCoy[937] and Milner[938] but contain mainly chemical colorimetric methods and little on ES. A book by Pinta,[939] originally written in French but now available in English, contains extensive chapters and many references to emission methods, including petroleum industry applications. Noar[940] included 97 references in a 1955 review on the uses of emission spectrography in the petroleum industry.

Maintenance programs for analyzing wear metals in lubricating oils are now numerous and have evolved into extensive use of many ESD methods.[906,912,913,918-925,927,941-943]

2.9.4.2 Emission Spectroscopic Techniques

Various techniques employing different types of electrodes have been used for oil analysis. The most popular technique seems to be an ESD method using the Rotrode, which will be discussed first. All the techniques used will be discussed and will include the following: (1) Rotrode, (2) vacuum cup electrode, (3) porous cup, (4) quenched electrode, (5) platform or flat top electrodes, (6) electrode coking, (7) residue methods–dry and wet ashing, and (8) plasma excitation sources.

2.9.4.2a Emission Spectroscopic Methods Using the Rotrode. Among all the varied techniques using ES or ESD methods, the disc electrode rotating directly in the oil sample (Rotrode) seems to be more universally used. Some workers, however, prefer the porous cup or other techniques.

As cited previously, Burt[897] and Pagliasotti and Porsche[898,899] were among the first to use the Rotrode. In 1954 Gillette *et al.*[912] described an ESD method using the Rotrode, providing precisions in the range 1-8% (3-6% in many cases) for Ca, Cu, Cr, Fe, Pb, and Si and 11% for B but at the low level of 7 ppm. A heavy aluminum boat was used to remove heat from the oil to minimize the chance of the sample catching fire. Reversing sample polarity to negative also helped cool the sample and improved precision.

Fry[944] reviewed the literature on different excitation techniques and compared some of them to an improved ES-Rotrode method. They used a spark excitation with a 60-s prespark, lithium as a buffer to reduce matrix errors, and obtained an average repeatability of 8.7%. Detailed discussion of matrix errors and particle-size influences is presented in Sections 2.9.4.7 and 2.9.4.8.

Cummins and Mason[942] made a thorough study of interelement effects and used NaCl-impregnated Rotrodes and special sets of standards to achieve good accuracy compared to chemical methods. Rappold and Ramsay,[927] using a Rotrode-ESD method, show standard deviations (expressed as relative %) in the range 1.4–2.7% for determining Ca, Zn, and P in lubricating oil additives and achieved good accuracy compared to chemical methods. Perkins *et al.*[926] evaluated various parameters influencing an ES–Rotrode method. They reported on cooperative results obtained in two laboratories showing some deviations. One laboratory used a Ca buffer (most oils they analyzed contained Ca as additive) and the other used Li. The use of Li was cited as better for a variety of oil types and reduced but did not eliminated matrix effects of various additives as they show in a table. The Rotrode has also been compared to other methods, including AA, and found adequate by Golden.[945] Hauptmann and Jager[946] compared the Platrode with Rotrode methods and found the latter to be superior. Woods[947] developed a Rotrode technique for small samples (0.5 ml) using samples floating on water. The Rotrode was also used by Jantzen[948] to analyze eight metals in oils, by Lakatos[949] studying hydrocarbon matrix effects, and by Westcott and Seifert[950] who studied the effects and the nature of iron particulates.

The ASTM has also studied use of the Rotrode and discussed results of a cooperative program.[928,929]

At Union Oil prior to 1958, an electrode coking method was used. Our participation in the ASTM program indicated that the Rotrode method would be faster and would give better precision and sensitivity. A method was developed specifically for determining 15 elements in railroad diesel lubricating oils. The method was patterned after that of Fry[944] and emphasized the need for good P determination. Precisions for the electrode coking and Rotrode method, averaged for various calibrations over extended periods of time, are combined in Table 2.19. A Bausch & Lomb dual grating spectrograph with a 30,000-line/inch grating blazed at 3000 Å and a high-voltage spark excitation were used. The precision of our photographic–spectrographic (ES) method varied from 4% to 10%, with that for P in the range 50–150 ppm being 8%. A special antimony buffer was used to level matrix effects and enhance emission. Special studies of matrix errors and particle-size effects are discussed in Sections 2.9.4.7 and 2.9.4.8.

The use of ESD compared to ES provides faster results and usually better precision. Pickles and Washbrook,[951] however, cite precisions using the Rotrode of 15% for Ba and 10% for Ca and Zn for ESD and 3% using ES. They attributed this to the use of more internal standard (I.S.) lines closer to analyte lines for ES compared to a single Ni I.S. line for the ESD method. Contrast this relatively poor precision to that of Rappold and Ramsay[927] in the range 1.5-2.7% for the same elements using ESD. This indicates that considerable care should be used when selecting the I.S. (Co in most cases), the line, and the conditions controlling possible background radiation.

2.9.4.2b Vacuum Cup Electrodes. The vacuum cup electrode was developed by Zink[952] for handling solutions. This electrode does not require special attachments such as that for the Rotrode, thus making changes from its use to other techniques easier. It also requires much less prespark time for signals to reach peak values. A modification to a flat top (Section 2.9.3) is used in the author's laboratory for acid solutions of ashed crude oils and residual oils and can also be applied to used lubricating oils. This provides better precision and is easier to use than the Rotrode.

All versions of the vacuum cup electrode tested by the author failed for applications directly in organic solutions. Since then a note was observed in the

Table 2.19 Comparison of Precision for Electrode Coking and Rotating Disc Methods Averaged for Various Calibrations

	Relative % standard deviation			Relative % standard deviation	
Element	Coking	Rotrode	Element	Coking	Rotrode
Ag	11	9	Ni	—	7
B	—	10	P	—	8
Ca	13	4	Pb	12	4
Cr	12	6	Si	16	7
Cu	13	8	Sn	11	4
Fe	10	6	V	11	8
Na	15	9	Zn	10	6

literature by McGowan,[953] who with a modified vacuum cup electrode was able to detect 0.05% Co, 0.07% Mg, and 0.23% Pb directly in oils. No further data were given.

2.9.4.2c Porous Cup Electrodes. The porous cup electrode was developed by Feldman[954] in 1949, who reviewed other solution methods currently used and applied the technique to the analysis of catalysts. This is an upper, drilled-out electrode that allows sample to seep through a thin bottom as excitation is applied. Gassman and O'Neill[902,903,955] used the technique, as did Gunn,[941,956] who compared the porous cup to the Rotrode. Gunn[941] made a thorough matrix study, finding that both techniques were influenced about equally, and although the Rotrode gave equivalent precision, he preferred the porous cup electrode combined with ESD as being faster and more flexible. The method is unsuited for samples containing suspended material.

2.9.4.2d Quenched Electrodes. The earliest direct ES analysis of oils was by Calkins and White,[900] who used a quenched electrode technique. The technique uses a preliminary heating of a graphite electrode, which is then inserted immediately into the sample, quenching it and thus drawing sample into the graphite. Carlson and Gunn[957] also used the technique in 1950, Hamm *et al.*[958] in 1952, and it was thoroughly studied by an API Committee on Analytical Research (COAR) and found to have limited applications.[959]

Biktimirova and Mashireva[960] have modified a quenched electrode method developed by Biktimirova and Baibazarov[961] to obtain increased detection limits (up to 10-fold) by using selected buffers. The use of 1.1% NaCl enhanced Cr and V from 10- to 11-fold, and 0.8% Ga_2O_3 enhanced Ni and Mo.

2.9.4.2e Platform Electrodes. Platform electrodes of two types have been used in the literature. One type is simply a flat tip or slightly dished graphite electrode. Solution evaporated on the tips of these electrodes and then sparked provides exceptional sensitivity (Section 2.6.8.9.3). This technique is called the graphite-spark or, when using Cu electrodes, the Cu-spark method. It has been utilized by Hoggan *et al.*[962] for application primarily to gas oils and to a limited extent in the author's laboratory.

Another type is the $\frac{1}{2}$-inch-diameter rotating platform electrode (Platrode) developed by Rozsa and Zeeb[911] They reviewed other methods and claimed that coking samples directly on Platrodes provided the advantage of exciting all the sample, including particulates. They obtained good accuracy and precision compared to other methods, within 10%, and claimed better precision for values below 10 ppm. The method was also tested by Fry,[944] who compared it to his improved Rotrode method. He found the Platrode excellent and that it provided better detectability than the Rotrode. The method does not appear to be used much currently. Hauptmann and Jager,[946] however, compared the methods and found the Rotrode superior.

In the author's laboratory the Platrode was found to be no better than a $\frac{1}{4}$-inch-diameter flat tip electrode (slightly dished), which is easier to use.

2.9.4.2f Electrode Coking and Excitation Directly in Cup Electrodes. Coking samples in cup electrodes is very similar to the Platrode technique except that an arc excitation is generally required. Veldhuis *et al.*[963] described a special technique using spark excitation. A $\frac{1}{8}$-inch-thick pellet punched from a Whatman ashless tablet is placed in a shallow $\frac{3}{16}$-inch-diameter cup electrode. The sample is heated, homogenized, and then applied to the pellet, after which spark excitation is employed. For 20 determinations of an oil, the relative standard deviation was 6.3% for Ba and 1.7% for a level of 0.04% P in a new oil. The technique was also applied to the wear metals Ag, Al, Cu, Fe, Ni, Pb, and Sn. Despite this early use in 1952, using a simple, rapid procedure that appeared to provide good precision, the method did not seem to catch on. Gunn[964] also used a spark excitation of lubricating oils mixed to make a paste with graphite and obtained precisions of ±5% for Ba and ±8% for Ca.

Coking the sample in an electrode followed by arc excitation has been applied to Fe determinations by Hansen *et al.*[965] and to additional elements by Barney,[966] Barney and Kimball[967] and Kyuregyan and Marenova.[933] This method was tested by the ASTM[928,930] and in the author's laboratory as described in Section 2.9.4.2.1 and found to be less acceptable than the Rotrode.

2.9.4.2g Ashing or Residue Methods. The surest means of obtaining accuracy for samples containing particulates is to ash a significantly sized sample of homogeneous oil. The remaining residues may then be dissolved in acid solution and excited with conventional spark techniques or a plasma torch. This can provide the advantage of using inorganic standards and often leads to better precision. More often the sample ash is arced after mixing with graphite. Meeker and Pomatti[968] used such a technique, employing Ba and Ni as a buffer to minimize matrix errors. A table showing the matrix effects of various additives in different buffers was included. Work and Juliard[969] used alumina as an ashing aid and buffer with good success. Alumina has also been used as a buffer for arc methods in the author's laboratory and was the best for minimizing matrix variables and enhancing emission. The standards used were ideal for applications analyzing alumina-based catalyst. McEvoy *et al.*[970] used Al-Si for a catalytic ashing technique, mixed the residue with graphite, and used an arc excitation of a briquetted pellet to achieve a precision of ±5%. Soroka *et al.*[971] used MgO as a collector for Cr, Cu, and Fe.

Many others have used ashing-arc methods, including Refs 934 and 972-977. Childs and Kanehann[978,979] used the technique in conjunction with a log sector-ES procedure and claimed it to be faster since it eliminated density measurements on photographic plates. Polkanov and Sotnikov[980] obtained high reproducibility by using an improved arc generator and thin-walled carbon electrodes.

Some workers prefer to use wet-ash procedures, such as that of Hansen and Hodgkins,[981] to minimize the loss of metals. Such losses due to volatility of metals are usually not significant for lubricating oils (except for Pb). This is discussed in detail in Section 2.9.6 for feedstock and gas oils. Hansen and Hodgkins used Cu powder[981] as an ashing aid during the wet-ash procedure with sulfuric acid. They used a special feathered cup electrode to reduce wandering of the arc for both ES and ESD methods. Precision ranged from 8 to 12% for Ni and V and was poor for Fe (about 25%) in gas oils. Agrawal and Fish[982] compared three ashing methods and obtained the most reliable results with wet ashing.

2.9.4.2h Plasma Excitation Sources. The use of the plasma arc, and especially the plasma torch, was cited in Section 2.6 as providing exceptional detection limits and in Section 2.9.1 for water analysis. They should be especially good when used in conjunction with ashing methods. Karacki and Corcoran[983] determined nine elements in coal ash, including P and Si, using an argon stabilized plasma arc. Heemstra and Foster[984] studied the influence of solvents on V emission using a plasma arc. Such sources usually are less stable upon introducing organic solutions into them, but recent advances have minimized some of these instabilites. McElfresh and Parsons[985] used force feeding into a plasma arc to overcome lubricating oil sampling problems. Greenfield and Smith[986] used a plasma torch for the direct analysis of Al, Cr, and Cu in oils and trace metals in blood. They presented the method as a serious competitor to nonflame AA methods.

Varnes *et al.*,[987] Pforr and Aribot,[988] and Fassel *et al.*[989] also analyzed metals in oils using plasma torches. A 10-fold dilution of samples was used[989] and the results were compared to those from the U.S. Air Force (SOAP) program. Detection limits were established for 15 elements, but only 6 were analyzed in samples. A direct-reading spectrometer was used for the simultanious analysis of 5 of the elements. The results were good for jet engine oils, but some Cr and Fe values were high for reciprocating engine oils.

2.9.4.3 Atomic Absorption Methods

Although AA was introduced as an analytical technique in 1955, the first P.I. application was not reported until 1962-1963 by Barras and Helwig (Section 2.4) for the determination of metals in gas oils (Section 2.9.6). Although no publications occurred in the literature, AA was used for P.I. applications prior to 1962 by various oil companies. It is especially valuable for determining Zn used as an additive in lubricating oils. In the author's laboratory such determinations were first made using a Beckman DU monochromator and a simple dc arrangement.

The first AA method for lubricating oils was reported by Sprague and Slavin[990] in 1963, and Slavin also reviewed P.I. applications. In the period 1964-1967, other applications for lubricating oil analysis appeared,[991-995] and Slavin and Slavin[996] describe a fully automated AA procedure applied to aircraft lubricating oils. They compared results to many samples previously analyzed by an ESD method on a number of samples previously analyzed by the Army Air Force (SOAP) programs and an AA ashing method. There were many discrepancies, which indicate poor accuracy in some areas. Kittinger and Ellis[916] presented extensive data for AA and ESD methods obtained in a cooperative program by the Army Air Force involving many labs. It appeared that no positive conclusion was formed and that differences between the methods do occur.

It is interesting to note that the Navy Air Force in a similar application of lubricating oil analyses preferred ESD methods.[915] Although the Army Air Force originally embarked on a program using AA, they later made extensive comparisons with ESD methods. A more recent survey made by Prazak[997] indicated AA to be the best method.

Additional AA applications of lubricating oil analysis[951, 998–1007] include the determination of an unusual Sb additive,[1008] and a unique AA determination of sulfur in oil.[1009] Sanders[1010] and Skujins[1011] reviewed petroleum applications for Varian-Techtron. For Perkin-Elmer, Barnett *et al.*[1012] described the use of their instrumentation for the rapid sequential determination of six elements in a single lubricating oil sample using a nitrous oxide flame and a multielement lamp.

Lush[1013] discussed AA methods for lubricating oils in a chapter dealing with the applications of atomic absorption for trace element analysis in the petroleum industry. Welz[1014] surveyed the applications of AA spectroscopy in industrial analysis and included analysis of engine, hydraulic, and transmission oils of trucks for wear metals. Vigler and Gaylor[1015] discussed AA methods for analyzing 23 elements in petroleum products using magnesium sulfonate or potassium sulfate as an ashing aid. Kaegler[1016] reviewed the West German Standards Committee use of DIN standards for the AA analysis of lube oils. Tuzar *et al.*[1017] analyzed both wear metals and additives using AA. Jackson *et al.*[1018] simultaneously determined four wear metals in lubricating oils using AA and a vidicon detector. They compared results to a single-channel instrument, and although precision was less with the use of the vidicon detector, it was claimed to be adequate for the application.

Samples that are not homogeneous due to suspended material (i.e., iron metallic and inorganic particulates) influence the accuracy of both AA and FE methods (Sections 2.9.4.7)

2.9.4.3a Nonflame Methods. Nonflame methods used in AA provide significant sensitivity gains as compared to flame AA methods and have been studied for applications to lubricating oil analysis. Brodie and Matousek[1019] tested a carbon rod for the determination of Al, Cu, Cr, Mg, Ni, and Pb in oils. Using 0.5-μl samples they obtained precisions better than 4%. Reeves *et al.*[1020] used a graphite rod for determining Ag, Cr, Cu, Fe, Pb, and Sn in oils. They compared these results to those obtained in the Air Force SOAP program with flame

AA methods and obtained good agreement, except that higher results were obtained for Cr. The results for Cr were in better agreement with the results obtained by Barnett *et al.*,[1021] who used the nitrous oxide flame. Reeves *et al.*[1021] also applied the technique to a sequential determination of Ag and Cu by atomization at two different temperatures. Nonflame methods have also been used for the determination of Ag and Cu by Alder and West,[1022] Pb by Bratzel and Chakrabarti[1023] and for Ni and V by Omang.[1024] Hall *et al.*[1025] reviewed the development of nonflame methods and applications to oil analysis. They tested carbon rod and filament atomizers for analyzing five metals in crude and lubricating oils and suggested design changes for the carbon rod. Winefordner and coworkers made further studies of carbon rod methods for oil analysis[1026-1028] and used aqueous standards for analyzing oils.[1028] Through the selection of appropriate compounds, for example iron sulfate in preference to more volatile iron compounds, they were able to obtain good results compared to SOAP values. Prevot and Gente[1029] observed that the graphite furnace gave better sensitivity but poorer precision than flame AA methods.

2.9.4.3b Additives and New Oils. In addition to wear metals in oils, the control of lubricating oil blending by analyzing for the metallic components of various compounds employed as additives to improve oils is important to the P.I. Lush[1013] reviewed these applications for various Ba, Ca, and Zn additive types. They cited difficulty in the use of NBS metalloorganic standards because of solubilization and stability problems; a similar observation was also made in the author's laboratory. They also stated that there were no interferences caused by various additive types but showed considerable experimental error deviating from straight-line calibration curves for Zn. Lukasiewicz and Buell[1030] discussed errors in the Zn analysis and showed up to 19% potential errors for the various additive types. An excess of viscosity-improving additive was found to decrease Zn signals by 15%. An addition technique and an alternative direct procedure using the corresponding additive as standards were proposed. Concentrated additives can be analyzed by AA using a wet oxidation in an Erlenmeyer flask. The wet oxidation proceeds quite rapidly, since small samples (0.1 g) can be used but caution should be used in the initial charring step to eliminate volatility losses. Since surfuric acid depresses Zn signals, acid concentrations in sample and standard solutions should be matched. This technique is a convenient and accurate way to establish values for the alternative direct method proposed.[1030]

Kashiki *et al.*[1031] used the addition of iodine to level differences in AA response for the different compounds containing Ba, Ca, Cu, Pb, and Zn. Although this procedure works well with Pb, it did not level differences found for Zn,[1030] nor did the addition of dithizone (to complex with Zn), even though color developed, indicating the formation of a complex.

Holding *et al.*[1032,1033] and Guttenberger and Marold[1034] used a mixed solvent system to analyze for Ca and Ba and for Ca, Ba, and Zn, respectively, in order to use inorganic compounds as standards. A similar approach was tested in the author's laboratory, but again the errors reported in Ref. 1030 were not eliminated.

2.9.4.4 Atomic Fluorescence Methods

Atomic fluorescence is relatively new and is still developing technically. There have been few applications to oil analysis, and they do not offer anything exceptional. The AF technique does offer the possibility of simpler designs of multichannel instruments, and Miller *et al.*[(1035)] used such an instrument with combined AF and FE measurements. Demers *et al.*[(1036)] describe a similar instrument for application to water and oil. Davis[(1037)] and Winefordner and co-workers[(1026,1038,1039)] also applied AF to oil analysis. A special burner was designed for AF applications by Cotton and Jenkins[(1040)] that utilizes hydrocarbon samples as the fuel, with oxygen added as oxidant after starting the flame as a nitrogen-supported diffusion flame. It is claimed that there is no luminosity for kerosene or benzene, and detection limits for determining Cu, Fe, and Pb in kerosene were 4, 40, and 60 ppb, respectively.

2.9.4.5 Flame Emission Methods

The first commercial flame spectrometers appeared in the United States in about 1950, at a time when the ES applications were expanding in the P.I. Conrad and Johnson[(1041)] in 1950 were among the first to apply FE to oil analysis for the elements Ba, Ca, Li, and Na. In his book, Milner[(938)] reviewed applications for the analysis of Na and K (Section 2.9.3) with regard to Na analysis being of interest mainly in residual fuel oils. Buell[(1042)] reviewed the earlier applications of Fe in the P.I. The compilation by Aleksandrov *et al.*[(935)] included information on FE applications for petroleum. Additional references for the determination of alkali and alkaline earth elements in oil by FE include Refs. 1043-1050.

The determination of other elements in oils has been reported by Muntean and Badea for Cu, Fe, Ni, and V[(1051)] and by Sarkovic for Cu, Ba, Mo, Ni, and V.[(1052)] Qualitative flame spectrograms for elements present in lubricating oil have been recorded by Gleason and Hold[(1053)] and Whisman and Eccleston.[(1054)] Direct determination of B in oil via bandhead emission with a precision in the range 1-2% was developed by Buell.[(1055)] This method has subsequently been applied to the analysis of boron in oils down to about 5 ppm using a Perkin-Elmer Model 306 and either a nitrous oxide-hydrogen or a nitrous oxide-acetylene flame.[(1056)] These flames provided increased sensitivity for the BO_2 emission as compared to the oxygen-hydrogen flame used in the earlier report.[(1055)] Both flames provided precision in the range 1-2% at the 20-ppm calibration level in special naphtholite solvent and detection limits of about

0.3 ppm. With either flame, Ca, K, and Na interferences are serious but are considerably less (3- to 6-fold lower) for K and Na using nitrous oxide–acetylene. With the Perkin-Elmer instrument, overlapping second-order OH radiation and stray light affect the boron detection limits, which can only be reduced about 2-fold with a plain glass filter at 5180 Å.

For the determination of Ba in new lubricating oils, ac FE is used in the author's laboratory since interference from up to a 100-fold excess of Ca is eliminated by the technique. Although applications to elements other than the alkali and alkaline earths have been limited, the use of ac FE with the nitrous oxide–acetylene flame should provide increased applications (see applications to crude oil, Section 2.9.3.9). Lush[(1013)] commented on the use of a good monochromator and nitrous oxide–acetylene flame to provide detection limits better than those obtained by AA. Comments such as those by Jackson *et al.*[(1018)] that AA and AF, compared to FE, provide less matrix effects and better precision are misleading and, hopefully, will be recognized as such. For example, the better detection limits for many elements, and those cited in Section 2.9.3.9 for metals determination directly in crude oils plus precision below 1%, which is consistently equal or even better than that for AA, definitely show the latter statement[(1018)] to be in error. Matrix effects using nitrous oxide–acetylene are about equal for AA and FE, and as measured in the author's laboratory were equally bad for Mo and V. Using ac FE, direct spectral interference from other elements is seldom encountered; thus, AA provides only an occasional advantage. Using FE with an instrument such as a Perkin-Elmer Model 306, however, spectral interferences from excess matrix elements such as Ca and Na and from flame background may be up to 1000-fold greater (measured at a variety of wavelengths in the author's laboratory).

Thus, with the best instrumentation and techniques, modern FE must be considered an equal partner to AA for elements such as Ag, Ca, Cu, Fe, Ni, and Pb and is best for most Al, Ba, Cr, Mo, and V determinations.

2.9.4.6 Relative Merits of Most Popular Methods

As of 1973, the two most popular methods of analyzing lubricating oils for metals were AA and ESD using the Rotrode. Modern instrumentation and techniques should also make FE equally applicable, especially if ac FE is used or if multichannel instrumentation is used in conjunction with FE (Ref. 179). Both FE and AA use flames to produce free atoms. Oil samples are diluted with solvent, which is then introduced into the flame through a nebulizing chamber. Only about 10% of the sample passes through the chamber, and larger particles, including particulates in oil samples, are selectively discarded. There has been much debate on the accuracy of AA flame methods because of this fact, and procedures to minimize the errors involved and matrix errors are still being evaluated (Sections 2.9.4.7 and 2.9.4.8).

Because sudden increases of metal content (relative amounts) are of more importance than the absolute amount present, absolute accuracy is not mandatory for effective analyses in a maintenance program. For this reason AA and direct ESD methods continue to be used primarily because of their speed and simplicity, although both techniques can be influenced by particulates and matrix errors.

The ESD Rotrode technique may be influenced by particle-size effects but not to the extent that AA is influenced. The Rotrode can carry suspended material into the spark gap, where it is at least partially excited. The weakest part of the ESD method is the instability of the spark excitation discharge. Through the use of integration with time and the I.S. technique, the fluctuations involved are minimized. However, the stable flame used with AA and FE is an advantage. Reliability and flexibility for immediate calibration response and better precision result. On the other hand, spark excitation provides better sensitivity for certain elements, such as B, P, and Si. In the future we should see more use of the ac FE method, especially the plasma torch with direct readers.

2.9.4.6a Speed of Analysis. The largest advantage of ESD methods is speed of analysis. With a channel for each element and computerized calibration and calculation, a sample can be analyzed in 2 min for a large number of elements. This speed cannot be matched using AA, even with fully automated sequential analysis. An equivalent turnaround time for AA is more like 30 min. This is the reason two of the large commercial laboratories use ESD.[910] Plasma torch direct-reader methods should be a strong competitor of ESD methods in the future.

2.9.4.7 Particle-Size Influence

In any discussion of particle-size influence, the type of oil, engine, and history of the used oil must be considered. A well-used automobile engine oil with a high sludge and lead content (oil A) and a railroad diesel engine oil from an engine equipped with a flow-through clay-type filter (oil B) probably represents two extremes.

2.9.4.7a Rotrode Methods. In a study conducted at Union Oil[1057] using an ES–Rotrode method, particle-size influence in the two oil types A and B were studied. Oil A contained a heavy sludge and lead deposit and was analyzed for Ba, Cu, Fe, Pb, and Zn before (in the supernatant after a long settling period) and after shaking. After shaking, the emission intensity for Fe and Pb, respectively, were over 2-fold and 4-fold greater. The change for the other elements was much smaller or insignificant. Precision of measurements were also determined after shaking and after homogenizing with a blender. A marked increase in precision after homogenizing was obtained for Ba, Fe, and Zn, as shown in Table 2.20.

Table 2.20 Precision of Spectrographic Analysis as a Function of Homogeneity

	Element				
	Fe	Pb	Ba	Cu	Zn
Average percent standard deviation, hand-mixed samples	15	6.9	10	2.4	37
Average percent standard deviation, homogenized samples	9.3	8.0	3.3	4.0	15

For a type B oil and also another diesel railroad oil obtained from an engine not equipped with a clay-type flow-through filter (oil C), microscopic examinations and precision measurements were made before and after homogenizing. In this case a Virtis high-speed homogenizer operated at about 35,000 rpm was used. Conclusions were as follows: (1) homogenizing has little, if any, influence on oil B, and particle size is small and distributed uniformly without homogenizing; (2) particle size in oil C was reduced somewhat, but only caused small improvements (less than 20% relative); and (3) adding a dispersant and shaking oil samples with a Wig-L-Bug dental amalgamator was simpler and gave essentially equivalent results to homogenizing oil samples.

2.9.4.7b Atomic Absorption. In a thorough study of metal particle behavior for AA results, Taylor, Bartels and Crump[1058] evaluated variables including particle size, flame condition, and burner–flame combinations. Synthetic samples were prepared from iron particles having a mean diameter of 1.2 and 3.0 μm and the atomization efficiency (ppm of Fe recovered compared to organometallic standards) and aspiration efficiency (how much Fe actually reached the flame) were determined. The results are summarized below and show that large errors (greater than 10-fold) can occur if samples contain iron metal particles.

Sample	Atomization efficiency, (% recovery compared to an organometallic standard)	Aspiration efficiency (flow spoiler out) (%)
3.0 μm synthetic	2	7.8
1.5 μm synthetic	6	11.4
Organometallic standard	100	31.3

Their results gave lower values with a nitrous oxide–acetylene flame. They were not conclusive and were attributed possibly to burner configuration. Bartels and Slater[1059] also prepared suspensions of micrometer-sized particles in oil and achieved low recoveries by both AA and FE.

In another study for the determination of Fe in used oils, Golden[945] indicated that the nitrous oxide–acetylene flame (compared to other AA flames) gave the highest and most accurate AA results. He also indicated these results were reasonably accurate compared to chemical, ES-Rotrode, and x-ray fluorescence results (one sample, however, appeared to give results 37% high compared to the chemical method). Samples were selected from operating gas turbine engines and were of a synthetic ester type. Effects of iron particle size were determined on samples filtered through various Millipore filters. Most of the iron was retained on a 0.45-μm filter, and electron microprobe measurements verified that particles were of the order 1 μm or less in diameter. Even on the filtrate from a 0.25-μm filtrate, air–acetylene gave low results compared to nitrous oxide–acetylene and ES–Rotrode techniques, indicating that some type of matrix error was encountered. For this filtrate the results for ES on all three samples tested were higher (20–45%), indicating perhaps that similar but smaller matrix errors might exist even using the nitrous oxide flame. Golden's conclusions were that all direct methods may be influenced to some extent by particle size.

Westcott and Seifert,[950] in comparing ferrograph and ESD–Rotrode measurements for Fe in used oils, found discrepancies. They concurred that nonmagnetic particulates such as oxides and chlorides were present that are not precipitated on the ferrograph and caused low results by Rotrode methods. In connection with Fe-particulate studies, Freegarde and Barnes[1060] developed a novel separation method with chromatography on a metaldehyde column. The metaldehyde can then be sublimed away from the particulates at 112°C, after which they can be analyzed. They found that Fe occurred in three distinct fractions.

Some comments on observations in the author's laboratory may be worthwhile. It was observed for certain gas oils that upon dilution with hydrocarbon solvents a purple precipitate formed. The Fe content was high in this precipitate; thus, it was apparently a sulfur-containing complex that formed in the oil. To eliminate errors caused by such compounds and particulates, a mixed solvent system containing 1% HC1 and ammonium thiocyanate in an isopropanol–cyclohenanone-special naphtholite mixture was developed. Warming samples in this mixture and using direct FE or AA measurements gave results equivalent to ashing methods. The procedure, however, may not be suitable for certain samples, such as shale oils, wherein Fe may be contained in suspended, complex shale components.

Kriss and Bartels[1061] also proposed the use of an acidified diluent to solubilize the iron before analysis. Holding and Noar[1032] propose a mixed solvent system containing acid to minimize a peculiar difference in results for Ca compared to naphthenate standards in cyclohexanone. Golden,[945] however, upon testing the acid solubilizer technique, obtained unexplainably high results.

Various reports of discrepancies such as those discussed indicate that particle size does indeed influence AA determinations and also that we need to learn more about matrix errors. In the final analysis, it would seem that for reliable, accurate results a preliminary ashing of the sample would be wisest.

2.9.4.8 Matrix Errors Caused by Oil Composition

Various types of oils compounded by different oil companies may contain P and S compounds and associated cations, such as Ca, Ba, Mg, K, Zn, and even Sb.[1008] These substances are generally higher in concentration than trace amounts of wear metals that occur in used oils and thus may cause matrix errors during analyses. There may also be mutual interferences for the analyses of these additive elements. See Section 2.9.4.3b for effects on AA methods with new oils.

2.9.4.8a Matrix Errors for Emission Spectrometric Methods. It is well established that lubricating oil composition can influence relative emission intensities when using ES methods. This matrix type of error, caused by the types of compounds present and interelement effects, should not be confused with spectral interference, which is a completely different and easier type of interference to correct for.

In this respect the Rotrode method appears to be no better than other ES methods. Gunn[941] made a statistical study of elemental interactions using the porous cup electrode method and developed equations to correct for such interferences. He compared such interferences to those using the Rotrode method and concluded that matrix errors influenced both methods, and that precision of measurements was about the same. Data were also presented showing that viscosity in the range 200-800 (SSU at 100° F) caused only small errors, if any.

In another study using the Rotrode method, Cummins and Mason[942] showed significant differences due to viscosity influence in the range SAE 10-50. They found that there differences were less using high inductance (more arclike) excitation. They also studied interelement effects and found that these were less when using graphite electrode discs impregnated with aqueous solutions of magnesium and sodium chloride.

Some interesting matrix influences were studied in the author's laboratory,[1057] which reveal complex situations that can occur using the Rotrode method. It was noted that calcium-containing additives and other additives caused a strong matrix error for certain elements. Buffers were tested to minimize these errors and also to enhance emission, particularly for phosphorus. Adding about 0.2% of an organometallic antimony compound enhanced emission markedly and nullified the influence of most additives tested, including up to 0.1% of calcium. At this time an unusually large enhancing effect was discovered upon the addition of a Stan-Add additive with a high sulfur content and no other inorganic elements. An enhancing effect of slightly different magnitude was noted upon addition of the compound to standards prepared from metal naphthenates. These observations are cited to stress that the type of oil and additive should be known and their influence on signals compensated for in order to provide proper calibrations and accurate determinations. If changes in concentration are more important to know and absolute amounts are of lesser importance, many of these matrix influences can be tolerated.

2.9.4.8b Matrix Errors for Atomic Absorption Methods. Although many papers have been published on the use of AA methods for oil analysis, the possibility of matrix errors has been somewhat neglected in these studies. Some matrix effects in new oils were discussed in Section 2.9.4.3b. Golden[945] cites a suspected matrix influence for FE determination in oils using low-temperature flames. Holding and Noar[1032] found that the addition of acid in a mixed solvent system eliminated the matrix error for Ca analysis, yet Golden appeared to obtain erroneously high results using a similar mixed and acidified solvent system.

Chemical interferences are generally greater in lower temperature flames such as air–acetylene and less in the higher temperature nitrous oxide–acetylene flame. (Interferences were thoroughly discussed in Section 2.7.) Bowman and Willis[1062] showed that although interferences were less using the latter flame, some chemical interferences still existed in the AA determinations of Ba and Sr as applied to rock analysis. With the development of the nitrous oxide flame, the determination of V in oil became practical.[1062] Sensitivity for other elements, such as Al, Ba, Ca, Cr, and Si, was also increased. Bowman and Willis, in their study of the AA determination of V in fuel oil, did not study matrix errors except for the influence of oil but did find that low results were obtained using an N.B.S. standard dissolved in solvent. It was necessary to use a well-analyzed sample as a standard, but even then some discrepancies occurred for certain samples comparing AA-ashing and direct methods (up to 30%). Oil, in addition to depressing absorption at levels over 20%, enhanced absorption.

Ionization and certain other types of interferences are greater with the nitrous oxide flame, although large, condensed phase interferences for alkaline earth elements so prevalent in low-temperature flames are much less. Ionization interference can occur for alkaline earth elements, for Al, and to a smaller extent for Cr and V. The determination of Ba using either AA or FE is especially influenced by ionization. The addition of K as a buffer is very effective in eliminating such interferences, as shown by Capacho-Delgado and Manning[1063] in the determination of Ca and Ba in lubricating oils and in Ref. 1062 for aqueous solutions. The FE response is similar, and for the determination of Ca and Ba in oils, ionization occurs even in the oxyhydrogen flame. As measured for the Ba ion line using FE in the author's laboratory, the addition of K suppressed ionization to less than 2% of that without K. The addition of about 50 ppm K was effective for up to 30 ppm of Ca and Ba.

Other types of interferences in the nitrous oxide flame seem to involve refractory elements and lateral diffusion effects (Section 2.7). Barnett[1064] studied interferences of acids on Cu, Cr, Mn, and Ni absorption and found them least using air and greater using nitrous oxide flames. Maruta *et al.*[1065] found the opposite effects for Cr, Cu, and Fe, finding no interferences using a nitrous oxide-acetylene flame. Marks and Welcher,[1066] using nitrous oxide flames, recorded the mutual interferences between Al, Cr, Ni, and Ti and also additional interferences caused by Al, Cr, Ni, and Ti. Similar interferences occur for FE but appear to be far less for nitrous oxide burners used for AF as measured for elements such as Al, Mo, and V by Dagnall *et al.*[1067] Miller *et al.*[1035] also studied interferences in their application of AF to lube oil analysis. Ferris *et al.*[1068] evaluated AA correction coefficients for mutual interferences of Al, Si, and Fe. A nitrous oxide flame was used for Al and Si and for determinations of Fe made using an air-acetylene flame, and phosphate was found to be a good buffer for leveling depressing interferences caused by Al and Si. Other interference studies for Fe determinations using air flames[1069–1071] reveal many interferences, including a depressing effect caused by thiocyanate even in organic solvent solutions. In Ref. 1069 La was used as a buffer and in Ref. 1070 oxine was an effective buffer for Co, Cu, and Ni interferences.

Yanagisawa *et al.*[1072] compared the use of air and nitrous oxide flames for the AA determination of Cr, finding many interferences using air but none using the latter flame. This is contrary to the interferences recorded by Marks and Welcher,[1066] although interferences were found in Ref. 1066 to be much less using a lean flame. Hurlburt and Chriswell[1073] overcame Cr interferences in an air-acetylene flame by the use of sodium sulfate, which also enhanced absorption.

Additional references to interference studies are Refs. 1074–1078 and those included in Table 2.21. The table summarizes potential interferences given in

Table 2.21 Interference References for AA Determinations of Metals That May Occur in Lubricating Oils

		References	
Element	Interfering elements	Air flame	N_2O flame
Al	Many elements, Na[b]		1066–1068, 1074
Ba	Al, K,[b] NH_4Cl,[b] oxine[b]		1062, 1063
Ca	K[b]		1062, 1076
Cr	Acids, many elements	1064, 1065, 1072	1064–1066, 1072[a]
Cu	Acids, thiocyanate	1064, 1065, 1071	1064–1065,[a] 1071
	Na_2SO_4	1073	
Fe	Acids, thiocyanate, many elements	1065, 1068, 1069	1065,[a] 1071
	La,[b] P,[b] oxine[b]	1070, 1071	
Ni	Acids, many elements	1064	1064, 1066
Si	Al, Fe		1068
Sn	Many elements, $FeCl_3$[b]	1077 (air–H)	
Sr	Al, K[b]		1062
V	Oil, many elements		1062, 1067
	(Al + diethylene glycol diethyl ether)[b]		1062, 1067

[a]Interferences tested but none found.
[b]Tested for use as buffers.

references for interference studies concerning elements that occur in lubricating oils.

Matrix interferences appear to be greater for nonflame, AA methods studied in Refs. 1079–1081 and include a type of interference for Pb not found in flames.[1079]

Utilization of information given in these references may lead to better methods and may resolve some of the discrepancies between direct and ashing methods mentioned in Section 2.9.3.7e and discrepancies between AA and other methods of analysis shown in various references.[916, 996, etc.]

Vanadium and also molybdenum are particularly susceptible to interferences in the nitrous oxide flame. For aqueous solutions a combination of some of the buffers used in various references has been found in the author's lab to be effective as a universal buffer which levels interferences and enhances AA and FE. The buffer consists of a mixture of 0.03% oxine, 0.8% phosphoric acid, 0.1% K, and 4% isopropanol. Matrix errors in organic systems are being investigated.

2.9.5 Gasoline and Light Petroleum Distillates (LPD)

Gasoline is the premium product of refining operations and is produced in tremendous volumes. As a result, methods for quality control of finished gasolines, gasoline blending stocks, and feedstocks is of considerable importance. Techniques of analysis for gasolines can be applied to other LPD, such as kerosenes, naphthas, and other light hydrocarbon products. Ashing procedures are generally unsatisfactory for such materials, but extraction methods can readily be applied. Direct methods are naturally preferred, and AA and FE are well suited for such approaches.

2.9.5.1 Lead in Gasoline

The determination of lead in gasolines is still important. For many years lead has been added to gasolines in the form of tetraethyllead (TEL) as a means of increasing the octane rating. In more recent years mixed alkyl lead compounds (Tetramix), including tetramethyllead (TML), have been used. Through the years automobiles in America gradually became more powerful. This increase in power (and compression ratio) required higher-octane gasoline if a given high-power car was not to knock or rumble (caused by too low an octane rating). The most effective and economical means of raising the octane rating is to add lead compounds. The trend is now changing and less powerful cars (also lower in cost) are more popular. In addition, there was an increasing awareness of pollution problems. Discharges from automobile exhausts were evaluated as a major contribution to smog. Pollution, including smog, has become a serious problem. As a consequence, federal regulations have been passed which require the eventual elimination of lead in gasoline. This will require monitoring gasoline to contain no more than 0.05 g lead/gal (about 13 ppm).

In addition, at certain refining stages, treatment of naphtha and other LPD may involve the use of lead oxide to remove sulfur. If such feedstocks are processed through any sort of catalyst bed, it is desirable to keep lead concentrations below 0.1 ppm so as not to poison the catalyst bed.

As a consequence, the analysis of Pb in gasoline, gasoline feedstocks, and other LPD is of considerable importance to the P.I.

2.9.5.1a Flame Emission Methods for Lead in Gasoline. In early days, before 1950, lead in gasoline was determined primarily by chemical methods; Lykken[1082] reviewed and studied such methods in 1945. Probably the earliest application of ES to studies of TEL in gasoline was by Clark and Smith[1083] in 1929. Spectrographic methods such as the porous cup electrode used by Gassman and O'Neill in 1949[1084] were tried but are generally not precise enough. Emission methods did not become competitive with chemical methods until the early 1950s. When flame emission spectroscopy (FE) became popular and commercial instrumentation became available in America, FE methods for determining TEL began to appear. Gilbert investigated such methods in 1951.[1085] Numerous methods were published in the following years,[1086–1090] including a detailed study by Buell[1091, 1092] of the problems involved in providing the best results. Buell[1092] investigated the problems involved with differences in signals provided by various lead compounds and calibrating methods. Extremes are observed in the responses of TEL and TML, which is more volatile and dissociates faster in flames, thereby giving a much larger response. With a variable-height total consumption oxygen–hydrogen burner a special procedure can be used to compensate for the differences.[1092] At one particular height in the flame, TEL and TML give equal signals and at that height Tetramix can be analyzed accurately. A better procedure is to react the samples with iodine (discussed under AA methods) and use a nitrous oxide–acetylene flame. Using ac FE in this manner, the detection limit for Pb is 50 ppb.

2.9.5.1b Atomic Absorption Methods for Lead in Gasoline. Atomic absorption methods are now more popular and offer some advantages but are influenced by differences in alkyllead forms in the same way as FE methods.

Early champions of AA methods appear to have paid little attention to the previous FE literature revealing this fact[1093] and additional data. Dagnall and West[1094] in 1964 made a detailed study of Pb and noted different responses. Currently, the best AA methods recognize the problems and have devised techniques for rapidly converting alkyllead compounds to a common form.[1095, 1096] Since the lead additives in gasoline are nonionic compounds, special procedures are used to decompose them. Kashiki *et al.*[1095] reacted samples with iodine diluted in methyl isobutyl ketone and verified that the procedure provided equivalent responses for five alkyllead compounds. The iodine also removed "memory" effects, which can be quite persistent for Pb compounds in laminar flow burners. In the author's laboratory it was verified that reaction *in situ*, by adding a small amount of solvent containing iodine, converted TML and TEL to forms giving equivalent AA and FE response. The magnitude of response was lower than that provided by TML and nearly equal to that provided by TEL. A DuPont method[1096] also utilizes reaction with iodine and Aliquot 336 to increase sensitivity and storage stability. Using a 5-fold dilution in methyl isobutyl ketone, they obtained a determination limit as low as 0.001 g Pb/gal. Nishishita *et al.*[1097] investigated a report that olefins interfered with the iodine-reaction in certain gasolines but found no interference up to a concentration of olefins 254 times that of the iodine. Lukasiewicz *et al.*[1098] used a nitrogen oxide–hydrogen flame for the direct analysis of lead in undiluted gasoline samples containing less than 0.1 g/gal. Parameters for the iodine reaction and the addition of Aliquot 336 (to stabilize Pb) were studied and optimized. McCorriston and Ritchie[1099] found that a total consumption burner using air–hydrogen gave better results for Pb and cited advantages, which included equivalent responses for TML and TEL without prior iodine reaction. For gasoline samples well below the "no" Pb limit (0.05 g/gal) a direct analysis without the iodine reaction is acceptable and faster.

The first application of AA for determining lead in gasoline was by Robinson,[1100] who found that his results were comparable to those obtained by other methods. He investigated interference possibilities from various cations and N and S compounds, finding none. He did not, however, study possible errors caused by variations in the hydrocarbon composition of gasolines, but the 10-fold dilution he used probably made these differences insignificant. Chakrabarti,[1101] Trent,[1102] and Mostyn and Cunningham[1103] also applied AA to the determination of Pb in gasoline. Nagypataki and Tamasi[1104] reviewed U.S. and European methods for Pb in gasoline. Sachdev and West[1105] studied the AA response of Pb as applied to solvent extractions. Moore *et al.*[1106] analyzed Pb in simple dilute acid extractions as a means of evaluating gasoline stability by determining the amount of ionic Pb resulting from the natural decomposition of the nonextractable alkyllead compounds. The carbon rod method has been applied to Pb in gasoline by Bratzel and Chakrabarti.[1107]

2.9.5.2 Miscellaneous Elements

Various approaches have been used for analyzing Na in hydrocarbons, although a direct determination is relatively simple. Nelson and Grimes[1108] used water extraction followed by FE for determining Na in liquid hydrocarbons. Kapitaniak[1109] also used a FE determination of Na but first ashed the samples. Kaegler[1110] applied AA directly to the determination of Na, Mn, and Pb in liquid hydrocarbons.

Jordan[1111] applied FE to the determination of Cu, and Goux[1112] applied AA and FE for determining Cu and Pb directly in petroleum products, including kerosine and light oil. Manjarrez and Pereda[1113] applied AA directly to the determination of Cu, Ni, Fe, Pb, and V in petroleum products, including kerosine, and Moore *et al.*[1106] determined Cu, Ni, and Zn in gasolines using AA.

Boron in gasoline has been determined by Buell[1114, 1115] and Dubois and Barieau[1116] using FE, and by Vigler and Failoni[1117] using a plasma arc.

Manganese has been determined in JP-4 fuel using AA by Bartels and Wilson[1118] and Kaegler,[1110] and by Smith and Palmby[1090] using FE. Nickel as a gasoline additive has been determined by Schoer and Pontious[1119] using FE.

Atomic fluorescence has been applied to the determination of Cu, Fe, and Pb in kerosine by Cotton and Jenkins[1120] and for Ni in petroleum fractions diluted in *n*-heptane and xylene by Sychra *et al.*[1121–1123] Cotton and Jenkins[1120] developed a special burner which used a nitrogen-supported flame with liquid hydrocarbon as fuel. Kerosine and even benzene were quoted as burning with a nonluminous flame, and detection limits were 4, 40, and 60 ppb for Cu, Fe, and Pb in kerosine, respectively.

Griffing *et al.*[1124] used a Rotrode–ES method for the determination of 5–75 mg/liter of P in gasoline with a standard deviation of 0.8 mg/liter. Special techniques using an extension of the Rotrode armature and an inert gas stream were used to prevent flaming of the sample.

For an EPA study, von Lehmden *et al.*[1125] investigated various methods, including AA and ES, for determining 28 elements in coal, fly ash, fuel oil, and gasoline. Agreement among methods and laboratories was poor. No details of the methods for gasoline were given. Most of the results for gasoline were less than the detection limits, which were often below 0.1 ppm.

Gasoline analyses were included in methods developed for 13 trace elements using wet ashing and nonflame AA in a consortium approach by a group of oil companies.[836]

2.9.6 Diesel Fuels, Turbine Fuels, Gas Oils, and Feedstocks for Refinery Processing Units

In the American petroleum industry gasoline is the premium product. To produce higher yields of gasoline, crude oil distillates are subjected to catalytic cracking processes, which on a volume basis can produce greater than 100% yields. Modern procedures have been slanted toward greater capabilities of processing heavier feedstocks into gasolines and jet fuels.[1126,1127]

This diversification of feeds creates problems for the analytical chemist. As boiling ranges increase from light gas oils to heavier feedstocks, direct methods for analyzing their metal contents become more difficult. The control of trace metal content (Cu, Fe, Ni, V) is important because certain metals (especially nickel and vanadium, which are indigenous to crude oils) can poison solid, cracking catalyst, thus reducing gasoline production and increasing coke production.[1128,1129] Delays in production caused by regenerating catalyst or replacing them (an expensive, major task) are very costly. For this reason, close control of the metal content of feeds is desired. Because of the large volumes of feeds processed in a given unit, even concentrations below 1 ppm can gradually accumulate on a catalyst bed, leading to poisoning. Because the metal contents are so low, methods employing concentration procedures are generally required (see Section 2.9.6.3 for a direct ac FE method).

Diesel and turbine fuels are distillate fuels ranging from kerosine to gas oils that have low-metal-content specifications (Ref. 1160, p. 170). The methods of analysis used for feedstocks, ranging from very light gas oils to the heavier, waxy gas oils, may also be used for these distillate-type fuels.

2.9.6.1 Emission Spectrographic and Spectrometric Methods

Spectrographic methods were the first emission techniques applied to such determinations. Murray and Plagge[1130] in 1949 ashed samples in the presence of silica as an ashing aid and determined about a dozen metals with a semiquantitative ES procedure. Additional early applications of ES using dry ashing and arc excitation of sample ash mixed with lithium carbonate buffer include those discussed in Refs. (1131-1133).

Karchmer and Gunn[1134] reviewed colorimetric and spectrographic methods currently in use at the time and made a detailed study of analyzing oil samples. They analyzed Na by Fe and Na, Al, Ca, Cr, Fe, Mg, Ni, and V by the ES method of Gunn and Powers[1132] using lithium carbonate as the buffer. They made a detailed study of the variables involved for dry-ashing oil samples. The ashing technique they developed gave results within ±10% for ash values in the range 1–100 ppm and metal analyses comparable to colorimetric methods. Many of their data, however, were obtained on samples other than gas oils (overhead distillates). They ascertained that severe stratification of oil samples, concentrating metals in lower layers, is of considerable importance. Sampling must be done very carefully and they recommended using the entire bottle of sample once it has been obtained.

Work and Juliard[1135] used an ES method which employed alumina to collect sample ash and to minimize loss of elements due to volatility. This use of alumina should be quite good, since it can chromatographically adsorb metalloorganic compounds, including porphyrins. Alumina has been tested in the author's laboratory and found to be an excellent buffer for nullifying matrix errors, and it also enhances emission (discussed in Section 2.9.4). McEvoy,

Milliken, and Juliard[1136] used a catalytic ashing technique with Al–Si to minimize losses. The ashed residue was mixed with graphite, then pressed into a briquetted pellet which was arced. This procedure provided a precision of about ±5%.

2.9.6.1a Volatility of Metal Compounds in Gas Oils and Possible Losses. It was learned early that some metal compounds are volatile and can occur in distillates.[1137,1138] These volatile metals compounds include Ni and V porphyrin compounds, which incidentally are nonionic and thermally stable. This makes it difficult to extract such compounds from oil and also leads to complications when using dry-ashing or wet-ashing procedures designed to minimize loss due to volatility.

In a comparison of dry ashing and wet ashing with a preliminary charring of sample with sulfuric acid, Milner *et al.*[1139] showed there was no loss of Fe, Ni, and V for crude oils or residual fractions, but that losses do occur for overhead distillates. Losses for the latter using dry ashing, as determined by comparison with wet ash–chemical analysis, varied from 22 to 32% for Ni (67% in one case for a level of less than 0.1 ppm Ni) and from 2 to 38% for V, with one exception of a 59% loss. Gamble and Jones[1137] also compared dry and wet ashing. For the wet-ashing method they added magnesium nitrate as an ashing aid and spectrographic buffer and cobalt naphthenate as the internal standard dissolved in alcohol. The ash was packed into a cup electrode with a center post and excited with a 10-A dc arc. Standard deviations for the range 0.1–2 ppm were ±0.02 ppm for V and ±0.06 ppm for Mn and Ni. For dry-ashed samples, the Ni losses varied from 37 to 57% (70% in one case) when compared to the wet-ashing method. They also show curves of T versus Ni and V volatility for redistillation of heavy gas oils. They also demonstrated that the technique of dry ashing is very touchy and obtained losses of 34 to 59% by increasing the burning rate from 30 to 175 g/h with a larger ashing dish. This may be part of the reason they obtained greater losses for Ni than did Milner *et al.*[1139] and why Barney and Haight[1140] obtained no loss for Fe, Ni, and V and only loss for Cu and Pb. Barney and Haight,[1140] in a similar comparison, used an ES method and arc excitation of an acid solution of sample residue evaporated on a flat, chamfered $\frac{1}{4}$-inch electrode. They compared four extraction methods and dry ashing to sulfated and partially sulfated ashing. A gas-oil cut from a Venezuela crude contained the following elemental concentrations via the sulfated ash method:

Element	Cu	Fe	Pb	Ni	V
ppm	0.051	0.92	0.38	0.023	0.18

The sample contained V porphyrin compounds. They found that losses for Cu and Pb also occurred when a partial sulfating procedure was used, which indicated that a wet-ashing method can be touchy. The wet (sulfated)-ash method and extractions with hydrobromic–acetic acids and sulfuric–hydrochloric acids gave equivalent results. They preferred the latter extraction method, which they studied further.[1141]

Hansen and Hodgkins[1142] developed an ES and ESD method for feedstocks using a wet-ash procedure with copper powder as an ashing aid. They used an arc excitation of sample ash in a cup electrode with a feathered edge to reduce arc wandering. With 10-g samples they obtained precisions of 7.7–9.4% for Ni and 11.3–11.4% for V by the ES method and 8–10% for Ni and 8–11% for V using the ESD method. Precision for FE was poor possibly, owing to the non-homogeneous samples.

An ESD–platform electrode method developed by Hoggan *et al.*[1143] used a mixture of 20% sulfuric acid with 80% butanol as an ashing additive to reduce losses of volatile metals. Sample ash was dissolved in acid and evaporated to dryness on the top of a dished-platform electrode.

At Union Oil samples are ashed in the presence of benzenesulfonic acid to minimize volatility losses. Acid solutions of the ash are then analyzed for Cu, Fe, Ni, and V using an ESD method with a vacuum cup electrode. A modification[1144] of the usual vacuum cup electrode (a flat top instead of a hemispherical tip) was found to be more precise and easier to use than the well-known Rotrode technique. Precision for two calibration ranges is given in Table 2.22 and varies from 1.6 to 2.7% for the high range (8–20 ppm) and from 2.0 to 2.9% for the low range (2–5 ppm).

Other methods of analyzing gas oils have also used special procedures to minimize volatility losses. Agazzi *et al.*[1145] developed an ashing method in which elemental S was added to the oil. Results of colorimetric determinations, using a sample predetermined to lose metals upon dry ashing, compared favorably to wet oxidation procedures. Using x-ray fluorescence methods,[1146,1147] employed benzenesulfonic acid plus MgO and xylenesulfonic acid as ashing aids.

It is interesting to note that all these methods designed to minimize volatility losses, with the exception of the work by Barney and Haight,[1140] apparently utilized little from the chemical studies of Ni and V porphyrin compounds

Table 2.22 Calibration Precision for a Direct-Reading, Emission Spectrometric Method Using a Modified Vacuum Cup Electrode

	High range		Low range	
Element	ppm	Precision	ppm	Precision
Cu	8	1.6	2	2.9
Fe	20	2.7	5	2.8
Ni	10	2.2	2.5	2.1
V	16	2.4	4	2.7

presented in many references on crude oil studies (Section 2.9.3). Nor did they utilize testing with any natural concentrates of volatile porphyrin compounds from crude oils. Horeczy *et al.*[1148] did make a study using synthesized Cu, Fe, Ni, and V compounds and found considerable loss using the dry-ashing procedure in Ref. 1139. Copper and iron porphyrin compounds have never been found in crude oils (Section 2.9.3), however, so part of their study was not truly realistic. Because of the large differences between the studies cited on volatility losses and the particular nature of Ni and V, it would appear wisest to keep an open mind as to the true situation applied to oils from various sources. Unless a closed system is used, it might be possible even with wet-ashing methods for some loss of Ni and V porphyrin compounds to occur. If the efficiency and rate of decomposing these stable porphyrin compounds is not rapid enough and too rapid a heating rate is used, it would appear likely that losses could occur. Some of the studies applied to such compounds (Section 2.3) show that they can have exceptional thermal stability and strong resistance toward quantitative decomposition with chemical reactions.

A high-Ni, high-V (320-ppm Ni and 740-ppm V) solid asphaltene concentrate was obtained at Union Oil from heavy crude oil. Wet oxidation under a condenser gave 8% higher Ni and 4% higher V results than wet oxidation in an open beaker. Routine low-temperature ashing in a crucible with 0.5-g samples produced results about 5 to 8% low, probably due to the entrainment losses of fly ash even with the addition of benzenesulfonic acid. Ashing at an extremely slow rate in a larger high-form crucible cut losses 4–5%.

Recently, a consortium approach was used by a group of oil companies to develop methods for trace analysis for 13 elements in fuel oils and gasolines (Ref. 836). Various methods of sample preparation were tested. Ashing with a preliminary charring with sulfuric acid followed by furnace ignition at 540°C was selected. Small samples (0.5–2 g) were ashed in 50-ml Vycor crucibles for nonflame AA methods. In addition to sulfuric acid, magnesium nitrate was added as an ashing aid to process 100 g of oil for spectrographic analysis. The resulting ash was mixed with graphite, pressed into a pellet, and sparked using a graphite counter electrode. Results were compared to those for activation analysis, but most results were reported for standard addition studies.

Shmulyakovskii *et al.*[1149] used a spectrographic method for determining As, Cu, and Pb in feedstocks and substituted for the ashing step an evaporation of samples directly on the electrodes. A similar method, showing some promise, has been tested by the author using platform electrodes and ESD.

2.9.6.2 Atomic Absorption Methods

The first use of atomic absorption for such applications was reported by Barras and Helwig[1150,1151] in 1962 and 1963. Their technique was not without problems, such as absorption interference from organic solvents for nickel and a lack of sensitivity. The addition of 10 ppm of Ni to various organic solvents and obtaining the net response for Ni gave a variation of −15 to +43% absorption. For determinations at low levels a blank was encountered which was equivalent to 3.75 ppm of nickel, not a favorable ratio when analyzing 0.2 to 5 ppm of nickel. Despite these problems, they claimed that the results were good. Later, in 1967, Barras and Smith[1152] reviewed the use of atomic absorption in the petroleum industry and described applications to the analysis of gas oils for Cu, Fe, Ni, and V using a multichannel, atomic absorption spectrometer. They cited the need for making corrections for molecular absorption which was worse for certain sample types. Their earlier success for V at levels less than 1 ppm, for which they used a total consumption burner, is questionable. They claimed verification by activation analysis, but a cross comparison of their results in cooperative testing with oil companies using other methods was very poor.[1153]

Trent and Slavin[1154] made a detailed study for the application of atomic absorption to Ni in feedstocks. They used the method of successive additions of standards and made careful measurements at 2320 Å and at a nonabsorbing wavelength, 2316 Å, to correct for molecular absorption. They used a 5-fold dilution in xylene, which they claimed absorbed slightly less than the bare flame. They reduced sample flow and studied parameters influencing flame characteristics for various solvents. The precision and detection limit was 0.05 ppm of Ni. Kerber[1155] reviewed the use of this technique applied to Ni analysis.

Capacho-Delgado and Manning[1156] and Bowman and Willis[1157] applied atomic absorption to the determination of V in gas oils. The application was made possible by using a 5-cm laminar flow nitrous oxide-acetylene burner. Prior to the development of this burner, sensitivity for V was not adequate for such an application. In Ref. 1156 the same technique used for Ni was employed except that background corrections were not required. The detection limit for samples was about 0.05 ppm, and results for two samples were equivalent to emission results. For a third sample, 1.05 mg/liter was consistently obtained versus 1.8 mg/liter by the emission method. The discrepancy was not explained but may be due to an unknown matrix effect. Severe interferences are known to occur for the measurements of vanadium in nitrous oxide flames (Section 2.7) and comments were made in Section 2.9.3.7.5 concerning erratic results for direct flame methods. Bowman and Willis[1157] obtained fair agreement for some samples but found discrepancies (up to 30%) for others upon comparing ashing to the direct method. They used a well-analyzed sample as a standard because low results were obtained with an N.B.S. standard compound.

Obidinski and Johnson[1158] used an ashing method to analyze for Ca, Mg, and Pb in turbine fuels. FE was used for K and Na, and V was determined colorimetrically. Smith *et al.*[1159] discussed x-ray fluorescence and AA methods for Fe, Ni, Cu, and V in feedstocks using a standard addition technique.

In a 1972 symposium on turbine fuels,[1160] H. A. Braier presented general considerations using different analytical methods, including AA. J. B. Harrell discussed the paper and an AA standard addition method used by the Petrolite

Corp. At the same symposium C. C. Ward presented a survey of trace metals in distillate fuels, followed by a discussion by J. A. Vincent and D. L. Beers citing a Standard Oil of California spectrographic method for Ca, Cu, Na, Pb, and V, and an AA method for K.

2.9.6.3 Flame Emission Methods

An ashing procedure is generally used to concentrate the analyte as described in Section 2.9.6.1 or as used in Section 2.9.6.2 (Ref. 1158) for the analysis of K and Na by FE.

Metals in gas oils have been analyzed directly without ashing in the author's laboratory using ac FE as described in Section 2.9.3.9. The better detection limits obtained (1.8 ppb of Ni and 0.7 ppb of V in solvent) made this technique superior to AA without the need of making a correction for molecular absorption (see Section 2.9.3.9 for AA detection limits and for problems involved with direct methods). For gas oils and turbine fuels, the method is capable of analyzing down to the limits listed in Table 2.23. A good reference solvent extremely low in metal content and similar to turbine fuels is *n*-hexadecane. For such samples burner spraying rates should be increased.

2.9.6.4 Atomic Fluorescence Methods

Sychra and Matousek[1161] applied AF to the determination of Ni in gas oils. Any of the techniques using AF analysis of oils discussed in Section 2.9.4.4 might also be applicable to gas oils.

2.9.6.5 Comments on Methods

For some time atomic flame spectroscopy was not in general use in the P.I. for feedstock analysis, but its use is now increasing. ES or ESD methods are often used. Cooperative comparison of methods using similar ES techniques by different oil companies for low metal concentrations in gas oils too often produce results that do not agree (various participations from 1955 to the present). At times it appeared more difficult to produce results which agree when cross-comparing different techniques. These comments are added to emphasize that determinations of metal contents below 1 ppm in feedstocks appear to be quite touchy. Extreme care should be observed, regardless of the method used, and

Table 2.23 Detection Limits for Direct ac FE Method

	Gas oils		Turbine fuels			
	Ni	V	Ca	K	Na	Pb
Detection limit (ppm in sample)	0.02	0.01	0.005	0.03	0.005	0.05

one should carefully avoid jumping to conclusions regarding the accuracy of such methods.

2.9.7 Catalysts Used in Petroleum Refining Operations

Refining operations producing gasoline and other LPD use solid alumina and silica-based catalysts (Section 2.9.6). With the demands during and after World War II to produce more gasoline and jet fuel by processing higher-boiling oil fractions, poisoning of catalyst by accumulation of undesirable metals became more of a problem. Mills[1162,1163] and Duffy and Hart[1164] presented data and discussed the importance of this problem to the P.I.

In addition to the analysis of trace metals (Fe, Ni, V, etc.) that poison catalysts, the principal and minor components of catalysts are also monitored. In addition to Al and Si, catalysts may contain such elements as Co, Cr, Mo, Pd, Pt, and Re and possibly other noble metals and even rare earths. A research group may be continually compounding catalysts which could contain other elements.

2.9.7.1 Emission Methods

Among the first applications in the United States of ES for P.I. applications was the determination by Marling[1165] in 1948 of Na in catalysts. Among other early applications of ES to catalyst analysis were those by Gunn[1166,1167] and Gamble.[1168] In addition to arc methods, spark techniques have also been applied to the analysis of catalysts. Key and Hoggan[1169] developed a special high-voltage-spark technique. A mixture of the sample mixed with graphite and lithium carbonate buffer was pressed into a rotating disc electrode on a special mandril and sparked. The elements Cr, Fe, Na, Ni, and V were analyzed with a maximum standard deviation of 10% and a better average. Pagliassotti and Porsche[1170] dissolved samples in acid solution and used a Rotrode-spark technique to determine Ca, Cu, Fe, Mg, Ni, Pb, and V. Since the catalyst composition was 2% magnesia, 15% alumina, and 80% silica, a preliminary treatment with HF was used first to remove the silica, then the residue was dissolved in HCl. The sodium content was determined by a FE method.

In Japan, Okada *et al.*[1171] used an interrupted arc and a unique electrode arrangement for analyzing catalysts. A glass rod was inserted through a hole in the electrode so that with a gear arrangement to drive the glass rod, the sample is kept even with the electrode tip during excitation. They achieved relative standard deviations of ±4 to 7% for Cu, Fe, Ni, and V and ±15% for Na. In Russia, ES and FE techniques applied to catalyst were discussed in the compilation by Aleksandrov *et al.*[1172]

There appear to be few recent references using ES for catalyst analysis. One of the main reasons is probably due to the extensive use of x-ray fluorescence, which has considerable advantage for such applications. Raburn[1173] briefly reviewed the methods in use and described an x-ray method using fusion with lithium tetraborate.

Some ES methods, however, are still being used in the P.I. and often it is wise to have an alternative method. At Union Oil we have used arc excitation of a modified Linde Company method. Catalysts are often prepared by specialized organizations for the P.I. In such cases specification limits for various elements and the methods for their analysis are part of the negotiations which must be agreed upon. In testing this method,[1174] a Stallwood jet with oxygen was used. Some advantages were found for a few elements (Cr, Ni, and Mo), but other disadvantages occurred. Sample is consumed faster using the jet, and the background from swan bands is reduced. The background for alumina and silica is increased, and sensitivity is not as good for Ti and Zr. The method selected used a 6-fold dilution with a simple arc excitation. For determining nine elements in catalyst, the latter method gave better emission intensity for all elements except Cr, Ni, and Mo, and precision was markedly better for Fe, Mo, Pb, and Ti (7-9%).

A special ESD technique has also been used in the author's lab for determining moderate amounts of Mo and Ni in alumina-based catalyst. A 50-mg catalyst sample is dissolved readily in concentrated pyrophosphoric acid (see Section 2.9.8.4 for preparation) at high hot-plate temperatures. A high-voltage excitation with a modified vacuum cup electrode (Section 2.9.8.4) and 240 ppm of Co I.S. were used. For duplicate determinations of 11 different samples the average percent deviation from the mean was 1.2 for Mo and 1.3 for Ni. Comparison to chemical results was good as shown in Table 2.24, and accuracy of the method is estimated to be within ±3%.

Table 2.24 Comparison of Emission and Chemical Results (wt. %)

	MoO_3			NiO		
Sample No.	Chem. av.	Em. sp. av.	Dev. from chem. value	Chem. av.	Em. sp. av.	Dev. from chem. value
1	16.5	15.8	0.7	2.61	2.65	0.04
2	20.4	19.8	0.6	2.79	2.75	0.04
3	18.9	18.3	0.6	3.73	3.65	0.08
4	18.1	17.9	0.2	2.78	2.68	0.10
5	17.9	17.4	0.5	2.945	2.805	0.14
6	19.5	18.8	0.7	3.09	3.03	0.06
7	14.8	15.1	0.3	3.255	3.14	0.11
8	16.8	16.95	0.15	2.775	2.63	0.14
9	16.3	16.95	0.65	2.84	2.94	0.10
10	16.7	16.1	0.6	3.815	3.78	0.03
11	15.3[a]	14.95[a]	0.45	–	3.72	–
Av. dev.:			0.5			0.08

[a]American Cyanamid–average of six determinations gave 15.1%.

2.9.7.2 Flame Emission Methods

The development of FE in the United States was of immense value for catalyst analyses, particularly for Na and Ca. In the 1950s, many oil companies in an API Committee on Analytical Research participated in a cooperative program to evaluate FE for determining Na in catalyst. It was found to give rapid and accurate results as long as the silica was first eliminated. As described in Ref. 1170, preliminary treatment with HF was required. One particular contaminated catalyst distributed during the cooperative study appeared to tie-up Na in the silica structure. Rigorous leaching with acid mixtures such as $HClO_4$-HNO_3-H_2SO_4 gave results which were about 3-fold lower when compared to methods using HF to eliminate silica.

The application of FE to catalyst analysis was also discussed in the compilation cited in Section 2.9.7.1,[1172] and by Zaidman and Orechkin[1175] for the determination of K, Na, and other elements.

In the author's laboratory, in addition to Na, determinations of Ca at levels under 0.1% using a nitrous oxide flame to minimize interferences have been invaluable for certain catalysts. In conjunction with the use of ac FE (Section 2.6), many other elements can also be determined in catalysts. After dissolution of the catalyst using the rapid pyrophosphoric acid technique for alumina-based catalysts cited in Section 2.9.7.1, Mo, Ni, and Co can be determined by FE as an alternative to the ESD method, and other elements can be analyzed as well. Determinations of Al, Ca, Co, Cr, Mo, Ni, Pd, and V with excellent sensitivity are provided by use of ac FE with a nitrous oxide flame. For the Mo, Co, and Ni determinations, either ac FE or AA can be used, but the precision for Ni is better with ac FE. Long-term testing of the method and comparison with chemical methods has shown it to be accurate within 2% relative for Mo, with a precision of less than 1% RSD. For Co and Ni, accuracy was within 3% relative and precision approached 1%.

2.9.7.3 Atomic Absorption Methods

The development of AA also created a valuable tool for the analysis of catalysts. It can be applied to Ca, K, and Na determinations but with less sensitivity than was provided by FE. For Mg the AA method provides outstanding detection limits and as a consequence has been extremely valuable where it is desired to keep the Mg content in certain catalyst types as low as possible (less than 0.1%). Magnesium is also employed as a moderate component of other catalysts. When applied to the determination of low amounts of alkaline earth elements with an air–acetylene flame, La is used as a buffer to eliminate interference from alumina. Since the ratio of alumina to Mg is high and unfavorable for this application, using the nitrous oxide flame (for which interferences are far less) is more expedient. La can be omitted, but K must be added as an ionization buffer in this case.

Sato *et al.*[1176] have applied AA to the determination of Ni and V in studies of hydrodesulfurization processes. Coudert and Vergnaud[1177] applied an AA method directly to the determination Pd in catalyst mixed with calcium carbonate. A direct consumption arrangement was used with a sample hopper and mechanized rod to force solid sample into the burner. Kashiki and Oshima[1178] also used a direct-solid technique for determining Co and Mo. Their arrangement used 10–50 mg of finely pulverized catalyst dispersed in 100 ml of methanol and an electromagnetically vibrated steel tubing for introducing sample through the burner capillary. They showed results with relative errors ranging from 1.2 to 7.2%. More recently, Labrecque[1179] determined Mo, Co, and Ni in hydrodesulfurization catalysts, and Potter[1180] determined Pd and Pt in oxidation catalysts for automotive exhausts.

In the author's laboratory, samples for Si in catalysts are first fluxed with KOH at hot-plate temperatures of 450–500°C in vitreous carbon crucibles. After leaching the flux with water, the caustic solution is transferred to a large volume of water in a 250-ml Nalgene volumetric flask containing a slight excess of acid and 1000 ppm of Mo. This prevents occlusion of Si in residues, prevents formation of insoluble Si polymers, and buffers interferences. See Section 2.10 for standard silicate minerals.

2.9.8 Miscellaneous Heavy Petroleum Products

Petroleum products which are heavy and viscous include greases, residual oils, asphalts, and solid cokes. Residuum oil fractions are used as energy-fuel sources. The nomenclature used for these in the P.I. includes fuel oils, residual fuel oils, bunker fuel oils, and vacuum tower bottoms. Asphalts are deliberately made heavier yet, for better durability as paving asphalts. Roofing asphalts require special specifications to have the proper qualities and durability. Certain crude oils, for reasons as yet not completely known, produce better roofing asphalts. Greases are specially compounded, but because of their similar viscosity (semisolid), present similar analytical problems. Cokes are the final residual product of certain petroleum operations and constitute a special product.

2.9.8.1 Analytical Problems Imposed by Viscosity

Heavy petroleum samples impose severe restrictions for direct atomic spectrometric methods of analysis. Different types of fuel oils may vary considerably in viscosity and may even be semisolid at room temperature or quite fluid for certain distillate cuts. Problems involved are similar to those for some crude oils (discussed in Section 2.9.3) cited as perhaps best analyzed by ashing methods.

If an ashing method is used, the thickness of the sample only presents a problem of obtaining a representative sample. Gentle heating of such samples

Table 2.25 Precision of Analyzing a Residual Fuel Oil by a Direct-Reading Emission Spectrometric Method Using a Modified Vacuum Cup Electrode

	ppm				
	Cu	Ni	Fe	V	Ash
	0.25	2.06	11.4	1.25	328
	0.27	2.12	12.1	1.30	348
	0.23	2.28	12.5	1.22	328
	0.20	2.16	11.0	1.17	324
	0.27	1.90	10.4	1.24	366
	0.28	1.87	11.3	1.29	338
	0.28	2.35	12.5	1.25	338
	0.27	2.50	12.5	1.18	—
	0.30	2.10	10.5	1.26	—
	0.20	2.30	10.4	1.30	—
	0.19	2.00	13.0	1.20	—
	—	2.08	—	1.16	
	—	2.25	—	1.40	361
	—	2.23	—	1.40	—
Av.:	0.25	2.16	11.6	1.26	341
Av. % S.D. from mean:	13	6.0	5.6	4.5	3.4

is generally required to obtain proper sample mixing. If a direct method is used, more problems are involved. Whereas a simple 10-fold dilution in a hydrocarbon solvent may suffice for lubricating oils (Section 2.9.4), there are greater complications for residuum samples. In addition to errors caused by viscosity influences, the composition of heavy samples may cause matrix errors influencing analyte and background response. Erratic results by comparing direct to ashing methods (discussed in Section 2.9.3.7e) may thus be caused. After dilution these influences may still occur; Perry and Keyworth,[1181] using AA to determine Ni, found that even adding Ni in mineral oil to α-methylnaphthalene caused a difference in blank and molecular absorption. One other effect also occurs in laminar flow burners. Heavy samples diluted in lower-boiling hydrocarbon solvents selectively coat nebulizing chambers and the insides of burner-head jaws. This leads to burner clogging, greater fluctuations, and odd memory effects. Even after a 25-fold dilution for some heavy samples, these effects still appear to linger.

2.9.8.2 Analytical Methods—Emission Spectroscopy

Any of the methods discussed in Sections 2.9.3 through 2.9.5 could be applied to heavy petroleum samples, and some of the methods were applied to several types of products, including fuel oils. In 1951 Anderson and Hughes[1182] applied ES to the determination of V in residual fuel oils using an ashing procedure. Other early work included the ashing-ES methods applied to residual oils by Dryoff *et al.*,[1183] Work and Juliard,[1184] and to greases by Childs and Kanehann[1185] and by Gent *et al.*[1186] Key and Hoggan applied an ES–Rotrode method via ashing to bunker fuel oils[1187] and to greases dissolved in a mixture of mineral oil–heavy naphtha-amylacetate.[1188] Lakatos[1189] used an ES–Rotrode method for analyzing asphalts dissolved in $CHCl_3$.

In the author's laboratory, residual oils have been analyzed using the ESD–vacuum cup electrode discussed in Section 2.9.6. Precision using an ashing technique for a residual fuel oil is given in Table 2.25 and was 4.5% for 1.46 ppm of V and 6.0% for 2.16 ppm of Ni. Von Lehmden *et al.*[1190] made an interlaboratory comparison for the EPA using various methods, including ES and AA, for analyzing 28 elements in fuel oils and fly ash. Poor agreement was obtained in many cases.

2.9.8.3 Atomic Absorption and Flame Emission Methods

Prior to 1970 there had been few AA applications to the analysis of residual oils. In 1965 Burrell[1191] applied AA to the determination of Co, Fe, and Ni in a study of asphaltic fractions in recent sediments. In 1966 Thomas[1192] applied FE to the determination of Ca, K, and Na in fuel oils and obtained a precision of 3–5%. In 1967 Mostyn and Cunningham[1193] applied AA to the direct determination of Cr in fuel oil and analyzed Mo in grease with the use of a special ashing–extraction procedure. In 1967 Perry and Keyworth[1181] used AA for determining Ni in vacuum tower bottoms and crude oils as discussed in Section 2.9.8.1. Since that time applications have increased; the Institute of Petroleum in England has published a methods book.[1194] This book used AA for the determination of Na, Ni, and V in fuel oils and crudes, and FE was used for Li and Na in greases. An ACS publication based on a 1975 symposium included the development of acid-ashing and nonflame AA methods for 13 elements in fuel oils, previously discussed in Sections 2.9.3.8 and 2.9.6. Goux[1195] and Serbanescu *et al.*[1196] used both AA and FE for analyzing metals in fuel oils and cokes. AA was also used for analyzing metals in fuel oils in Refs. 1197–1199. Alder and West[1200] applied nonflame AA to the determination of Ni in residual fuel oil and crudes. May and Presley[1201] used the same technique for determining V in beach asphalts diluted in carbon tetrachloride but obtained poor comparisons to activation analysis. Subsequently, they first wet-ashed samples and obtained better results.[1202] Norwitz and Gordon[1203] used AA for the determination of Li in greases.

2.9.8.4 Analysis of Cokes

There has long been an interest in the metals in petroleum coke, and it appears that some of the early references on the subject may have used the terms "petroleum ash" and "coke" interchangeably. In 1936 Herman[1204]

Table 2.26 Precision for Ni and V in Cokes Using an ESD Method and Pyrophosphoric Acid Solution

		Ni			V		
Sample No.	Aliquot No.	ppm	Instrument precision[a]	Method precision[b]	ppm	Instrument precision[a]	Method precision[b]
A6-GR	1	72.7	1.51	5.22	40.3	0.82	2.83
	2	75.0	0.0	2.22	41.5	1.10	2.36
	3	81.5	0.61	6.28	45.0	2.22	5.70
	4	77.5	4.52	1.04	43.0	2.33	1.18
	Av.	76.7	1.66	2.52	42.5	1.64	3.06
A6-CAL	1	110	0.0		46.0	2.17	2.75
	2	100	0.0		45.5	1.10	3.81
	3	112	1.79		47.5	1.05	0.42
	4	110	0.0		50.0	4.00	5.70
	Av.	108	0.45	3.70	47.3	2.08	3.17
Grand av.		—	1.06	3.1	—	1.86	3.1

[a]Percent deviation from the mean of duplicates for each sample aliquot (precision for a given solution of sample).
[b]Percent deviation from the mean, including sample preparation for the four sample aliquots.

studied the metal content of petroleum coke and heavy-oil ash. In 1946 Wells[(1205)] presented analyses for a number of petroleum-coke ashes.

Coke produced by the P.I. has several uses, foremost of which is as electrodes used in the processing of Al ores. These uses of coke impose a limit on the amount of trace metals, especially Ni and V. Specification limits may be on the order of 200 ppm.

A more modern ES determination of Fe and V in coke is described by Vigler and Conrad.[(1206)] Using a spark-ignited arc directly on coke gave a relative standard deviation of 12.8% for Fe and 5.5% for V. Comparisons to an ash-colorimetric method were reasonable.

A direct ES-arc method used in the author's laboratory for determining Fe, Ni, and V in coke gave biased results compared to other methods when synthetic standards in graphite were used. Good results were obtained using coke samples which had been analyzed by other methods as standards.

More recently, a special ESD-vacuum cup electrode method similar to that used for gas oils (Section 2.9.6) and residual fuel oils has been used in the author's laboratory for the simultaneous determination of Al, Cu, Fe, Ni, Si, and V. Pyrophosphoric acid mixtures used to dissolve coke ash provide improved precision, especially for Si (see Section 2.9.1.4), and also act as a good buffer against matrix influences. Precision data for Ni and V are given in Table 2.26, both as an instrumental function and for method repeatability, including ashing. A brief method outline is as follows:

1. Weigh 3.5–4 g of coke into a Pt crucible and ash at 600°C in a muffle furnace.

2. Dissolve the ash in 1.5 ml of concentrated pyrophosphoric acid at high hot-plate temperature. High-purity acid can be prepared by bringing 200 g of solid P_2O_5 to a total weight of 300 g with distilled water (cautiously) after allowing the P_2O_5 to equilibriate overnight in an open Teflon beaker. Warming cautiously and stirring may be required to effect complete solution.

3. Transfer the sample solution (after cooling to a warm temperature) quantitatively into a 25-ml volumetric flask and dilute to the mark, adding Co as I.S.

4. Compare to appropriate standards using a high-voltage spark excitation and a vacuum cup electrode as modified to a flat tip by Buell.[1207]

This method is employed for the control of metal contents which are undesirable in cokes used for some purposes. For this application concentrations over 100 become more critical, especially for Ni and V, and concentrations at levels of 20 ppm are less important. If it is desired to determine lower concentrations and elements such as Ca with good precision, AA and ac FE methods of analysis can be used. Sample solutions prepared for the ESD method can be diluted (at least 5-fold more to reduce phosphoric acid content, which causes trouble in laminar flow burners) and appropriate buffers added such as that used in Section 2.9.4.8b.

2.9.9 Petrochemical and Agricultural Products

With the wide diversification of the modern P.I., most large oil companies are active in one or more phases of petrochemical production. The products may be entirely or partially derived from petroleum. Agricultural products (fertilizers, etc.) constitute the largest volume of these products. Although not petroleum products in the true sense, references pertaining to their analysis will be reviewed here. There is much literature on agricultural analysis, of which only selected references will be given.

2.9.9.1 Agricultural Petrochemicals

Fertilizer (nutrient) products such as ammonium nitrate and anhydrous liquid phosphates, and other products such as insecticide and defoliant sprays (perhaps containing added plant nutrients), are manufactured in large quantities by various petroleum companies. The great interest in the plant nutrient elements (which may be added to certain products) and other contaminating elements leads to the need for analytical control and research methods for agricultural products. Companies supplying such products may become involved in research or customer service projects in determining the effectiveness of plant nutrients. As a consequence, many plant and soil samples may be analyzed by petroleum laboratories.

2.9.9.2 History, Reviews, and Analytical Methods

Lundegardh, the Swedish plant physiologist, agricultural chemist, and father of modern flame spectroscopy, beginning in 1928 did much research on agricultural materials, resulting in his classic two-volume work.[1208] By 1936 other applications began to appear, including Vanselow and Laurance,[1209] and Mitchell,[1210] who applied the Lundegardh FE techniques to agricultural studies. These men have continued to work in this area. Vanselow and Bradford[1211] prepared a chapter on ES methods in a 1961 work published by the University of California Division of Agricultural Science on analysis for soils, plants and waters. In 1947 Mitchell published a review on spectrographic analysis of plants and soils, including 260 references.[1212] He and Scott have since become famous for their extensive work at the Macaulay Institute for Soil Research in Scotland. Their emission methods, which employ preconcentration with mixed organic precipitants, is well known and has seen much use by others. Of their many publications, those from 1955 to 1960 amply review their work and agricultural applications.[1213–1215] Their methods include FE techniques, including a three-channel spectrometer for Ca, K, and Na determinations,[1216] a porous cup spark-solution method, direct arc excitation of plant ash, and an arc method on precipitated concentrates. The latter method was developed to a high degree and gave exceptional precision for an arc–ES method: 1.4–3.5% for Co, Mo, and V; and 7.9 and 5.5% for Sn and Zn, respectively. Heggen and Strock[1217] applied this technique to a variety of samples, including leaves and oil-field brines. In addition to the techniques described above, Scott[1218] reported on very promising ESD results using plant ash mixed with graphite, cellulose powder, and buffers, which is briquetted into a Rotrode shape.

In the United States, in addition to Vanselow and coworkers, Schrenk[1219] at the Kansas Agricultural Experiment Station reviewed ES methods and states that FE is better for certain elements, especially for rubidium. He found unusually large amounts in soybeans, 220 ppm; made considerable agricultural and nutritional studies of the element; and obtained a relative standard deviation of less than 3%. Schrenk and coworkers developed many ES and FE applications. These included an ES determination of B in plants[1220] with a precision of ±4%, and a FE determination of Cu, Fe, and Mn in plants using a selective precipitation of Cu and Fe with oxine.[1221] Toth *et al.*[1222] also applied FE to the analysis of Ca, K, and Na in plants, as well as chemical methods for other elements after a wet ashing of plants.

Grant[1223] presented an excellent review on trace elements and analytical methods in agriculture. He also included a discussion of essential nutrients and a map of the United States showing known areas where mineral nutritional diseases of animals occur. The map shows goiter belts; areas of Co, Cu, and P deficiency; and regions of Mo and Se toxicity as plotted by the U.S. Plant, Soil and Nutrition Laboratory in Ithaca, New York. Allaway[1224] has prepared an excellent chapter on the trace elements in biological systems, including plants, soils, and animals. In one table he gives examples for certain elements of plant deficiencies and normal ranges, dietary requirements, toxic levels, and levels found in animals, including man.

2.9.9.3 Atomic Absorption and ESD Methods

The agricultural applications of AA have been reviewed at various times by Allan,[1225] David,[1226] and Slavin.[1227] The technique provided the means for improved Zn and Mg determination in plants. In comparison to ES methods, including an ESD method for Zn, Cu, and Pb in plants[1228] and a FE method for Mg in plants,[1229] the exceptional sensitivity of AA for Mg and Zn provided marked advantages. AA has now been applied to determinations of many additional elements in agricultural materials. Breck[1230] compared AA and ESD methods, concluding that both methods can provide the required good results, with ESD being faster. Chaplin and Dixon[1231] also used a Rotrode-ESD method and Jones[1232] reviewed a collaborative investigation by 11 laboratories using the same technique. As a result, the method has been adopted as official first action by the AOAC. In the author's laboratory a combination of three techniques is used for plant analysis: AA for Mg and Zn; FE for Ca, K, and Na; and ESD for the simultaneous determination of Al, B, Cu, Fe, and Mn. Basson and Bohmer[1233] applied AA to an extensive study of ashing and extraction procedures for determining Ca, Cu, Fe, Mg, Mn, and Zn in plants. They preferred a rapid, dry-ashing technique which gave results equivalent to wet oxidation procedures (there were no losses due to volatility for the elements determined). AA has also been used by Cary and Olson[1234] for Cr and by Boline and Schrenk[1235] for Cd. Pau *et al.*[1236] used FE with a nitrous oxide-hydrogen flame for determining B in plants after extraction with 2-ethylhexane-1, 3-diol in chloroform.

2.9.9.4 Soil Analysis by AA

Pawluk[1237] reported on the use of AA for soil analysis at the University of Alberta, Edmonton. He found the technique to be accurate and precise. At the Ohio State University Soil Testing Laboratory, which analyzes approximately 100,000 samples per year, Linville[1238] described the use of a fully automated AA method for soil analysis after standard sample preparations.

Lead has become particularly important in antipollution studies. Methods using AA and ES for the determination of Pb in soil were compared and used for studies of possible contamination along highways by Seeley *et al.*[1239] Although the AA method using an extraction with hot nitric acid was more precise, the ES was preferred because it was faster and provided spectra for additional elements. Kahn *et al.*[1240] used an AA-Delves cup and flame procedures for determining Pb and Cd (both of ecological interest) in soils and leaves. The Delves cup method was promising and when applied directly to leaves showed that the Pb distribution was not uniform.

2.9.9.5 Fertilizers

McBride[1241] compared AA to chemical methods for determining minor nutrients in fertilizer in a report on a collaborative study. Good agreement was shown for many results, but some discrepancies were found. The precision of the AA method was good and the method was accepted as an official AOAC method. Even B in fertilizers (which has poor AA sensitivity) has been determined by Weger *et al.*[1242] using AA and an extraction with 2-ethylhexane-1, 3-diol.

Brabson and Wilhide[1243] determined Ca in wet-process phosphoric acid by FE using cation exchange to separate Ca from phosphate interference and adding Sr to buffer against Al interference. For this application in the author's laboratory, ac FE with a nitrous oxide flame has been found to eliminate most of these interferences. Samples are diluted 25-fold, adding a small excess of phosphoric acid and oxine in isopropanol to eliminate residual matrix effects. The analysis of other elements in anhydrous liquid phosphates is also important. A combination of AA and ac FE is used by the author to determine Al, Ca, Cr, Fe, Mg, V, and Zn. A Rotrode-ES method has also been used with precisions in the range 3–8%. The ES method is faster but not as precise and perhaps slightly less accurate than the AA and FE methods.

2.9.9.6 Unusual Applications

The use of nitrogen 15 as a tracer in agricultural studies is increasing. Ferraris and Proksch[1244] discussed this unique use and an ES method using electrodeless discharge tubes and N_2-band measurements for determining N-isotope ratios. Although a commercial instrument was used, they tested several parameters which might aid in setting up one's own instrument. They showed promising results and cited the much smaller expense as compared to a mass spectrometer.

Another unique application is that by Gutsche *et al.*,[1245] who used a special FE method to detect down to 0.01 and 0.07 μg of P and S, respectively, in insecticide residues.

2.10 GEOCHEMICAL APPLICATIONS

There is considerable interest in the analysis of rocks and minerals by the P.I. In drilling for oil, the composition of core samples is of extreme interest, as are different geological formations as related to the exploration for oil and geological studies of oil fields. Petroleum companies that are diversified may also have interests in mining ventures. As a consequence, petrographic laboratories are indispensable to the P.I. and other methods of analysis are used, including instrumental methods.

2.10.1 Geochemistry in Mineral Exploration

In geochemical studies and in mineralogical methods of prospecting, analysis of the chemical elements, particularly trace elements, is used. The technique of utilizing trace analyses for mineral prospecting has received much attention. Such techniques received detailed study by Fersman and co-workers, which culminated in a book published in 1939. The book contained a chapter of such interest and value to geochemists that it was translated and in 1952 published as a USGS Circular.(1246) Special methods of prospecting include the study of dispersion halos, botanical prospecting, and the use of ES analytical procedures. Since then, such methods have become used worldwide, and analyses of trace elements are used extensively. General tests are those by Hawkes and Webb(1247) on geochemistry in mineral exploration and by Wainerdi and Uken(1248) on modern methods of geochemical analysis. The latter contains chapters on methods of analysis, including AA, emission spectroscopy, and x-ray techniques. A good text on the principles of geochemistry is that by Mason,(1249) and much information on studies of the composition of the earth's crust is given by Parker.(1250) An excellent encyclopedia of geochemistry and environmental sciences, edited by Fairbridge, has been published.(1251)

2.10.2 Analytical Methods—Emission Spectroscopy

In addition to the references already cited, ES methods are thoroughly treated for the analysis of silicates by Ahrens.(1252) Murata(1253) reviewed ES methods for analysis of trace elements in geological materials. Cruft and Giles(1254) discussed applications of automated ESD programs. With the aid of a computer they determined 13 elements in rock samples at the rate of about 3 min/sample, with precisions varying from 2 to 9% for 10 elements and from 10 to 13% for the remaining three elements. They mixed the sample with a phosphate buffer and used a carbon cup electrode with an 18-A arc in a Stallwood jet using an argon–oxygen mixture as the gas. Schwander(1255) devised a unique technique by pressing sample into a 6-mm-diameter cylinder. They first fused the samples to produce a uniform physical form of samples (thus lessening emission variabilities due to different sample compositions). After mixing with graphite and pressing the sample mixture into a pellet, they inserted the pellet into a special Stallwood jet device. Either an arc or spark excitation could be used, but they preferred the arc for better sensitivity. To reduce the greater fluctuations with the arc, they used a flow of CO_2 in the Stallwood jet.

In the author's laboratory, spark excitation using a similar ES method was found to give better precision and also better detection limits. Samples are mixed 1:5 in graphite and pressed into a $\frac{1}{4}$-inch-diameter cylinder which is inserted into a metal electrode holder similar to that used by Helz and Scribner(1256) for their pellet method.

2.10.3 Flame Emission Methods

Geological applications of FE, including some early history, have been reviewed by Buell in the encyclopedia edited by Fairbridge (Ref. 1251, p. 367). Camacho-Calderon and Vallecilla-Risscos[1257] compiled a bibliography on FE applications to rock and mineral analysis. The bibliography on flame spectroscopy by Mavrodineanu[1258] contains 170 references for FE applied to rocks, minerals, and ores.

The technique has been applied mainly to alkali and alkaline earth elements, but other elements, including Ga, In, Tl, Fe, Cu, Co, Mn, La, V, and PO_4, have been determined using FE. The development of improved instrumentation and ac FE (Section 2.6) in recent years should extend applications by the increased sensitivity and reduced interferences provided.

Selected applications include the following. In 1950 Lundegardh[1259] determined Ca, Fe, K, Mg, Mn, and Na in silicate rocks. During the 1950s many other applications were published. For the determination of alkali elements, Na and K in rocks and minerals were analyzed in Refs. 1260–1262; Li in rocks and spodumene was analyzed in Refs. 1263–1265; Li, Cs, and Rb in rocks were analyzed in Refs. 1266 and 1267; alkali and alkaline earth elements in rocks and minerals were analyzed in Refs. 1268 and 1269. Minerals and rocks were analyzed for Ca in Refs. 1270 and 1271 and Ref. 1270 used a Mg buffer to level Al interference. Phosphate in rocks was analyzed in Refs. 1272 and 1273 using indirect methods.

Care must be used in any FE (or AA) method applied to complex geological samples to avoid matrix interferences. Techniques and chemical interferences in the references given in Section 2.7 for AA using a nitrous oxide flame apply equally to FE methods using the same flame.

2.10.4 Atomic Absorption Methods

Geochemical applications of AA are numerous and only selected references will be given here. The technique is well suited to such applications and has been reviewed extensively by Billings as early as 1964 in the AA *Newsletter*[1274, 1275] and in a Varian-Techtron booklet.[1276] A book on atomic absorption spectrometry in geology has been authored by Angino and Billings[1277] Slavin[1278] in 1965 reviewed applications of AA to geochemical prospecting and mining. Ward *et al.*[1279] in 1969 reviewed AA methods used by the USGS in geochemical exploration, citing USGS publications dating back to 1966. Bernas[1280] briefly reviewed AA references, citing increased applications with the development of the nitrous oxide flame. He discussed problems of sample preparation for AA, and described a silicate decomposition procedure using HF in a special Teflon-lined bomb followed by the addition of boric acid. This "single matrix system" was found to diminish or eliminate interferences. The Parr Instrument Company now makes a bomb with a removable 25-ml Teflon cup for such applications.

In using AA methods for geochemical samples, careful consideration should be given to the possibility of interference. The elements Mo and V are particularly susceptible to interferences in the nitrous oxide flame. Van Loon[1281] studied interference for determining Mo in soils for geochemical exploration purposes and added Al as a buffer to suppress interference. Additional references concerning interference are given in Section 2.7.

The AA method used in the author's laboratory cited for Si in Section 2.9.7.3 works well for a variety of silicate minerals. For NBS flint clay 97a and soda feldspar 99, and the USGS standard reference samples of basalt, granite, and peridotite, the average deviation from accepted values was 0.88% relative and the average precision was 0.95% RSD. References to the USGS and the international geochemical reference samples are given in Refs. 1282 and 1283.

Additional references to AA and FE methods on this topic are tabulated in the *Analytical Chemistry* review.[1284] The review also cites a most interesting and unique application in the extensive analyses of moon rocks presented at the Lunar Science Conferences. Another interesting reference is by Hey,[1285] who presented a review on the history of methods development from 1770 to recent years and cited the usefulness of instrumental methods.

2.11 REFERENCES

References to Section 2.3

1. J. R. Donnel, *7th World Petrol. Congr. 9*, 699 (1967).
2. R. H. Dott, Sr., and M. J. Reynolds, *Sourcebook for Petroleum Geology*, American Association of Petroleum Geologists, Tulsa, Okla. (1969), p. 311.
3. E. D. Innes and J. V. D. Fear, *7th World Petrol. Congr. 3*, 633 (1967).
4. F. L. Hartley, Environmentalists and the Real World, paper presented to the Air Pollution Control Association, Lake Tahoe, Calif., October 6, 1972.
5. *Oil Gas J. 60*, 146 (January 29, 1962).
6. D. K. McIvor and A. W. Brown, *Amer. Chem. Soc. Div. Petrol. Chem. Paper 16* (May 1970).
7. L. G. Weeks, ed., *Habitat of Oil*, American Association of Petroleum Geologists, Tulsa, Okla. (1958), p. 59.
8. M. A. Bestougeff, Petroleum hydrocarbons, Chap. 3, and G. Costantinides and G. Arich, Non-hydrocarbon compounds in petroleum, Chap. 4, in: *Fundamental Aspects of Petroleum Geochemistry* (B. Nagy and U. Colombo, eds.,), American Elsevier Publishing Company, Inc., New York (1967).
9. L. R. Snyder and B. E. Buell, *Anal. Chem. 40*, 1295 (1968).
10. L. R. Snyder, B. E. Buell, and H. E. Howard, *Anal. Chem. 40*, 1303 (1968).
11. L. R. Snyder, *Accounts Chem. Res. 3*, 290 (1970).
12. L. R. Snyder, *Amer. Chem. Soc. Div. Petrol. Chem. Preprints 15*, C43 (1970).
13. E. C. Copelin, *Anal. Chem. 36*, 2274 (1964).
14. D. M. Jewell, J. H. Weber, J. W. Bunger, H. Plancher, and D. R. Latham, *Anal. Chem. 44*, 1391 (1972). Characterization of heavy ends of petroleum (various reports), *Amer. Petrol. Inst. Project 60.*
15. J. F. McKay and D. R. Latham, *Anal. Chem. 44*, 2132 (1972).
16. J. F. McKay and D. R. Latham, *Anal. Chem. 45*, 1050 (1973).
17. H. M. Smith, *Ind. Eng. Chem. 44*, 2577 (1952).

18. H. T. Rall, C. J. Thompson, H. J. Coleman, and R. L. Hopkins, *Sulfur Compounds in Crude Oil*, U.S. Department of the Interior, Bureau of Mines, Washington, D.C. (1972).
19. L. R. Aalund, *Oil Gas J. 74* (1976): series of eight articles on "Guide to World Crudes" from nos. 13, 98 (March 29) through 27, 98 (July 5).
20. D. H. Stormont, *Oil Gas J. 64*, 51 (October 24, 1966).
21. B. Peralta and J. Sosnowski, The Unicracking-JHC Family of Processes, Union Oil Company of California, Technical Sales Division, Brea, Calif.
22. P. C. Sylvester-Bradley and R. J. King, *Nature 198*, 728 (1963).
23. A. T. Wilson, *Nature 196*, 11 (1962).
24. R. Robinson, *Naturc 199*, 113 (1963); *Nature 212*, 1291 (1966).
25. P. C. Marx, *Amer. Chem. Soc. Div. Petrol. Chem. 9* (1), 25 (1964).
26. R. A. Friedel and A. G. Sharkley, Jr., *Science 139*, 1203 (1963).
27. P. H. Fan and T. L. Kister, *Gas. Assoc. Monthly 47*, 12 (1965).
28. P. N. Kropotkin, in: *Aspects of the Origin of Life* (M. Florkin, ed.), Pergamon Press, Inc., Elmsford, N.Y. (1960).
29. A. A. Wolsky and F. W. Chapman, Jr., *Proc. Amer. Petrol. Inst. Sect. III*, *40*, 423 (1960).
30. T. A. Lind, *Amer. Assoc. Petrol. Geol. Bull. 41*, 1387 (1957).
31. R. B. Cate, *Amer. Assoc. Petrol. Geol. Bull. 44*, 423 (1960).
32. K. O. Emery, *The Sea off Southern California: A Modern Habitat of Petroleum*, John Wiley & Sons, Inc., New York (1960).
33. O. K. Bordovsky, *Marine Geol. 3*, 3 (1965).
34. J. G. McNab, P. V. Smith, Jr., and R. L. Betts, *Ind. Eng. Chem.*, *44* (11), 2556 (1952).
35. P. V. Smith, Jr., *Amer. Petrol. Geol. 38* (3), 377 (1954).
36. N. P. Stevens, *Amer. Assoc. Petrol. Geol. Bull. 40* (1), 51 (1956).
37. B. J. Mair, *Geochim. Cosmochim. Acta 28*, 1303 (1964).
38. E. W. Baker, Phorphyrins in Deep Ocean Sources, Petrol. Res. Fund Rept., Amer. Chem. Soc. 15th Annual Meeting, 1970; also see Ref. 42, p. 498.
39. E. W. Baker, C. Dereppe, and J. R. Boal, Fossil Porphyrins and Chlorins in Deep Ocean Sediments, Petrol. Res. Fund Rept., Amer. Chem. Soc. Annual Meeting, 1970.
40. M. Blumer and G. S. Omenn, *Geochim. Cosmochim. Acta 25*, 81 (1961).
41. W. H. Bradley, *7th World Petrol. Congr. 3*, 695 (1967).
42. J. B. Davis and E. E. Bray, *Initial Reports of the Deep Sea Drilling Project*, Vol. 1, National Science Foundation, University of California (1968), p. 415.
43. M. L. Dunton and J. M. Hunt, *Amer. Assoc. Petrol. Geol. Bull. 46*, 2246 (1962).
44. G. W. Hodgson, B. Kitchon, K. Taguchi, B. L. Baker, and E. Peake, *Geochim. Cosmochim. Acta 32*, 737 (1968).
45. G. W. Hodgson and E. Peake, *Nature 191* 766 (1961).
46. K. B. Krauskopf, *Introduction to Geochemistry*, McGraw-Hill Book Company, New York (1967), Chap. 11.
47. J. A. Linebach and D. L. Gross, *Illinois State Geol. Surv. Environ. Geol. Note 58* (December 1972); contains a bibliography of additional studies of Lake Michigan sediments.
48. W. L. Orr, K. O. Emery, and J. R. Grady, *Amer. Assoc. Petrol. Geol. Bull. 42*, 925 (1958).
49. N. P. Stevens, E. E. Bray, and E. D. Evans, in: *Habitat of Oil* (G. Weeks, ed.), American Association of Petroleum Geologists, Tulsa, Okla. (1958), p. 779.
50. D. W. Thomas and M. Blumer, *Geochim. Cosmochim. Acta 48*, 1147 (1964).
51. P. D. Trask and C. C. Wu, *Amer. Assoc. Petrol. Geol. Bull. 14*, 1451 (1960).
52. National Science Foundation, *Initial Reports of the Deep Sea Drilling Project*. A project planned and carried out with the advice of the Joint Oceanographic Institu-

tions for Deep Earth Sampling. Over 10 volumes starting with Vol. 1 (August–September 1968), with more to come.
53. I. A. Breger, ed., *Organic Chemistry*, Macmillan Publishing Co., Inc., New York (1963).
54. U. Columbo and G. D. Hobson, eds., *Advances in Organic Geochemistry*, Macmillan Publishing Co., Inc., New York (1964).
55. R. H. Dott and M. J. Reynolds, *Sourcebook for Petroleum Geology* (Mem. 5), American Association of Petroleum Geologists, Tulsa, Okla. (1969).
56. G. Eglinton and M. T. J. Murphy, eds., *Organic Geochemistry*, Springer-Verlag, New York (1969).
57. W. A. Gruse and D. R. Stevens, *Chemical Technology of Petroleum*, McGraw-Hill Book Company, New York (1960).
58. K. B. Krauskopf, *Introduction to Geochemistry*, McGraw-Hill Book Company, New York (1967), Chap. 11, p. 301.
59. A. I. Levorsen, *Geology of Petroleum*, W. H. Freeman and Company, San Francisco (1954).
60. S. M. Manskaya and T. V. Drozdova, *Geochemistry of Organic Substances*, Pergamon Press, Inc., Elmsford, N.Y. (1968).
61. B. Mason, *Principles of Geochemistry*, 3rd ed., John Wiley & Sons, Inc., New York (1966).
62. M. Alexander, *Microbial Ecology*, John Wiley & Sons, Inc., New York (1971).
63. P. H. Abelson, *6th World Petrol. Congr. Sect. I* (1963), paper 41, p. 397.
64. E. G. Baker, *Science 129*, 871 (1959).
65. L. M. Banks, *Amer. Assoc. Petrol. Geol. Bull. 50*, 397 (1966).
66. H. N. Dunning and J. W. Moore, *Amer. Assoc. Petrol. Geol. Bull. 41*, 2403 (1957).
67. J. G. Erdman, in: *Amer. Assoc. Petrol. Geol. Mem. 4* (A. Young and J. E. Galley, eds.) (1965), p. 20.
68. H. D. Hedberg, *Amer. Assoc. Petrol. Geol. Bull. 48*, 1755 (1964).
69. G. W. Hodgson, *Amer. Assoc. Petrol. Geol. Bull. 38*, 2537 (1954). A study on Fe, Ni, and V in Canadian crudes that contains a good discussion of the origin of metals.
70. K. Krejci-Graf, *Geophys. Prospecting 11*, 244 (1963).
71. G. Murray, *Oil Gas J. 63*, 236 (April 5, 1965).
72. W. G. Meinschein, *Amer. Assoc. Petrol. Geol. Bull. 43*, 925 (1959).
73. C. E. Zobell, *Int. Sci. Technol. 1963*, 42 (August).
74. J. B. Davis, *Petroleum Microbiology*, American Elsevier Publishing Company, Inc., New York (1967).
75. E. Beerstecher, Jr., *Petroleum Microbiology*, American Elsevier Publishing Company, Inc., New York (1954).
76. M. B. Allen, ed., *Comparative Biochemistry of Photoreactive System*, Vol. 9, Academic Press, Inc., New York (1960).
77. J. H. Hedgpeth, ed., *Treatise on Marine Ecology and Paleoecology* (2 vols.), Geol. Soc. Amer. Mem. 67, Washington, D.C. (1957).
78. O. A. Radchenko, *Geochemical Regularities in the Distribution of the Oil-Bearing Regions of the World*, Israeli Program for Scientific Translations, 1965 (transl. 1968). Available from U.S. Department of Commerce, Washington, D.C.
79. R. L. Martin, J. C. Winters, and J. A. Williams, *Nature 199*, 110 (1963).
80. *Origin of Petroleum*, Amer. Assoc. Petrol. Geol. Reprint Ser. 1 (1971) and 9 (1974), Golden, Col.
81. C. Rimington and G. Y. Kennedy, in: *Comparative Biochemistry* (M. Florkin and H. S. Mason, eds.), Academic Press, Inc., New York (1962).
82. J. E. Falk, *Porphyrins and Metal Porphyrins*, Vols. 1 and 2, American Elsevier Publishing Company, Inc., New York (1964).
83. H. H. Inhoffen, J. W. Buchler, and P. Jaeger, *Fortschr. Chem. Org. Naturstoffe 26*, 284 (1968); in German.

84. J. H. M. Bowen, *Trace Elements in Biochemistry*, Academic Press, Inc., New York (1966).
85. A. P. Vinogradov, *The Elementary Chemical Composition of Marine Organisms*, Sears Foundation for Marine Research, New Haven, Conn. (1953).
86. E. D. Goldberg, in: *Treatise on Marine Ecology and Paleoecology*, Vol. 1 (J. H. Hedgpeth, ed.), Geol. Soc. Amer. Mem. 67, Washington, D.C. (1957), Chap. 12.
87. K. B. Krauskopf, *Geochim. Cosmochim. Acta 9*, 1 (1956).
88. D. F. Schutz and K. K. Turekian, *Geochim. Cosmochim. Acta 29*, 259 (1965).
89. G. M. Knebel and G. Rodriguez-Eraso, *Amer. Assoc. Petrol. Geol. Bull. 40*, 547 (1956).
90. D. I. Andrews and J. C. Stipe, *World Oil 1961*, 122 (June).
91. E. S. Barghoorn and J. W. Schopf, *Science 150*, 337 (1965). Excellent review on Precambrian organic material.
92. E. S. Barghoorn, W. G. Meinschein, and J. W. Schopf, *Science 148*, 461 (1965).
93. E. S. Barghoorn and S. A. Tyler, *Science 147*, 563 (1965).
94. A. L. Burlingame, P. Haug, T. Belsky, and M. Calvin, *Proc. Natl. Acad. Sci. U.S. 54*, 1406 (1965).
95. P. E. Cloud, Sr., *Science 148*, 1713 (1965).
96. A. E. J. Engel, B. Nagy, L. A. Nagy, C. G. Engel, G. O. W. Kremp, and C. M. Drew, *Science 161*, 1005 (1968).
97. B. Eglinton, Chap. 2, and L. R. Moore, Chap. 11, in: *Organic Geochemistry* (G. Eglinton and M. T. J. Murphy, eds.), Springer-Verlag, New York (1969). Good discussions on Precambrian organic material, including unpublished data by J. Han, Space Sciences Laboratory, University of California at Berkeley, Fig. 25, p. 62.
98. J. S. Harrington and C. R. Cilliers, *Geochim. Cosmochim. Acta 27*, 411 (1963).
99. G. E. Murray, *Amer. Assoc. Petrol. Geol. Bull. 49*, 3 (1965).
100. J. W. Schopf, E. S. Barghoorn, M. D. Maser, and R. O. Gordon, *Science 149*, 1385 (1965).
101. F. M. Swain, A. Blumentals, and N. Prokopovich, *Amer. Assoc. Petrol. Geol. Bull. 42*, 173 (1958).
102. J. G. Erdman and P. H. Harju, *Amer. Chem. Soc. Div. Petrol. Chem. Preprints*(1962), p. 43.
103. C. E. Dobbin, *Amer. Assoc. Petrol. Geol. Bull. 27*, 17 (1943).
104. J. M. Hunt, *Amer. Assoc. Petrol. Geol. Bull. 37*, 1837 (1953).
105. T. W. Todd, *Amer. Assoc. Petrol. Geol. Bull. 47*, 599 (1963).
106. J. M. Hunt, F. Stewart, and P. A. Dickey, *Amer. Assoc. Petrol. Geol. Bull. 38*, 1671 (1954).
107. A. G. Harrison and H. C. Thode, *Amer. Assoc. Petrol. Geol. Bull. 42*, 2642 (1958).
108. L. C. Bonham, *Amer. Assoc. Petrol. Geol. Bull. 40*, 897 (1956).
109. W. F. Meents, A. H. Bell, O. W. Rees, and W. G. Tilbury, *Illinois State Geol. Surv. Illinois Petrol. No. 66* (1952).
110. W. R. Eckelmann, W. S. Broecker, D. W. Whitlock, and J. R. Allsup, *Amer. Assoc. Petrol. Geol. Bull. 46*, 699 (1962).
111. S. R. Silverman and S. Epstein, *Amer. Assoc. Petrol. Geol. Bull. 42*, 998 (1958).
112. H. G. Thode and J. Monster, and H. B. Dunford, *Amer. Assoc. Petrol. Geol. Bull. 42*, 2619 (1958).
113. S. Epstein and T. Mayeda, *Geochim. Cosmochim. Acta 4*, 213 (1953).
114. J. M. Sugihara, J. F. Branthaven, G. Y. Wu, and C. Weatherbee, *Amer. Chem. Soc. Div. Petrol. Chem. Preprints* (February 1970), p. C65.
115. *Proc. Amer. Petrol. Inst. Sect. VIII, 42*, 17, 27, 33 (1962).
116. M. Carrales and R. W. Martin, *U.S. Bur. Mines Inform. Circ. 8676* (1975).

References to Section 2.4

117. L. E. Calkins and M. M. White, *Natl. Petrol. News 38*(27) 519 (1946).
118. W. H. Thomas, *Inst. Petrol. Technol. 10*, 216 (1924).
119. W. B. Shirey, *Ind. Eng. Chem. 23*, 1151 (1931).
120. A. Triebs, *Ann. Chem. 510*, 42 (1934).
121. W. H. Thomas, *Inorganic Constituents of Petroleum*, Vol. 2, Oxford University Press, London (1938).
122. L. Lykken, K. R. Fitzsimmins, S. A. Pibbits, and G. Wyld, *Petrol. Refiner 24*(10), 405 (1945). Refer to ASTM methods D810-44T and D811-44T.
123. F. M. Wrightson, *Anal. Chem. 21*, 1543 (1949).
124. R. G. Russell, *Anal. Chem. 20*, 296 (1948).
125. A. G. Gassman and W. R. O'Neill, *Amer. Petrol. Inst. Sect. III, Proc. 29M*, 79 (1949); *Anal. Chem. 21*, 417 (1949).
126. M. J. Murray and H. A. Plagge, *Amer. Petrol. Inst. Sect. III, 29M*, 84 (1949).
127. H. K. Hughes, J. W. Anderson, R. W. Murphy, and J. B. Rather, *Amer. Petrol. Inst. Sect. III, 29M* 89 (1949).
128. Suggested methods for spectrochemical analysis of used diesel lubricating oils, in: *Symposium on Lubricating Oils*, Amer. Soc. Testing Mater. 1956 National Meeting, American Society for Testing and Materials, Philadelphia (1956); F. R. Bryan, *Amer. Soc. Testing Mater. Spec. Tech. Publ. 208*, 24 (1955).
129. A. L. Conrad and W. C. Johnson, *Anal. Chem. 22*, 1530 (1950).
130. P. T. Gilbert, Jr., *Amer. Soc. Testing Mater. Spec. Tech. Publ. 116*, 67 (1951).
131. H. Lundegardh, *Die Quantitative Spektralanalyse der Elemente,* Gustav Fisher, Jena (Part I, 1929; Part II, 1934).
132. J. W. Robinson, *Anal. Chim. Acta 24*, 451 (1961).
133. C. R. Barras and J. D. Helwig, *Amer. Petrol. Inst. Midyear Meeting Div. Ref. Preprint 20-63*, 1 (1963); C. R. Barras, *Jarrell-Ash Newsletter 13*, 1–4 (June 1962).
134. S. Sprague and W. Slavin, *At. Absorption Newsletter*, No. 12, 4 (April 1963).
135. W. Walsh, *Appl. Optics 7*, 1259 (1968), Fig. 1.
136. M. S. Cresser and P. N. Keliher, *Amer. Lab. 4*, 8 (August 1970).
137. V. Sychra, J. Matousek, and S. Masek. *Chem. Listy 63* 177 (1969).

References to Section 2.5

138. *Anal. Chem.*, Annual Reviews, 1949 on.
139. Soc. Anal. Chem. (England), *Annual Reports on Analytical Atomic Spectroscopy*, Vol. 1, London (1971).
140. E. V. Il'ina, *Zavodsk. Lab. 29*, 1317 (1963).
141. B. E. Buell, *Encyclopedia of Spectroscopy* (G. L. Clark, ed.), Reinhold, Publishing Corporation, New York (1960), p. 365.
142. J. Noar, *Inst. Petrol. Rev. 9*, 187, 209 (1955).
143. Perkin-Elmer Corp., *At. Absorption Newsletter* (1962 and following years).
144. R. Mavrodineanu, *Bibliography on Flame Spectroscopy: Analytical Applications 1800–1966*, Natl. Bur. Std. Misc. Publ. 281, Washington, D.C. (1967).
145. H. K. Hughes, J. W. Anderson, R. W. Murphy, and J. B. Rather, *Proc. Amer. Petrol. Inst. Sect. III, 29M*, 89 (1949).
146. R. C. Barras and H. W. Smith, *7th World Petrol. Congr. 9*, 65 (1967).
147. S. K. Kyuregyan, *Emission Spectral Analysis of Petroleum Products*, Izdatelstvo Khimiya, Moscow (1969); *Chem. Abstr. 74*, 5285j (1971).
148. S. K. Kyuregyan, *Determination of Internal Combustion Engine Wear by Spectral*

Analysis, Izdatelstvo Mashinostroenie, Moscow (1966); *Chem. Abstr. 68*, 35621j (1968).
149. J. W. McCoy, *The Inorganic Analysis of Petroleum*, Chemical Publishing Company, New York (1962).
150. O. I. Milner, *Analysis of Petroleum for Trace Elements*, Pergamon Press, Inc., Elmsford, N.Y. (1963).
151. A. N. Aleksandrov, M. I. Dement'eva, and Ya. E. Shmulyakovskii, eds., *Metody Issled. Produktov Neftepereabotki i Neftekhim. Sinteza*, Gostoptekhizdat, Leningrad (1962); *Chem. Abstr. 59*, 3688H (1963).
152. M. Pinta, *Detection and Determination of Trace Elements*, Ann Arbor Humphrey Science Publishers Ltd., London (1971).
153. L. W. Gamble, *Anal. Chem. 23*, 1817 (1951).
154. M. Margoshes and B. F. Scribner, *Anal. Chem. 38*, 297R (1966).

References to Section 2.6

155. R. L. Mitchell, in: *Trace Analysis* (J. H. Yoe and H. J. Koch, Jr., eds.), John Wiley & Sons, Inc., New York (1957), Chap. 14, p. 398.
156. P. N. Keliher and C. C. Wohlers, *Anal. Chem. 45*, 111 (1973).
157. W. G. Elliott, *Amer. Lab. 2*(3), 67 (1970).
158. G. G. Olson, *Amer. Lab. 4*, 57 (February 1972).
159. P. Schierer, Pittsburgh Conference on Analytical Chemistry and Applied Spectroscopy, 1972, paper 51.
160. K. Laqua and W. D. Hagenah, *Xth Colloquium Spectroscopicum International*, Spartan Books, New York (1963), pp. 91–123.
161. B. E. Buell, U.S. Patent 3,583,811 (June 8, 1971).
162. C. Feldman, *Anal. Chem. 21*, 1041 (1949).
163. E. L. Gunn, *Anal. Chem. 32*, 1449 (1960).
164. B. E. Buell, *Anal. Chem. 38*, 1376 (1966).
165. V. A. Fassel and D. W. Golightly, *Anal. Chem. 39*, 466 (1967).
166. R. Skogerboe, 1972 Pacific Conference on Chemistry and Spectroscopy.
167. F. E. Lichte and R. K. Skogerboe, *Anal. Chem. 45*, 399 (1973).
168. R. K. Skogerboe and G. N. Coleman, *Anal. Chem. 48*, 611A (1976).
169. R. K. Skogerboe and G. N. Coleman, *Appl. Spectrosc. 30*, 504 (1976).
170. *Environ. Protection Agency Newsletter* (*Anal. Quality Control Lab.*) (April 1976), p. 5, reference to report on AA and plasma comparison 27th Pittsburgh Conference, 1976.
171. G. W. Wooten, Proceedings of the Annual Symposium on Trace Analysis and Detection in the Environment, Edgewood Arsenal (1975). NTIS AD-A021-948 (January 1976).
172. S. Greenfield and P. B. Smith, *Anal. Chim. Acta 59*, 341 (1972).
173. L. R. Layman and G. M. Hieftje, *Anal. Chem. 47*, 194 (1975).
174. V. A. Fassel and R. N. Kniseley, *Anal. Chem. 46*, 1110A (1974).
175. V. A. Fassel and R. N. Kniseley, *Anal. Chem. 46*, 1155A (1974).
176. R. H. Scott, V. A. Fassel, R. N. Kniseley, and D. E. Nixon, *Anal. Chem. 46*, 75 (1974).
177. R. G. Schleicher and R. M. Barnes, *Anal. Chem. 47*, 724 (1975).
178. G. F. Larson, V. A. Fassel, R. H. Scott, and R. N. Kniseley, *Anal. Chem. 46*, 238 (1975).
179. R. K. Skogerboe, I. T. Urasa, and G. N. Coleman, *Appl. Spectrosc. 30*, 500 (1976).
180. W. E. Rippetoe and T. J. Vickers, *Anal. Chem. 47*, 436 (1975).
181. W. Snelleman, T. C. Rains, K. W. Lee, H. D. Cook, and O. Menis, *Anal. Chem. 42*, 394 (1970).

182. B. E. Buell, 1972 Pacific Conference on Chemistry and Spectroscopy.
183. G. D. Christian and F. J. Feldman, *Appl. Spectrosc. 25*, 660 (1971).
184. G. D. Christian and F. J. Feldman, *Atomic Absorption Spectroscopy: Applications in Agriculture, Biology and Medicine*, Wiley–Interscience, New York (1970).
185. E. E. Pickett and S. R. Koirtyohann, *Anal. Chem. 41*, 28A (1969).
186. R. C. Barras and H. W. Smith, *7th World Petrol. Congr. 9*, 65 (1967).
187. K. Fuwa and B. L. Vallee, *Anal. Biochem. 17*, 444 (1966).
188. F. Zweibaum and J. Moorhead, *At. Absorption Newsletter 6*(6), 134 (1967).
189. R. Mavrodineanu and R. C. Hughes, *Appl. Opt. 7*, 1281 (1968).
190. A. Syty, CRC critical reviews, in: *Anal. Chem. 46*, 155 (1974).
191. M. D. Amos, *Amer. Lab. 4*(8), 57 (1972).
192. G. F. Kirkbright, *Analyst 96*, 609 (1971).
193. J. W. Robinson and P. J. Slevin, *Amer. Lab. 4*(8), 10 (1972).
194. R. B. Cruz and J. C. van Loon, *Anal. Chim. Acta 72*, 231 (1974).
195. J. G. T. Regan and J. Warren, *Analyst 101*, 220 (1976).
196. R. Woodriff and D. Shrader, *Anal. Chem. 43*, 1918 (1971).
197. D. Alger, R. G. Anderson, I. S. Maines, and T. S. West, *Anal. Chim. Acta. 57*, 271 (1971).
198. F. J. M. J. Maessen and F. D. Posma, *Anal. Chem. 46*, 1439 (1974).
199. J. W. Robinson and D. K. Wolcott, *Anal. Chim. Acta 74*, 43 (1975).
200. A. Montaser and S. R. Crouch, *Anal. Chem. 46*, 1817 (1974).
201. H. V. Lagesson, *Mickrochim. Acta 1974*(3), 527.
202. A. F. Ward, D. G. Mitchell, and K. M. Aldous, *Anal. Chem. 47*, 1656 (1975).
203. T. Kantor, S. A. Clyburn, and C. Veillon, *Anal. Chem. 46*. 2213 (1974).
204. C. J. Molnar and J. D. Winefordner, *Anal. Chem. 46*, 1419 (1974).
205. K. I. Aspila, C. L. Chakrabarti, and M. P. Bratzel, Jr., *Anal. Chem. 44*, 1718 (1972).
206. D. D. Siemer, R. Woodriff, and B. Watne, *Appl. Spectrosc. 28*, 582 (1974).
207. Y. Talmi, *Anal. Chem. 46*, 1005 (1974).
208. G. H. Morrison and Y. Talmi, *Anal. Chem. 42*, 809 (1970).
209. M. R. McCullough and J. T. Vickers, *Anal. Chem. 48*, 1006 (1976).
210. K. Govindaraju, G. Mevelle, and C. Chouard, *Anal. Chem. 46*, 1672 (1974).
211. R. Belcher, S. L. Bogdanski, S. A. Ghonaim, and A. Townshend, *Anal. Letters 7*, 133 (1974).
212. R. Belcher, S. A. Ghonaim, and A. Townshend, *Anal. Chim. Acta 71*, 255 (1974).
213. R. Belcher, S. L. Bogdanski, D. J. Knowles, and A. Townshend, *Anal. Chim. Acta 77*, 53 (1975).
214. R. Belcher, S. L. Bogdanski, I. H. B. Rix, and A. Townshend, *Anal. Chim. Acta 84*, 226 (1976).
215. R. K. Skogerboe, D. L. Dick, D. A Pavlica, and F. E. Lichte, *Anal. Chem. 47*, 568 (1975).
216. J. D. Winefordner and R. C. Elser, *Anal. Chem. 43*(4), 24A (1971).
217. B. W. Bailey, *The Element*, Publ. 20, Aztec Instruments, Inc. 10K4568.
218. D. W. Ellis and D. R. Demers, *Advan. Chem. Ser. 73*, 326 (1968).
219. P. L. Larkins and J. B. Willis, *Spectrochim. Acta 26B*, 477, 491 (1971).
220. R. C. Elser and J. D. Winefordner, *Appl. Spectrosc. 45*(3), 345 (1971).
221. J. D. Norris and T. S. West, *Anal. Chem. 45*, 226 (1973).
222. W. B. Barnett and H. L. Kahn, *Anal. Chem. 44*(6), 935 (1972).
223. S. R. Koirtyohann and E. E. Pickett, *Appl. Spectrosc. 23*, 597 (1969).
224. M. D. Amos and J. B. Willis, *Spectrochim. Acta 22*, 1325 (1966).
225. F. W. Hofman and H. Kohn, *J. Opt. Soc. Amer. 51*, 512 (1961).
226. J. A. Fiorino, R. N. Kniseley, and V. A. Fassel, *Spectrochim. Acta 23B*, 413 (1968).
227. P. E. Thomas and M. D. Amos, *Resonance Lines* (*Cary Instru.*) *1*(1), 1 (1969).
228. G. F. Kirkbright, A. Semb, and T. S. West, *Talanta 14*, 1011 (1967).

229. D. N. Hingle, G. F. Kirkbright, and T. S. West, *Talanta 15*, 199 (1968).
230. G. F. Kirkbright, A. Semb, and T. S. West, *Talanta 15*, 441 (1968).
231. R. S. Hobbs, G. F. Kirkbright, M. Sargent, and T. S. West, *Talanta 15*, 997 (1968).
232. G. F. Kirkbright, M. Sargent, and T. S. West, *Talanta 16*, 245 (1969).
233. R. M. Dagnall, G. F. Kirkbright, T. S. West, and R. Wood, *Anal. Chim. Acta. 47*, 407 (1969).
234. G. F. Kirkbright, M. Sargent, and T. S. West, *At. Absorption Newsletter 8*(2), 34 (1969).
235. R. M. Dagnall, G. F. Kirkbright, T. S. West, and R. Wood, *Analyst 95*, 425 (1970).
236. G. F. Kirkbright and T. S. West, *Appl. Opt. 7*(7), 1305 (1968).
237. M. D. Amos, P. A. Bennett, and K. G. Brodie, *Resonance Lines 2*(1), 3 (1970).
238. A. Hell, W. F. Ulrich, N. Shrifrin, and J. Ramirez-Munoz, *Appl. Opt. 7*, 1317 (1968).
239. A. A. Venghiattis, *Appl. Opt. 7* 1313 (1968).
240. G. Ury, J. N'Guea Lotten, J. P. Tardiff, and J. Spitz, *Spectrochim. Acta 26B*, 151 (1971).
241. M. Aldous, R. F. Browner, R. M. Dagnall, and T. S. West, *Anal. Chem. 42*, 939 (1970).
242. M. S. Wang, *Appl. Spectrosc. 26*, 653 (1972).
243. I. E. A. Boling, *Spectrochim. Acta 22*, 425 (1966).
244. M. B. Denton and H. V. Malmstadt, *Anal. Chem. 44*, 241 (1972).
245. C. H. Perrin and P. A. Ferguson, *J. Assoc. Offic. Agri. Chemists 51*(3), 654 (1968).
246. B. E. Buell, *Anal. Chem. 34*, 635 (1962).
247. B. E. Buell, *Anal. Chem. 35*, 372 (1963).
248. J. E. Allan, *Spectrochim. Acta 17*, 459, 467 (1971).
249. C. L. Chakrabarti and S. P. Singhal, *Spectrochim. Acta 24B*, 663 (1969).
250. Instrumentation Laboratories, performance data on detection limits and sensitivity; personal communication
251. Fisher Scientific–Jarrell Ash, performance data on detection limits and sensitivity for Model 810 spectrometer; personal communication (1972).
252. W. Slavin, *Atomic Absorption Spectroscopy*, Wiley–Interscience, New York (1968).
253. W. Slavin, *At. Absorption Newsletter 11*, 37 (1972).
254. M. D. Amos and J. P. Matousek, 1972 Pittsburgh Conference on Analytical Chemistry and Applied Spectroscopy, paper 42; Carbon Rod Atomizer Model 63, Varian-Techtron brochure (1972).
255. T. W. Steele and B. D. Guerin, *Analyst 97*, 77 (1972).
256. P. W. J. M. Boumans and F. J. de Boer, *Spectrochim. Acta 27B*, 391 (1972).
257. G. W. Dickinson and V. A. Fassel, *Anal. Chem. 41*, 1021 (1969).
258. H. Kaiser, *7th World Petrol. Congr. 9*, 3 (1967).
259. M. Fred, N. H. Nachtrieb, and F. S. Tompkins, *J. Opt. Soc. Amer. 37*, 279 (1947).
260. J. A. Goleb, J. P. Faris, and B. H. Meng, *Appl. Spectrosc. 16*, 9 (1962).
261. J. M. Morris and F. X. Pink, *Amer. Soc. Testing Mater. Spec. Tech. Publ. 221*, 39 (1957).
262. D. V. L'Vov and A. D. Khartsyzov, *Zh. Analit. Khim. 24*, 799 (1969); in Russian.
263. G. F. Kirkbright, T. S. West, and P. J. Wilson, *At. Absorption Newsletter 11*, 53 (1972).
264. R. Herrmann and B. Gutsche, *Analyst 94*, 1034 (1969).
265. B. Gutsche and R. Herrmann, *Analyst 95*, 805 (1970).
266. D. F. Tomkins and C. W. Frank, *Anal. Chem. 44*, 1451 (1972).
267. H. E. Taylor, J. H. Sibson, and R. K. Skogerboe, *Anal. Chem. 42*, 1569 (1970).
268. K. M. Aldous, R. M. Dagnall, and T. S. West, *Analyst 95*, 417 (1970).
269. R. M. Dagnall, K. C. Thompson, and T. S. West, *Analyst 93*, 72, 78 (1968).
270. A. Syty and J. A. Dean, *Appl. Opt. 7*, 1331 (1968).
271. G. F. Kirkbright, M. Marshall, and T. S. West, *Anal. Chem. 44*, 2379 (1972).

272. R. Herrmann and C. T. J. Alkemade, *Flame Photometry*, 2nd rev. ed. (P. T. Gilbert, Jr., transl.) Wiley–Interscience, New York (1963).
273. H. E. Taylor, J. H. Gibson, and R. K. Skogerboe, *Anal. Chem. 42*, 876 (1970).
274. F. J. Fernandez and D. C. Manning, *At. Absorption Newsletter 10*, 86 (1971).
275. D. C. Manning, *At. Absorption Newsletter 10*, 123 (1971).
276. W. Reichel and L. Acs, *Anal. Chem. 41*, 1886 (1969).
277. R. Hurford and D. F. Boltz, *Anal. Chem. 40*, 379 (1968).
278. K. Fuwa and B. L. Vallee, *Anal. Chem. 35*, 942 (1963).
279. I. Rubeska and B. Moldan, *Analyst 93*, 148 (1969).
280. C. Iida and M. Nagura, *Spectrosc. Letters 3*(3), 63 (1970).

References to Section 2.7

281. P. T. Gilbert, "Analysis Instrumentation-1964," *Proceedings of the 10th National Analysis Instrumentation Symposium*, San Francisco, 1964, Plenum Press, New York (1964), pp. 193–223. (Beckman reprint, R-6217, "Advances in Emission Flame Photometry.")
282. V. A. Fassel, J. O. Rasmussen, and T. G. Cowley, *Spectrochim. Acta 23B*, 579 (1968).
283. E. E. Pickett and S. R. Koirtyohann, *Anal. Chem. 41*, 28A (1969).
284. R. J. Lovett, D. L. Welch, and M. L. Parsons, *Appl. Spectrosc. 29*, 470 (1975).
285. C. W. Frank, W. G. Schrenk, and C. E. Mcloan, *Anal. Chem. 38*, 1005 (1966).
286. J. D. Norris and T. S. West, *Anal. Chem. 46*, 1423 (1974).
287. V. K. Panday and A. K. Ganguly, *Spectrosc. Letters 9*, 73 (1976).
288. B. E. Buell, in *Flame Emission and Atomic Absorption Spectrometry* (J. A. Dean and T. C. Rains, eds.), Vol. 1, Marcel Dekker, New York (1969).
289. W. Snellman, T. C. Rains, K. W. Yee, H. D. Cook, and O. Menis, *Anal. Chem. 42*, 394 (1970).
290. B. E. Buell, 1972 Pacific Conference on Chemistry and Spectroscopy.
291. F. E. Lichte and R. K. Skogerboe, *Anal. Chem. 45*, 399 (1973).
292. G. F. Larson, V. A. Fassell, R. K. Winge, and R. N. Kniseley, *Appl. Spectrosc. 30*, 384 (1976).
293. S. R. Koirtyohann and E. E. Pickett, *Anal. Chem. 37*, 601 (1965).
294. H. L. Kahn and D. C. Manning, *Amer. Lab. 4*, 51 (August 1972).
295. J. B. Ezell, Jr., *At. Absorption Newsletter 5*(6), 122 (1966).
296. V. J. Smith and J. W. Robinson, *Anal. Chim. Acta 49*, 161 (1970).
297. V. J. Smith and J. W. Robinson, *Anal. Chim. Acta 49*, 417 (1970).
298. T. G. Cowley, V. A. Fassel, and R. N. Kniseley, *Spectrochim. Acta 23B*, 771 (1968).
299. C. J. Perry and D. A. Keyworth, *Can. Spectrosc. 12*, 47 (1967).
300. J. D. Kerber, *Appl. Spectrosc. 20*, 212 (1966).
301. L. Capacho-Delgado and S. Sprogue, *At. Absorption Newsletter 4*, 363 (1965).
302. G. K. Billings, *At. Absorption Newsletter 4*, 357 (1965).
303. H. L. Fleming, *Anal. Chim. Acta 59*, 197 (1972).
304. S. R. Koirtyohann and E. E. Pickett, *Anal. Chem. 38*, 585 (1966).
305. S. R. Koirtyohann and E. E. Pickett, *Anal. Chem. 38* 1087 (1966).
306. B. V. L'Vov, *Spectrochim. Acta. 24B*, 53 (1969).
307. H. L. Kahn and D. C. Manning, *Amer. Lab. 4*, 51 (August 1972); *At. Absorption Newsletter 7*, 40 (1968); *11*, 112 (1972).
308. T. Hadeishi and R. D. McLaughlin, *Science 174*, 404 (1971).
309. H. Koizumi and K. Yasuda, *Anal. Chem. 47*, 1679 (1975).
310. R. D. McLaughlin, 1972 Pacific Conference on Chemistry and Spectroscopy.
311. J. Y. Marks, R. J. Spellman, and B. Wysocki, *Anal. Chem. 48*, 1474 (1976).
312. P. B. Zeeman and J. A. Brink, *Analyst 93*, 388 (1968).

313. S. R. Koirtyohann, in: *Flame Emission and Atomic Absorption Spectroscopy* (J. A. Dean and T. C. Rains, eds.), Marcel Dekker, New York (1969).
314. R. Mavrodineanu and H. Boiteux, *Flame Spectroscopy*, John Wiley & Sons, Inc., New York, 1965.
315. A. C. West, V. A. Fassel, and R. N. Kniseley, A Study of Lateral Diffusion Interferences in Flame Atomic Absorption and Emission Spectroscopy, National Applied Spectroscopy Meeting, 1972.
316. A. C. West, V. A. Fassel, and R. N. Kniseley, *Anal Chem. 45*, 2420 (1973).
317. S. R. Koirtyohann and E. E. Pickett, *Anal. Chem. 40* 2068 (1968).
318. J. Y. Marks and G. G. Welcher, *Anal. Chem. 42*, 1033 (1970).
319. D. C. Manning and L. Capacho-Delgado, *Anal. Chim. Acta 36*, 312, (1966).
320. G. R. Kornblum and L. DeGalan, *Spectrochim. Acta. 28B*, 139 (1973).
321. I. Rubeska, in: *Flame Emission and Atomic Absorption Spectrometry* (J. A. Dean and T. C. Rains, eds.), Marcel Dekker, New York (1969).
322. T. C. Rains, *in: Flame Emission and Atomic Absorption Spectrometry* (J. A. Dean and T. C. Rains, eds.), Marcel Dekker, New York (1969).
323. C. T. J. Alkemade, *Anal. Chem. 38*, 1252 (1966).
324. I. Rubeska and B. Moldan, *Anal. Chim. Acta. 37*, 421 (1967).
325. P. B. Adams and W. O. Passmore, *Anal. Chem. 38*, 633 (1966).
326. T. V. Ramakrishna, P. W. West, and J. W. Robinson, *Anal. Chim. Acta. 40*, 347 (1968).
327. P. E. Thomas and W. F. Pickering, *Talanta 18*, 127 (1971).
328. V. S. Sastri and C. L. Chakrabarti, *Talanta 16*, 1093 (1969).
329. E. M. Bulewicz and P. J. Padley, *Spectrochim. Acta. 28B*, 125 (1973).
330. K. Fujiwara, H. Haraguchi, and K. Fuwa, *Anal. Chem. 47*. 743 (1975).
331. I. Rubeska, *Anal. Chem. 48*, 1640 (1976).
332. J. A. Dean, *Flame Photometry*, McGraw-Hill Book Company, New York (1960).
333. R. Herrmann and C. T. J. Alkemade, *Flame Photometry* (P. T. Gilbert, Jr., transl.) Wiley-Interscience Pub., New York (1963).
334. J. E. Hicks, R. T. McPherson, and J. W. Salyer, *Anal. Chim. Acta. 61*, 441 (1972).
335. W. Slavin, *Atomic Absorption Spectroscopy*, Wiley-Interscience, New York (1968).
336. A. Strasheim and G. J. Wessels, *Appl. Spectrosc. 17*, 65 (1963).
337. M. D. Amos and J. B. Willis, *Spectrochim. Acta 22*, 1325 (1966).
338. W. B. Barnett, *Anal. Chem. 44*, 695 (1972).
339. A. M. Bond, *Anal. Chem. 42*, 932 (1970).
340. A. M. Bond and J. B. Willis, *Anal. Chem. 40*, 2087 (1968).
341. C. R. Belcher and K. Kinson, *Anal. Chim. Acta 30*, 483 (1964).
342. W. D. Cobb, W. W. Foster, and T. S. Harrison, *Anal. Chim. Acta 60*, 430 (1972).
343. R. M. Dagnall, G. F. Kirkbright, T. S. West, and R. Wood, *Anal. Chem. 42*, 1032 (1970).
344. A. P. Ferris, W. P. Jepson, and R. C. Shapland, *Analyst 95*, 574 (1970).
345. B. Fleet, K. V. Liberty, and T. S. West, *Talanta 17*, 203 (1970).
346. H. D. Fleming, *Spectrochim. Acta 23B*, 207 (1967); also see Ref. 303.
347. W. W. Harrison and W. H. Wadlin, *Anal. Chem. 41*, 374 (1969).
348. J. B. Headridge and D. P. Hubbard, *Anal. Chim. Acta 37* 151 (1967).
349. J. Husler, *At. Absorption Newsletter 10*, 205 (1971).
350. M. E. Hofton and D. P. Hubbard, *Anal. Chim. Acta 52*, 425 (1970).
351. R. J. Jaworowski, R. P. Weberling, and D. J. Bracco, *Anal. Chim. Acta. 37*, 284 (1967).
352. G. F. Kirkbright, M. K. Peters, and T. S. West, *Analyst 91*, 705 (1966).
353. M. J. Martin, *Chem. Ind. (London) 1971*, 514 (May 8).
354. T. Maruta, M. Suzuki, and T. Takeuchi, *Anal. Chim. Acta 51*, 381 (1970).

355. J. M. Ottoway, D. T. Coker, W. B. Rawston, and D. R. Bhattarai, *Analyst 95*, 567 (1970).
356. D. C. G. Pearton, J. B. Taylor, P. K. Faure, and T. W. Steele, *Anal. Chim. Acta 44*, 353 (1969).
357. T. V. Ramakrishna, P. W. West, and J. W. Robinson, *Anal. Chim. Acta. 44*, 437 (1969).
358. T. V. Ramakrishna, P. W. West, and J. W. Robinson, *Anal. Chim. Acta. 37*, 81 (1967).
359. T. V. Ramakrishna, P. W. West, and J. W. Robinson, *Anal. Chim. Acta 37*, 20 (1967).
360. S. L. Sachdev and P. W. West, *Invest. Sci. Technol. 4*, 749 (1970).
361. S. L. Sachdev, J. W. Robinson, and P. W. West, *Anal. Chim. Acta. 37*, 12 (1967).
362. S. J. Sachdev, J. W. Robinson, and P. W. West, *Anal. Chim. Acta. 37*, 156 (1967).
363. J. C. van Loon, *At. Absorption Newsletter 11*, 60 (1972).
364. K. Wada, *Japan Analyst 21*, 221 (1972).
365. M. Yanagisawa, M. Suzuki, and T. Takeuchi, *Anal. Chim. Acta 52*, 386 (1970).
366. M. Yanagisawa, H. Kihara, M. Suzuki, and T. Takeuchi, *Talanta 17*, 888 (1970).
367. S. Kallmavn and E. W. Hobart, *Anal. Chim. Acta 51*, 120 (1970).
368. C. A. Baker and F. W. J. Garton, *U.K. At. Energy Authority, Res. Group Rept. AERE R-3490* (1961). A study of interferences in emission and absorption flame photometry.
369. G. E. Janauer, F. R. Smith, and J. Mangan, *At. Absorption Newsletter 6*, 3 (1967).
370. D. C. Manning, *At. Absorption Newsletter 6*, 35 (1967).
371. R. A. Mostyn and A. F. Cunningham, *Anal. Chem. 38*, 121 (1966).
372. J. A. Platte and V. M. March, *At. Absorption Newsletter 4*, 289 (1965).
373. W. G. Shrenk, D. A. Lehman, and L. Neufeld, *Appl. Spectrosc. 20*, 389 (1966).
374. R. Smith and A. E. Lawson, *Analyst 96*, 631 (1971).
375. D. F. Tomkins and C. W. Frank. *Appl. Spectrosc. 25*, 539 (1971).
376. M. Atwell and J. Herbert, *Appl. Spectrosc. 23*, 480 (1969).
377. S. Kallmann and E. W. Hobart, *Anal. Chim. Acta 51*, 120 (1970).
378. R. C. Mallet, D. C. G. Pearton, E. J. Ring, and T. W. Steele, *Talanta 19*, 181 (1972).
379. A. A. G. Houze, J. S. *African Chem. Inst. 23*, 115 (1970).
380. W. B. Barrett, *Anal. Chem. 44*, 695 (1972).
381. J. A. Hurlburt and C. D. Chriswell, *Anal. Chem. 43*, 465 (1971).
382. J. A. Bowman and J. B. Willis, *Anal. Chem. 39*, 1210 (1967).
383. I. Janousek and M. Malat, *Anal. Chim. Acta 58*, 448 (1972).
384. S. S. Krishnan, K. A. Gillispie, and D. R. Crapper, *Anal. Chem. 44*, 1469 (1972).
385. T. Maruta, T. Takeuchi, and M. Suzuki, *Anal. Chim. Acta 58*, 452 (1972).
386. R. Dickson and C. M. Johnson, *Appl. Spectrosc. 20*, 214 (1966).
387. J. M. Ottaway and N. K. Pradhan, *Talanta 20*, 927 (1973).
388. W. W. Brachaczek, J. W. Butler, and W. R. Pierson, *Appl. Spectrosc. 28*, 585 (1974).
389. E. G. Bradfield, *Analyst 99*, 403 (1974).
390. S. Dilli, K. M. Gawne, and G. W. Ocago, *Anal. Chim. Acta 69*, 287 (1974).
391. L. L. Sundberg, *Anal. Chem. 45*, 1460 (1973).
392. J. P. Macquet and T. Theophanides, *Spectrochim. Acta 29B* 241 (1974).
393. D. Carter, J. G. T. Regan, and J. Warren, *Analyst 100*, 721 (1975).
394. V. Yofe and R. Finkelstein, *Anal. Chim. Acta 19*, 88 (1957).
395. J. I. Dinnin, *Anal. Chem. 32*, 1475 (1960); *U.S. Geol. Surv. Profess. Paper 424D*, 391, 393 (1961).
396. A. Shatkay, *Anal. Chem. 40*, 2097 (1968).
397. G. Lindstedt, *Analyst 95*, 264 (1970).
398. I. Rubeska and M. Miksovsky, *At. Absorption Newsletter 11*, 57 (1972).
399. T. Nakahara, M. Munemori, and S. Musha, *Anal. Chim. Acta 62*, 267 (1972).
400. H. L. Kahn and J. E. Schallis, *At. Absorption Newsletter 7*, 5 (1968).

401. G. F. Kirkbright, M. Sargent, and T. S. West, *At. Absorption Newsletter 8*, 34 (1969).
402. G. F. Kirkbright and L. Ranson, *Anal. Chem. 43*, 1238. 1241 (1971).
403. K. E. Smith and C. W. Frank, *Applied. Spectrosc. 22*, 765 (1968).
404. A. E. Smith, *Analyst 100*, 300 (1975).
405. M. D. Amos, P. A. Bennett, K. G. Brodie, P. W. Y. Lung, and J. P. Matousek, *Anal. Chem. 43*, 211 (1971).
406. D. Alger, R. G. Anderson, I. S. Maines, and T. S. West, *Anal. Chim. Acta 57*, 271 (1971).
407. K. W. Jackson, T. S. West, and L. Balchin, *Anal. Chem. 45*, 249 (1973).
408. T. Maruta and T. Takeuchi, *Anal. Chim. Acta. 62* (1972).
409. Y. Talmi and G. H. Morrison, *Anal. Chem. 44*, 1455 (1972).
410. T. Hadeishi and R. D. McLaughlin, *Anal. Chem. 48*, 1009 (1976).
411. M. S. Epstein, T. C. Rains, and T. C. O'Haver, *Appl. Spectrosc. 30*, 324 (1976).
412. D. Clark, R. M. Dagnall, and T. S. West, *Anal. Chim. Acta 58*, 339 (1972).

References to Section 2.8

413. Committee on Public Affairs, *Lead in the Environment*, American Petroleum Institute, New York (1967).
414. *Oil Gas J. 70* (47), 30 (1972).
415. J. G. Dean, F. L. Bosqui, and K. H. Lanuette, *Environ. Sci. Technol. 6*, 518 (1972).
416. D. G. Argo and G. L. Culp, *Water & Sewage Works 119* (8), 62 (1972).
417. P. N. Cheremisinoff and Y. H. Habib, *Water & Sewage Works 119* (8), 46 (1972).
418. S. Miller and C. Lewicke, *Environ. Sci. Technol. 6* (12); 974 (1972).
419. R. G. Bond and C. P. Straub, *Air Pollution*, CRC Press, Cleveland (1972).
420. G. V. James, *Water Treatment*, CRC Press, Cleveland (1971).
421. W. Summer, *Odor Pollution of Air*, CRC Press, Cleveland (1971).
422. L. S. Goodfriend, *Noise Pollution*, CRC Press, Cleveland (1973).
423. F. L. Parker and P. A. Krenkel, *Physical and Engineering Aspects of Thermal Pollution*, CRC Press, Cleveland (1973).
424. L. T. Friberg, M. Piscator, and G. F. Nordberg, *Cadmium in the Environment*, CRC Press, Cleveland (1971).
425. L. T. Friberg and J. J. Vostal, *Mercury in the Environment*, CRC Press, Cleveland (1972).
426. F. M. Ditri, *The Environmental Mercury Problem*, CRC Press, Cleveland (1972).
427. M. S. Shuman and W. W. Fogleman, *J. Water Pollution Control Federation 48*, 998 (1976).
428. H. A. Laitinin, *Anal. Chem. 44*(11), 1721 (1972).
429. H. M. Wilson, *Oil Gas J. 70* (30), 22 (1972).
430. Anon. *Chem. Eng. News 49*, 29 (July 1971).
431. P. N. Cheremisinoff and Y. H. Habib, *Water & Sewage Works 119* (7), 73 (1972).
432. E. D. Goldberg, Natl. Sci. Found. Int. Decade Ocean Exploration Baseline Conf., (New York, May 24–26, 1972, *U.S. Dept. Commer. Natl. Tech. Inform. Serv. PB-233*, 959 (1972).
433. D. C. Burrell, *Atomic Spectrometric Analysis of Heavy Metal Pollutants in Water*, Ann Arbor Science Publishers, Ann Arbor, Mich. (1974).
434. W. Gallay, H. Egan, J. L. Monkman, R. Truhaut, P. W. West, and G. Widmark, *Environmental Pollutants; Selected Analytical Methods*, Ann Arbor Science Publishers, Ann Arbor, Mich. (1975).

435. W. T. Pecora, *Water & Sewage Works 119* (7), 174 (1972).
436. S. S. Miller, *Environ. Sci. Technol. 9*, 194 (1975).
437. W. H. Durum and J. Haffty, *U.S. Geol. Surv. Cir. 445* (1961).
438. B. Mason, *Principles of Geochemistry*, 3rd ed., John Wiley & Sons, Inc., New York (1966), p. 45.
439. B. E. Buell, 1972 Pacific Conference on Chemistry and Spectroscopy.
440. V. A. Fassel, R. W. Slack, and R. N. Kniseley, *Anal. Chem. 43*, 186 (1971).
441. F. J. Johnson, T. C. Woodis, and J. M. Cummings, Jr., *At. Absorption Newsletter 11* (6), 118 (1972).
442. P. Strohal, K. Molnar, and I. Bacic, *Mikrochim. Acta 1972*, 586.
443. E. B. Sandell, *Colorimetric Determination of Trace Metals*, Wiley–Interscience, New York (1959), pp. 388–389.
444. A compilation of papers on mercury in the environment, *U.S. Geol. Surv. Profess. Paper 713* (1970).
445. R. A. Wallace, *Mercury in the Environment–The Human Element*, ORNL-NSF-EP-1, Oak Ridge, Tenn. (1971).
446. Study Group on Mercury Hazards, *Environ. Res. 4* (1); 1 (1971).
447. H. Burnes, *Oceanogr. Marine Biol. 8*, 203 (1971).
448. C. E. Billings, *Eng. Sci. 176*, 1232 (1972).
449. J. Friedman and N. Peterson, *Science 172*, 1027 (1971).
450. D. H. Klein, *Environ. Sci. Technol. 6* (6), 560 (1972).
451. W. H. Smith, *Science 176*, 1237 (1972).
452. W. Buller, *U.S. Geol. Surv. Prof. Paper 800-C*, 233 (1972).
453. M. E. Hinkle, *U.S. Geol. Surv. Profess. Paper 750-B*, B171 (1972).
454. R. H. Filby, An Investigation of Trace Elements in Petroleum Using Activation Analysis, Ph.D. thesis, Washington State University (1971).
455. E. H. Bailey, P. D. Snavely, Jr., and D. E. White, *U.S. Geol. Surv. Profess. Paper 424D*, D306 (1961).
456. D. C. Manning, *At. Absorption Newsletter 9* (5); 97 (1970).
457. C. Huffman, R. L. Rahill, V. E. Shaw, and D. R. Norton, *U.S. Geol. Surv. Profess. Paper 800-C*, C203 (1972).
458. R. V. Coyne and J. A. Collins, *Anal. Chem. 44*, 1093 (1972).
459. G. Lindstedt, *Analyst 95*, 264 (1970).
460. H. L. Kahn and D. C. Manning, *Amer. Lab. 4*, 51 (August 1972).
461. H. L. Kahn and D. C. Manning, *At. Absorption Newsletter 7*, 40 (1968); *11*, 112 (1972).
462. B. V. L'Vov, *Spectrochim. Acta 24B*, 53 (1969).
463. R. L. Windham, *Anal. Chem. 44*, 1334 (1972).
464. T. Hadeiski and R. D. McLaughlin, *Science 174*, 404 (1971).
465. F. E. Lichte and R. K. Skogerboe, *Anal. Chem. 44*, 1321 (1972).
466. R. W. April and D. N. Hume, *Science 170*, 849 (1970).
467. G. W. Dickinson and V. A. Fassel, *Anal. Chem. 41*, 1021 (1969).
468. J. J. McNerney, P. R. Buseck, and R. C. Hanson, *Science 178*, 612 (1972).
469. G. K. Billings, *At. Absorption Newsletter 4*, 357 (1965).
470. M. P. Bratzel, Jr., and C. L. Chakrabarti, *Anal. Chim. Acta 61*, 25 (1972).
471. J. L. Seeley, D. Dick, J. H. Arvik, R. L. Zimdahl, and R. K. Skogerboe, *Appl. Spectrosc. 26*, 456 (1972).
472. *Varian Instr. Appl. 6* (3); 8 (1972).
473. T. J. Chow and J. L. Earl. *Science 176*, 510 (1972).
474. Y. K. Chau, International Symposium on Identification and Measurement of Environmental Pollutants, National Research Councils of Canada (1971), p. 354.

475. J. H. Carpenter, in: *Analytical Chemistry: Key to Progress on National Problems* W. W. Meinke and J. K. Taylor, eds.), Chap. 7, Natl. Bur. Std. Spec. Publ. 351, Washington, D.C. (1972).
476. P. G. Brewer and D. W. Spencer, Trace Element Intercalibration Study, Red. 70-62, unpublished manuscript (1970).
477. J. V. Brunnock, D. F. Duckworth, and G. G. Stevens, *J. Inst. Petrol. 54*, 310 (1968).
478. E. R. Adlard, *J. Inst. Petrol. 58*, 63 (1972).
479. R. A. Stuart and R. D. Branch, *Instr. News 21* (2) (1970).
480. A. D. Thruston, Jr., and R. W. Knight, *Environ. Sci. Technol. 5*, 64 (1971).
481. Amer. Petrol. Inst. Abstr. 19-31598, *Chem. Week 111* (1), 24 (1972); *Chem. Eng. News 50* (27); 2 (1972).
482. D. C. Grey and M. L. Jensen, *Science 177*, 1099 (1972).
483. B. E. Saltzman and J. E. Cuddeback, *Anal. Chem. 47*, 1R (1975).
484. J. Y. Hwang, *Anal. Chem. 44* (14), 20A (1972).
485. A. C. Stern, *Air Pollution*, 2nd ed., Academic Press, Inc., New York (1968).
486. R. Woodriff and J. F. Lech, *Anal. Chem. 44* (7), 1323 (1972).
487. H. P. Loftin, C. M. Christian, and J. W. Robinson, *Spectrosc. Letters 3* (7), 161 (1970).
488. C. M. Christian and J. W. Robinson, *Anal. Chim. Acta 56*, 466 (1971).
489. D. R. Scott, W. A. Loseke, L. E. Holbroke, and R. J. Thompson, *Appl. Spectrosc. 30*, 392 (1976).
490. *At. Absorption Emission Spectrometry Abstr. 7B* (1), 50 (1975).

References to Section 2.9.1

491. A. G. Collins, *J. Water Pollution Control Federation 43*, 2383 (1971).
492. A. A. Coburn and C. E. Bowlin, Jr., Water Use by the Petroleum Industry, 1967 Annual Meeting of Interstate Oil Compact Commission, Oklahoma City, Okl.
493. E. R. Smith and F. A. Olson, *Amer. Petrol. Inst. Meeting, Div. Ref. 39* (4), 15 (May 1959).
494. M. J. Fishman and D. E. Erdmann, *Anal. Chem. 49*, 139R (1977); *47*, 334R (1975); *45*, 361R (1973); *43*, 356R (1971); reviews on water have appeared biennially in *Anal. Chem.* since 1949.
495. Res. Comm. Water Pollution Control Federation, Annual reviews on water analysis, *J. Water Pollution Control Federation 48*, 998 (1976), and previous years.
496. *Methods for Chemical Analysis of Water and Wastes*, Environmental Protection Agency, Water Quality Office, Cincinnati, Ohio (1971, 1974).
497. *Standard Methods for the Examination of Water and Wastewater*, 13th ed., American Public Health Association–American Water Works Association–Water Pollution Control Federation, New York (1971).
498. Anal. Ref. Serv., *J. Amer. Water Works Assoc. 60*, 739 (1968).
499. R. F. Gould, ed., *Trace Inorganics in Water*, Advan. Chem. Ser. 73, American Chemical Society, Washington, D.C. (1968); symposium sponsored by the Div. Air Waste Chem., 153rd Amer. Chem. Soc. Meeting, April 1967.
500. E. F. McFarren and R. J. Lishka, in: *Trace Inorganics in Water* (R. F. Gould, ed.), Advan. Chem. Ser. 73, American Chemical Society, Washington, D.C. (1968), p. 253.
501. A. E. Greenberg, N. Moskowitz, B. R. Tomplin, and J. Thomas, *J. Amer. Water Works Assoc. 61*, 599 (1969). Calif. Dept. of Public Health Study.
502. A. E. Greenberg, *Public Health Rept.* (*U.S.*) *76*, 783 (1961).
503. E. Brown, M. W. Skougstad, and M. J. Fishman, *U.S. Geol. Surv. Resources Invest. 5*, Chap. A-1 (1970).

504. W. H. Durum and J. Haffty, *U.S. Geol. Surv. Cir. 445* (1961).
505. J. Haffty, *U.S. Geol. Surv. Water Supply Paper 1540A* (1960).
506. W. D. Silvey and R. Brennan, *Anal. Chem. 34*, 784 (1962).
507. W. Silvey, *U.S. Geol. Surv. Water Supply Paper 1540B* (1961).
508. W. Silvey, *U.S. Geol. Surv. Water Supply Paper 1535-L* (1967).
509. J. H. Hem, *U.S. Geol. Surv. Water Supply Paper 1473* (1959).
510. F. H. Rainwater and L. L. Thatcher, *U.S. Geol. Surv. Water Supply Paper 1454* (1960).
511. D. A. Livingstone, *U.S. Geol. Surv. Profess. Paper 440-G* (1963).
512. Anon., *Water & Sewage Works 118*, 174 (1971).
513. U.S. Geol. Surv. National Water Resources Data System, Chief Hydrologist, Water Resources Division Department of the Interior, Washington D.C.
514. D. N. Hume, *Advan. Chem. Ser. 67*, 30 (1967).
515. J. F. Kopp and R. C. Kroner, *Appl. Spectrosc. 19*, 155 (1965).
516. D. C. Burrell, *Atomic Spectrometric Analysis of Heavy Metal Pollutants in Water*, Ann Arbor Science Publishers, Inc., Ann Arbor, Mich. (1974).
517. O. T. Høgdahl, *The Trace Elements in the Ocean: A Bibliographic Compilation*, Central Institute of Industrial Research, Oslo (1963).
518. H. Barnes, *Apparatus and Methods of Oceanography–Chemical*, George Allen & Unwin Ltd., London (1959).
519. D. F. Martin, *Marine Chemistry*, Vol. 1: *Analytical Methods*, 2nd ed., Marcel Dekker, New York (1972).
520. J. P. Riley and G. Skirrow, eds., *Chemical Oceanography*, Vol. 2, Academic Press, Inc., New York (1965).
521. F. A. Richards, *Geochim. Cosmochin. Acta 10*, 241 (1956).
522. D. Nehring, K. H. Rohde, and H. Berge, *Math.-Naturwiss. Reich 16*, 1085 (1967); *Chem. Abstr. 71*, 53344n (1969).
523. R. D. Hitchcock and W. L. Starr, *Appl. Spectrosc. 8*, 5 (1954).
524. J. R. Newton and M. E. Atkins, *U.S. Office Saline Water Res. Develop. Progr. Rept. 450* (1969).
525. D. F. Schutz and K. K. Turekian, *Geochim. Cosmochin. Acta 29*, 259 (1965).
526. J. P. Riley and D. Taylor, *Deep-Sea Res. 19*, 307 (1972).
527. Analytical Methods in Oceanography, (T. R. P. Gibb, Jr., ed.), *Advan. Chem. Ser. 147*; 1974 Amer. Chem. Soc. Symposium, Washington, D.C., 1975.
528. D. F. Martin, *Marine Chemistry*, Vol. 1: *Analytical Methods*, 2nd ed., Marcel Dekker, New York (1972).
529. E. D. Goldberg, Natl. Sci. Found. Int. Decade Ocean Exploration Baseline Conf. New York, Hay 24–26, 1972, *U.S. Dept. Commer. Natl. Tech. Inform. Serv. PB-233*, 959 (1972).
530. H. Khalil, *Instrumental Analysis for Water Pollution Control*, Ann Arbor Science Publishers, Inc., Ann Arbor, Mich. (1971).
531. W. J. Traversy, *Methods for Chemical Analysis of Waters and Wastewaters*, Canadian Water Quality Division, Inland Waters Branch (1971).
532. L. L. Ciaccio, ed., *Water and Water Pollution Handbook*, Vols. 1–4, Marcel Dekker, New York (1972).
533. I. M. Kolthoff, P. J. Elving, and F. H. Stross, eds., *Treatise on Analytical Chemistry*, Part III: *Analytical Chemistry in Industry*, Vol. 2, Wiley-Interscience, New York (1971); see chapters by R. C. Kroner and D. G. Ballinger and by K. H. Mancy and W. J. Weber.
534. M. W. Skougstad, *Anal. Chem. 46*, 982A (1974).
535. H. A. Laitinen, *Anal. Chem. 44*, 1721 (1972).

536. R. Mavrodineanu, *Bibliography on Flame Spectroscopy: Analytical Applications 1800–1966,* Natl. Bur. Std. Misc. Publ. 281, Washington D.C. (1967).
537. Perkin-Elmer Corp., *At. Absorption Newsletter* yearly bibliographies since 1962.
538. C. E. Roberson, *U.S. Geol. Surv. Res. Rept. 1961*, p. D335.
539. G. Rittenhouse, R. B. Fulton, R. J. Grabowski, and J. L. Bernard, *Chem. Geol. 4* (1/2) (1969); special issue on geochemistry of subsurface brines from a 1968 symposium.
540. D. M. Gullikson, W. H. Caraway, and G. L. Gates, *U.S. Bur. Mines Rept. Invest. 5737*, 45 (1961).
541. E. H. Bailey, P. D. Snavely, Jr., and D. E. White, *U.S. Geol. Surv. Profess. Paper 424D*, D306 (1961).
542. J. D. Bredehoeft, C. R. Blyth, W. A. White, and G. B. Maxey, *Amer. Assoc. Petrol. Geol. Bull. 47*, 257 (1963).
543. R. N. Clayton, I. Friedman, D. L. Graf, T. K. Mayeda, W. F. Meents, and N. F. Shimp, *J. Geophys. Res. 71*, 3869 (1966).
544. D. L. Graf, W. F. Meents, I. Friedman, and N. F. Shimp, *Illinois Geol. Surv. Circ. 393* (1966).
545. W. F. Meents, A. H. Bell, O. W. Rees, and W. G. Tilbury, *Illinois Geol. Surv. Illinois Petrol. No. 66* (1952).
546. L. C. Bonham, *Amer. Assoc. Petrol. Geol. Bull. 40*, 897 (1956).
547. A. E. Dong, *U.S. Geol. Surv. Profess. Paper 700B*, B242 (1970).
548. L. W. Strock and S. Drexler, *J. Opt. Soc. Amer. 31*, 167 (1941).
549. J. H. Carpenter, in: *Analytical Chemistry: Key to Progress on National Problems* (W. W. Meinke and J. K. Taylor, eds.), Chap. 7, Natl. Bur. Std. Spec. Publ. 351, Washington, D.C. (1972).
550. K. B. Krauskopf, *Geochim. Cosmochin. Acta 9*, 1 (1956).
551. B. Mason, *Principles of Geochemistry*, 3rd ed., John Wiley & Sons, Inc., New York (1966).
552. J. R. Vallentyne, in: *Comparative Biochemistry of Photoreactive Systems* (M. B. Allen, ed.), Academic Press, Inc. New York (1960), Chap. 7.
553. M. M. Braideck and F. H. Emery, *J. Amer. Water Works Assoc. 27*, 557 (1935); *Ohio Conf. Water Purification, 13th Ann. Rept.* (1933), pp. 68–77.
554. G. A. Uman, *J. Water Pollution Control Federation 40*, 484 (1968).
555. L. G. Young, *Analyst 87*, 6 (1962).
556. J. P. Pagliassotti and F. W. Porsche, *Anal. Chem. 28*, 1774 (1956).
557. T. H. Zink, *Appl. Spectrosc. 13*, 94 (1959).
558. B. E. Buell, U.S. Patent 3,583,811 (June 8, 1971).
559. C. Feldman, *Anal. Chem. 21*, 1041 (1949).
560. E. L. Gunn, *Anal. Chem. 32*, 1449 (1960).
561. M. Fred, N. H. Nachtrieb, and F. S. Tompkins, *J. Opt. Soc. Amer. 37*, 279 (1947).
562. J. A. Goleb, J. P. Faris, and B. H. Meng, *Appl. Spectrosc. 16*, 9 (1962).
563. J. M. Morris and F. X. Pink, *Amer. Soc. Testing Mater. Spec. Tech. Publ. 221*, 39 (1957).
564. R. Ko, *U.S. At. Energy Comm. Paper HW-48770* (1957).
565. M. W. Skougstad and C. A. Horr, *U.S. Geol. Surv. Circ. 420* (1960).
566. D. Hoggan, C. E. Marquart, and W. R. Battles, *Anal. Chem. 36*, 1955 (1964).
567. A. G. Collins, *Appl. Spectrosc. 21*, 16 (1967).
568. A. G. Collins and C. A. Pearson, *Anal. Chem. 36*, 787 (1964).
569. P. W. J. M. Boumans and F. J. de Boer, *Spectrochim Acta 27B*, 391 (1972).
570. G. W. Dickinson and V. A. Fassel, *Anal. Chem. 41*, 1021 (1969).
571. F. E. Lichte and R. K. Skogerboe, *Anal. Chem. 45*, 399 (1973).

572. F. E. Lichte and R. K. Skogerboe, *Anal. Chem. 44*, 1480 (1972).
573. F. E. Lichte and R. K. Skogerboe, *Anal. Chem. 44*, 1322 (1972).
574. F. L. Fricke, *Anal. Chem. 47*, 2018 (1975).
575. L. R. Layman and G. M. Hieftje, *Anal. Chem. 194* (1975).
576. C. S. Ling and R. D. Sacks, *Anal. Chem. 194*, 2074 (1975).
577. R. H. Scott, *Anal. Chem. 46*, 75 (1974).
578. W. R. Seitz, *Environ. Protection Technol. Ser. EPA-660/2-73-009* (1974); *Chem. Abstr. 82*, 102946 (1975).
579. R. J. Watling, *Anal. Chim. Acta 75*, 281 (1975).
580. A. W. Helz and B. F. Scribner, *J. Res. Natl. Bur. Std. 38*, 439 (1947).
581. C. W. Key and G. D. Hoggan, *Anal. Chem. 24*, 1921 (1952).
582. V. M. LeRoy and A. J. Lincoln, *Anal. Chem. 46*, 369 (1974).
583. G. E. Heggan and L. W. Strock, *Anal. Chem. 25*, 859 (1953).
584. R. F. Gould, ed., *Trace Inorganics in Water*, Advan. Chem. Ser. 73, Amer. Chem. Soc., Washington, D.C. (1968).
585. *Atomic Absorption Spectroscopy*, Amer. Soc. Testing Mater. Spec. Tech. Publ. 443, Philadelphia (1968).
586. J. P. Riley and D. Taylor, *Anal. Chem. Acta 40*, 479 (1968).
587. D. C. Burrell, *At. Absorption Newsletter 7*, 65 (1968).
588. K. Kuwata, K. Hisatomi, and T. Hasegawa, *At. Absorption Newsletter 10*, 111 (1971).
589. L. H. Windom and R. G. Smith, Deep-Sea Res. *19*, 727 (1972).
590. R. M. Burd, *Anal. Instru. 4*, 199 (1966).
591. C. G. Farnsworth, *Water & Sewage Works 119*, 52 (1972).
592. J. A. Platte and V. M. Marcy, *At. Absorption Newsletter 4*, 289 (1965).
593. P. E. Paus, *At. Absorption Newsletter 10*, 69 (1971).
594. C. R. Parker, Water Analysis by AA Spectroscopy, Varian-Techtron brochure, Springvale, Australia (1972).
595. O. K. Galle, *Appl. Spectrosc. 25*, 664 (1971).
596. R. E. Mansell and H. W. Emmel, *At. Absorption Newsletter 4*, 365 (1965).
597. International Symposium on Identification and Measurement of Environmental Pollutants, National Research Council of Canada (1971).
598. T. Surles, J. R. Tuschall, and T. T. Collins, *Environ. Sci. Technol. 9*, 1073 (1975).
599. K. M. Aldous, D. G. Mitchell, and K. W. Jackson, *Anal. Chem. 47*, 1034 (1975).
600. W. M. Edmunds, D. R. Giddings, and M. Morgan-Jones, *At. Absorption Newsletter 12*, 45 (1973).
601. F. J. Fernandez and D. C. Manning, *At. Absorption Newsletter 10*, 65 (1971).
602. T. H. Donnelly, J. Ferguson, and A. J. Eccleston, *Appl. Spectrosc. 29*, 158 (1975).
603. G. M. Shkolnik and D. E. Shrader, *Water Resource Instr. Int. Semin. Expo. 1*, 169 (1975).
604. C. J. Pickford and G. Rossi, *Analyst 97*, 647 (1972).
605. A. M. Ure, *Anal. Chim. Acta 76*, 1 (1975).
606. Y. Kimura and V. L. Miller, *Anal. Chim. Acta. 27*, 325 (1962).
607. D. C. Manning, *At. Absorption Newsletter 9*, 97 (1970).
608. W. R. Hatch and W. L. Ott, *Anal. Chem. 40*, 2085 (1968).
609. G. Lindstedt, *Analyst 95*, 264 (1970).
610. J. Y. Hwang, P. A. Ullucci, C. J. Mokeler, and S. B. Smith, *Amer. Lab. 5*, 43 (1973).
611. D. R. Demers, P. Heneage, and M. Gardels, Pittsburgh Conference on Analytical Chemistry and Applied Spectroscopy, 1971.
612. P. W. West, P. Folse, and D. Montgomery, *Anal. Chem. 22*, 667 (1950).
613. R. K. Scott, V. M. Marcy, and J. J. Hronas, *Symposium on Flame Photometry*, Amer. Soc. Testing Mater. Spec. Tech. Publ. 116, p. 105, Philadelphia (1951).

614. M. W. Skougstad and C. A. Horr, *U.S. Geol. Surv. Water Supply Paper 1456D*, 55 (1963).
615. C. A. Horr, "A Survey of Analytical Methods for the Determination of Strontium in Natural Waters," *U.S. Geol. Surv. Water Supply Paper 1486-A*, 1 (1959).
616. C. A. Horr, *U.S. Geol. Surv. Water Supply Paper 1496C*, 33 (1962).
617. A. R. Agg, N. T. Mitchell, and G. E. Mitchell, *Inst. Sewage Purification J. Proc. 1961*, 240–245; *Chem. Abstr. 55*, 25110 (1961).
618. T. R. Folsom, N. Hansen, G. J. Parks, and W. E. Weitz, *Appl. Spectrosc. 28*, 345 (1974).
619. G. E. Marsh, *Appl. Spectrosc. 10*, 8 (1956).
620. M. J. Prager and W. R. Seitz, *Anal. Chem. 47*, 148 (1975).
621. M. J. Prager, *Opt. Spectra 5* (8), 28 (1971).
622. C. R. Belcher, S. L. Bogdanski, D. L. Knowles, and A. Townshend, *Anal. Chim. Acta 77*, 53 (1975).
623. S. A. Ghonaim, *Proc. Soc. Anal. Chem. 11*, 138 (1974).
624. M. J. Milano, H. L. Pardue, T. E. Cook, R. E. Santini, D. W. Margerum, and J. M. T. Raycheba, *Anal. Chem. 46*, 374 (1974).
625. D. M. Gullikson, W. H. Caraway, and G. L. Gates, *U.S. Bur. Mines Rept. Invest. 5737* (1961).
626. G. L. Gates and W. H. Caraway, *U.S. Bur. Mines Rept. Invest. 6602* (1965).
627. A. G. Collins, *U.S. Bur. Mines Rept. Invest. 6047* (1962); *Chem. Abstr. 57*, 16337 (1962).
628. A. G. Collins, *Producers Monthly 26*, 22 (1962); *Chem. Abstr. 58*, 391 (1963).
629. A. G. Collins, *Producers Monthly 26*, 28 (1962); *Chem. Abstr. 59*, 354 (1963).
630. A. G. Collins, *Producers Monthly 27*, 2 (1963); *Chem. Abstr. 59*, 354 (1963).
631. A. G. Collins, *Anal. Chem. 35*, 1258 (1963).
632. A. G. Collins, *Air Water Pollution 9*, 145 (1965); *Chem. Abstr. 63*, 5415 (1965).
633. A. G. Collins, *U.S. Bur. Mines Rept. Invest. 6641* (1965); *Chem. Abstr. 63*, 12874 (1965).
634. R. G. Mihram and K. A. Catto, Jr., *Oil Gas. J. 59*, 126 (1961).
635. T. S. Bissett, J. Y. Hwang, and W. H. Hahn, Jr., Pittsburgh Conference on Analytical Chemistry and Applied Spectroscopy, 1971.
636. T. J. Chow, *Anal. Chim. Acta 31*, 58 (1964).
637. T. J. Chow and T. G. Thompson, *Anal. Chem. 27*, 910 (1955).
638. T. J. Chow and T. G. Thompson, *Anal. Chem. 27*, 18 (1955).
639. G. Joensson, *U.S. At. Energy Comm. AE-105*, 9 (1963).
640. J. P. Riley and M. Tongudai, *Deep-Sea Res. Oceanogr. Abstr. 11*, 563 (1964).
641. M. Honma, *Anal. Chem. 27*, 1656 (1955).
642. E. E. Pickett and S. R. Koirtyohann, *Anal. Chem. 41*, 28A (1969).
643. V. A. Fassel, R. W. Slack, and R. N. Kniseley, *Anal. Chem. 43*, 186 (1971).
644. B. E. Buell, 1972 Pacific Conference on Chemistry and Spectroscopy.

References to Section 2.9.2

645. A. Mizuiki, in: *Trace Analysis: Physical Methods* (G. H. Morrison, ed.), Wiley-Interscience, New York (1965), Chap. 4.
646. J. M. Rottschafer, R. J. Boczkowski, and H. B. Mark, Jr., *Talanta 19*, 163 (1972).
647. C. G. Farnsworth, *Water & Sewage Works 119*, 52 (1972).
648. J. F. Kopp and R. C. Kroner, *Appl. Spectrosc. 19*, 155 (1965).
649. E. F. Joy, N. A. Kershner, and A. J. Bernard, Jr., *Spex Speaker 16*, 1 (1971).
650. V. M. LeRoy and A. J. Lincoln, *Anal. Chem. 46*, 469 (1974).

651. M. Fred, N. H. Nachtrieb, and F. S. Tomkins, *J. Opt. Soc. Amer. 37*, 279 (1947).
652. J. A. Goleb, J. P. Faris, and B. H. Meng, *Appl. Spectrosc. 16*, 9 (1962).
653. J. M. Morris and F. X. Pink, *Amer. Soc. Testing Mater. Spec. Tech. Publ. 221*, 39 (1957).
654. A. K. De, S. M. Khopkar, and R. A. Chalmers, *Solvent Extraction of Metals*, Reinhold Publishing Corporation, New York (1970).
655. G. H. Morrison and H. Frieser, *Solvent Extraction in Analytical Chemistry*, John Wiley & Sons, Inc., New York (1957).
656. J. Stary, *The Solvent Extraction of Metal Chelates*, Pergamon Press, Inc., Elmsford, N.Y. (1964).
657. H. Frieser, *Chemist-Analyst 50*, 62 (1961).
658. H. Frieser, *Chemist-Analyst 50*, 94 (1961).
659. H. Frieser, *Chemist-Analyst 51*, 62 (1962).
660. H. Frieser, *Chemist-Analyst 52*, 55 (1963).
661. S. A. Katz, *Amer. Lab. 4*, 19 (1972).
662. S. L. Sachdev and P. W. West, *Environ. Sci. Technol. 4*, 749 (1970).
663. E. A. Kirichenko, A. V. Fokin, and S. M. Ivanova, *Tr. Mosk. Khim. Tekhnol. Inst. 3*, 259 (1969); *Chem. Abstr. 75*, 91154r (1971).
664. J. A. Dean, *Flame Photometry*, McGraw-Hill Book Company, New York (1960).
665. C. M. Stander, *Anal. Chem. 32*, 1296 (1960).
666. B. E. Buell, Pacific Conference on Chemistry and Spectroscopy, 1972.
667. O. G. Jeffery and G. O. Kerr, *Analyst 92*, 763 (1967).
668. H. Goto and E. Sudo, *Japan Analyst 10*, 1213 (1961) for V; pp. 171, 175, 456, 463 for other elements.
669. C. E. Mulford, *At. Absorption Newsletter 5*, 88 (1966).
670. R. R. Brooks, B. J. Presley, and I. R. Kaplan, *Talanta 14*, 809 (1967).
671. M. J. Orren, *J. S. African Chem. Inst. 24*, 96 (1971); *Chem. Abstr. 75*, 52642d (1971).
672. R. E. Mansell and H. W. Emmel, *At. Absorption Newsletter 4*, 365 (1965).
673. M. J. Fishman, *At. Absorption Newsletter 11*, 46 (1972).
674. M. Yanagisawa, M. Suzuki, and T. Takeuchi, *Talanta 14*, 933 (1967).
675. R. Woodruff, B. R. Culver, D. Shrader, and A. B. Super, *Anal. Chem. 45*, 230 (1973).
676. Y. K. Chau and K. Lum-Shue-Chon, *Anal. Chim. Acta. 5*, 201 (1970).
677. H. J. Crump-Weisner and H. R. Feltz, *Anal. Chim. Acta 55*, 29 (1971).
678. D. C. G. Pearton, J. D. Taylor, P. K. Faure, and T. W. Steele, *Anal. Chim. Acta 44*, 353 (1969).
679. D. Satyanarayana, N. Kurmajah, and V. P. R. Rao, *Chemist-Analyst 53*, 78 (1964).
680. R. J. Jacubiec and D. F. Boltz, *Anal. Letters 1*, 347 (1968).
681. H. N. Johnson, G. K. Kirkbright, and T. S. West, *Analyst 97*, 696 (1972).
682. G. K. Kirkbright, A. M. Smith, and T. S. West, *Analyst 92*, 411 (1967).
683. T. V. Ramakrishna, J. W. Robinson, and P. T. West, *Anal. Chim. Acta 45*, 43 (1969).
684. R. S. Danchik and D. F. Boltz, *Anal. Letters 1*, 901 (1968).
685. J. C. M. Pau, E. E. Pickett, and S. R. Koirtyohann, *Analyst 97*, 860 (1972).
686. W. Holak, *Anal. Chem. 54*, 1138 (1971).
687. E. J. Agazzi, *Anal. Chem. 39*, 233 (1969).
688. F. L. Moore, *Environ. Sci. Technol. 6*, 525 (1972).
689. M. D. Morris and L. R. Whitlock, *Anal. Chem. 39*, 1180 (1967).
690. H. A. Flaschka, R. Barnes, and D. Paschal, *Anal. Letters 5*, 253 (1972).
691. J. P. Riley and G. Toppin, *Anal. Chem. Acta. 44*, 234 (1969).
692. O. K. Galle, *Appl. Spectrosc. 25*, 664, (1971).
693. D. G. Biechler, *Anal. Chem. 37*, 1054 (1965).
694. T. W. Freudiger and C. T. Kenner, *Appl. Spectrosc. 26*(2), 302 (1972).

695. J. P. Riley and D. Taylor, *Deep-Sea Res. 19*, 307 (1972).
696. J. P. Riley and D. Taylor, *Anal. Chim. Acta. 40*, 479 (1968).
697. H. H. Le Riche, *Geochim. Acta. 32*, 791 (1968).
698. J. A. Brabson and W. D. Wilhide, *Anal. Chem. 26*, 1060 (1954).
699. W. H. Hinson, *Spectrochim. Acta. 18*, 427 (1962).
700. Reeves Angel, *Matter of Facts, 2*(3), 6 (1969).
701. O. Samuelson, *Ion Exchange Separation in Analytical Chemistry*, John Wiley & Sons, Inc., New York (1963).
702. H. Green, *Talanta 20*, 139 (1973). A review with 119 references.
703. W. D. Silvey, *U.S. Geol. Surv. Water Supply Paper 154-B*, 1961.
704. J. A. Buono, J. C. Buono, and J. L. Fasching, *Anal. Chem. 47*, 1926 (1975).
705. B. M. Vanderborght and R. E. Van Grieken, *Anal. Chem. 49*, 311 (1971).
706. J. I. Hoffman, *Chemist-Analyst 50*, 30 (1961).
707. J. I. Hoffman, *Chemist-Analyst 49*, 94, (1960).
708. K. L. Cheng, *Chemist-Analyst 50*, 126 (1961).
709. J. I. Hoffman, *Chemist-Analyst 49*, 126 (1960).
710. R. O. Scott and R. L. Mitchell, *J. Soc. Chem. Ind. 62*, 48 (1943).
711. R. L. Mitchell and R. O. Scott, *Spectrochim. Acta 3*, 367 (1948).
712. R. L. Mitchell and R. O. Scott, *Appl. Spectrosc. 11*, 6 (1957).
713. R. L. Mitchell, in: *Trace Analysis* J. H. Yoe and H. J. Koch, Jr., eds.), John Wiley & Sons, Inc., New York (1957).
714. M. C. Farquar, J. A. Hill, and M. M. English, *Anal. Chem. 38*, 208 (1966).
715. G. E. Heggan and L. W. Strock, *Anal. Chem. 25*, 859 (1953).
716. W. H. Durum and J. Haffty, *U.S. Geol. Surv. Cir. 445* (1961).
717. W. D. Silvey and R. Brennan, *Anal. Chem. 34*, 784 (1962).
718. J. W. Husler and E. F. Cruft, *Anal. Chem. 41*, 1688 (1969).
719. E. F. Cruft and J. W. Husler, *Anal. Chem. 41*, 175 (1969).
720. G. B. Marshall and T. S. West, *Talanta 14*, 823 (1967).
721. R. Ko and P. Anderson, *Anal. Chem. 41*, 177 (1969).
722. B. C. Severne and R. R. Brooks, *Anal. Chim. Acta 58*, (1972).
723. K. V. Krishnamurty and M. M. Reddy, *Anal. Chem. 49*, 222 (1977).
724. P. Strohal, K. Molnar, and I. Bacic, *Mikrochim. Acta. 1972*, 586.
725. M. Ishibashi and T. Shigematsu, *Bull. Inst. Chem. Res. Kyoto Univ. 23*, 59 (1950).
726. E. B. Sandell, *Colorimetric Determination of Trace Metals*, Wiley–Interscience, New York (1959), pp. 388–389.
727. R. Fukai, *Nature 213*, 901 (1967).

References to Section 2.9.3

728. W. E. Garwood, W. I. Denton, R. B. Bishop, S. J. Lukasiewicz, and J. N. Maile, *Ind. Eng. Chem. 51*, 1377 (1959).
729. J. E. H. Barney and J. Haight, Jr., *Anal. Chem. 27*. 1285 (1955).
730. G. Constantinides and G. Arich, *6th World Petrol. Congr. Sect. V, 11*, 1 (1963).
731. R. A. Dean and E. V. Whitehead, *6th World Petrol. Congr. Sect. V, 9*, 1 (1963).
732. G. W. Hodgson, B. L. Baker, and E. Peake, *7th World Petrol Congr. 9*, 117 (1967).
733. H. Bieber, H. M. Hartzband, and E. C. Kause, *J. Chem. Eng. Data 5*, 540 (1960).
734. O. A. Larson and H. Beuther, *Amer. Chem. Soc. Div. Petrol. Chem. Preprints* (1966), p. B-95.
735. H. J. Hyden, *U.S. Organic Chemistry Geol. Surv. Bull. 1100*, 1961.
736. H. N. Dunning, in: *Organic Geochemistry* (I. A. Breger, ed.), Pergamon Press, Inc., Elmsford, N.Y. (1963), Chap. 16.

737. J. M. Sugihara, *Proc. Amer. Petrol. Inst. Sect. VIII, 42*, 30 (1962).
738. W. L. Nelson, *Oil Gas J. 74*(45), 191 (1976).
739. J. M. Sugihara, J. F. Branthaven, G. Y. Wu, and C. Weatherbee, *Amer. Chem. Soc. Div. Petrol. Chem. Preprints* (1970), p. C-5.
740. D. S. Skinner, *Ind. Eng. Chem. 44*, 1159 (1952).
741. R. H. Filby, Trace Elements in Petroleum, Ph.D. thesis, Washington State University (1971).
742. K. R. Shah, R. H. Filby, and W. A. Heller, *J. Radioanal. Chem. 6*, 185 (1970).
743. K. R. Shah, R. H. Filby, and W. A. Heller, *J. Radioanal. Chem. 6*, 413 (1970).
744. R. H. Filby, in: *The Role of Trace Elements in Petroleum*, (T. F. Yen, ed.), Ann Arbor Science Publishers, Inc., Ann Arbor, Mich. (1975), Chap. 2.
745. B. Hitchon, R. H. Filby, and K. R. Shah, *The Role of Trace Elements in Petroleum* (T. F. Yen, ed.), Ann Arbor Science Publishers, Ann Arbor, Mich. (1975), p. 111.
746. T. F. Yen, J. G. Erdman and A. J. Saraceno, *Amer. Chem. Soc. Div. Petrol. Chem. Preprints 6*(3), B53 (1963).
747. J. G. Erdman and P. H. Harju, *Amer. Chem. Soc. Div. Petrol. Chem. Gen. Papers* 7(1), 43 (1962).
748. J. W. Dodd, J. W. Moore, and M. O. Denekas, *Ind. Eng. Chem. 44*, 2000 (1954).
749. J. S. Ball, W. J. Wenger, J. H. Hyden, C. A. Horr, and A. T. Myers, *J. Chem. Eng. Data 5*, 553 (1960).
750. J. W. Moore and H. N. Dunning, *Ind. Eng. Chem. 47*, 1440 (1955).
751. W. W. Howe and A. R. Williams, *J. Chem. Eng. Data 5*, 106 (1960).
752. A. J. Saraceno, D. T. Fanale, and N. D. Coggeshall, *Anal. Chem. 33*, 500 (1961).
753. F. E. Dickson and L. Petrakis, *Anal. Chem. 46*, 1129 (1974).
754. T. F. Yen, Chap. 1, and R. H. Filby, Chap. 2, in: *The Role of Trace Elements in Petroleum* (T. F. Yen, ed.), Ann Arbor Science Publishers, Inc., Ann Arbor, Mich. (1975).
755. J. W. Moore and H. N. Dunning, *U.S. Bur. Mines Rept. Invest. 5370*, (1957).
756. H. N. Dunning and J. W. Moore, *Petrol. Refiner 36*(5) 247 (1957).
757. K. H. Altgelt, *Amer. Chem. Soc. Div. Petrol. Chem. Gen. Papers 10*(3), 29 (1965).
758. I. A. Eldib, H. N. Dunning, and R. J. Bolen, *J. Chem. Eng. Data 5*, 550 (1960).
759. B. R. Ray, P. A. Witherspoon, and R. E. Grim, *J. Phys. Chem. 61*, 1296 (1957).
760. A. Triebs, *Ann. Chem. 510*, 42 (1934).
761. G. W. Hodgson, B. L. Baker, and E. Peake, in *Fundamental Aspects of Petroleum Geochemistry* (B. Nagy and U. Colombo, eds.), American Elsevier Publishing Company, Inc., New York (1967).
762. H. N. Dunning, J. W. Moore, H. Bieber, and R. B. Williams, *J. Chem. Eng. Data 5*, 546 (1960).
763. J. M. Sugihara and R. M. Bean, *J. Chem. Eng. Data 7*, 269 (1962).
764. A. A. Wolsky and F. W. Chapman, Jr., *Proc. Amer. Petrol. Inst. Sect. III 40*, 423 (1960).
765. S. Groennings, *Anal. Chem. 25*, 938 (1953).
766. J. M. Sugihara and R. G. Garvey, *Anal. Chem. 36*, 2374 (1964).
767. H. N. Dunning, J. W. Moore, and A. T. Myers, *Ind. Eng. Chem. 46*, 2000 (1954).
768. G. Constantinides, G. Arich, and C. Lomi, *5th World Petrol. Congr. Sect. V* (1959), paper 11.
769. C. N. Kinherler, Jr., U. S. Patent 3,203,892; *Chem. Abstr. 63*, 12948H (1965).
770. R. A. Dean and R. B. Girdler, *Chem. Ind. (London) 1960*, 100.
771. J. G. Erdman, U. S. Patent 3,190,829 (1965).
772. G. W. Hodgson, E. Peterson, and B. L. Baker, *Mikrochim. Acta. 1969*, 805.
773. E. W. Baker, *J. Amer. Chem. Soc. 88*, 2311 (1966).
774. J. M. Sugihara, J. F. Branthaver, and K. W. Wilcox, in *The Role of Trace Elements*

in Petroleum, (T. F. Yen, ed.), Ann Arbor Science Publishers, Inc., Ann Arbor, Mich. (1975), p. 183.
775. P. A. Witherspoon and R. S. Winniford, in *Fundamental Aspects of Petroleum Geochemistry* (B. Nagy and U. Colombo, eds.), American Elsevier Publishing Company, Inc., New York (1967), Chap. 6.
776. G. W. Hodgson, E. Peake, and B. L. Baker, *Res. Council Alberta Inform. Ser. 45*, 75 (1963).
777. L. K. Beach and J. E. Shewmaker, *Ind. Eng. Chem. 49*, 1157 (1957).
778. M. F. Millson, D. S. Montgomery, and S. R. Brown, *Geochim. Cosmochim. Acta 30*, 207 (1966).
779. F. H. Garner, S. J. Green, and F. D. Harper, *J. Inst. Petrol. 39*, 278 (1953).
780. V. Berti, A. M. Ilardi, and M. Nuzzi, *7th World Petrol. Congr. 9*, 149 (1967).
781. T. F. Yen, L. J. Boucher, J. P. Dickie, E. C. Tynan, and G. B. Vaughan, *J. Inst. Petrol. 55*, 87 (1969).
782. H. N. Dunning, J. W. Moore and M. O. Denekas, *Ind. Eng. Chem. 45*, 1759 (1953).
783. W. Howe, *Anal. Chem. 33*, 255 (1961).
784. H. N. Dunning and N. A. Rabon, *Ind. Eng. Chem. 42*, 2000 (1954).
785. J. F. Branthaver, G. Y. Wu, and J. M. Sugihara, *Amer. Chem. Soc. Div. Petrol. Chem. Preprints 12*(2), A73 (1967).
786. A. H. Corwin and E. W. Baker, *Amer. Chem. Soc. Div. Petrol. Chem. Preprints 9*(1), 19 (1964).
787. D. W. Thomas and M. Blumer, *Geochim. Cosmochim. Acta. 28*, 1147 (1964).
788. Union Oil report ARS72-240M, 1972.
789. H. H. Oelert, D. R. Latham, and W. E. Haines, *Separation Sci. 5*, 657 (1970).
790. J. H. Weber and H. H. Oelert, *Separation Sci. 5*, 669 (1970).
791. H. Blumer and W. D. Snyder, *Chem. Geol. 2*, 35 (1967).
792. H. P. Pohlmann and R. J. Rosscup, *Amer. Chem. Soc. Div. Petrol. Chem Preprints 12*(2) A103 (1967).
793. H. J. Coleman, D. E. Hirsch, and J. E. Dooley, *Anal. Chem. 41*, 800 (1969).
794. T. E. Cogswell, J. F. McKay, and D. R. Latham, *Anal. Chem. 43*, 645 (1971).
795. W. B. Shirey, *Ind. Eng. Chem. 23*, 1151 (1931).
796. W. H. Thomas, in: *The Science of Petroleum*, Vol. 2, Oxford University Press, London (1938).
797. W. H. Thomas, *Inst. Petrol. Technol. 10*, 216 (1924).
798. S. H. Southwick, Inorganic Constituents of Crude Oil, Ph.D. thesis, Massachusetts Institute of Technology (1951).
799. W. L. Whitehead and I. A. Breger, in *Organic Geochemistry*, (I. A. Breger, ed.), Pergamon Press, Inc., Elmsford, N.Y., (1963), Chap. 7.
800. P. A. Witherspoon and K. Nagashima, *Illinois Geol. Surv. Circ. 239* (1957).
801. G. W. Hodgson and B. L. Baker, *Amer. Assoc. Petrol. Geol. Bull. 41*, 2413 (1957).
802. M. J. Murray and H. A. Plagge, *Proc. Amer. Petrol. Inst. Sect. III, 29M*, 84 (1949).
803. O. I. Milner, J. R. Glass, J. P. Kirchner, and A. N. Yurick, *Anal. Chem. 24*, 1728 (1952).
804. L. C. Bonham, *Amer. Assoc. Petrol. Geol. Bull. 40*, 897 (1956).
805. L. W. Strock, *Spectrum Analysis with the Carbon Arc Cathode Layer*, Adam Hilger Ltd., London (1936).
806. R. F. Mast, N. F. Shimp, and P. A. Witherspoon, *Illinois Geol. Surv. Circ. 421* (1968).
807. K. Nagashima and J. S. Machin, *Illinois Geol. Surv. Cir. 235* (1957).
808. J. Hansen and C. R. Hodkins, *Anal. Chem. 30*, 368 (1958).
809. R. L. Mitchell and R. O. Scott, *Appl. Spectrosc. 11*, 6 (1957).
810. N. F. Shimp, J. Conner, A. L. Prince, and F. E. Bear, *Soil Sci. 83*, 51 (1957).

811. R. L. Erickson, A. T. Myers, and C. A. Horr, *Amer. Assoc. Petrol. Geol. Bull. 38*, 2200 (1954).
812. M. C. K. Jones and R. L. Hardy, *Ind. Eng. Chem. 24*, 742 (1952).
813. U. Colombo and G. Sironi, *5th World Petrol. Congr., Sect. I* (1959), paper 10, p. 177.
814. L. W. Gamble and W. H. Jones, *Anal. Chem. 27*, 1456 (1955).
815. U. P. Colombo, G. Sironi, G. B. Fasola, and R. Malvano, *Anal. Chem. 36*, 805 (1964).
816. J. H. Karchmer and E. L. Gunn, *Anal. Chem. 24*, 1733 (1952).
817. E. L. Gunn and J. M. Powers, *Anal. Chem. 24* 742 (1952).
818. E. H. Bailey, P. D. Snaveley, Jr., and D. E. White, *U.S. Geol. Surv. Profess. Paper 424D*, D306 (1961).
819. *Amer. Petrol. Inst. Res. Project 60*, Annual Report No. 12, July 1, 1971 to June 30, 1972.
820. G. Sebor, P. Vavrecka, V. Sychra, and O. Weisser, *Anal. Chim. Acta 78*, 99 (1975).
821. A. J. Smith, J. O. Rice, W. C. Sharer, Jr., and C. C. Cerato, in *The Role of Trace Elements in Petroleum*, (T. F. Yen, ed.), Ann Arbor Science Publishers, Inc., Ann Arbor, Mich. (1975), Chap. 8.
822. C. J. Perry and D. A. Keyworth, *Can. J. Spectrosc. 12*, 47 (1967).
823. A. Manjarrez and C. Pereda, *Rev. Soc. Quim. Mex. 13*, 53A (1969).
824. P. Vavrecka, G. Sebor, *et al.*, *Riv. Combust.*, *29* (no. 9), 375 (1975) (abstract).
825. R. J. Ibrahim and S. Sabbah, *At. Absorption Newsletter 14*, 131 (1975).
826. J. F. Alder and T. S. West, *Anal. Chim. Acta 61*, 132 (1972).
827. M. P. Bratzel, Jr., and C. L. Chakrabarti, *Anal. Chim. Acta 61* 25 (1972).
828. S. H. Omang, *Anal. Chim. Acta 56*, 470 (1971).
829. Y. E. Araktingi, C. L. Chakrabarti, and I. S. Maines, *Spectrosc. Letters 7*, 97 (1974).
830. G. Hall, M. P. Bratzel, and C. L. Chakrabarti, *Talanta 20*, 755 (1973).
831. E. M. Magee, H. J. Hall, and G. M. Varga, Jr., Potential Pollutants in Fossil Fuels, EPA-R2-73249, Office of Research and Monitoring, U.S. Environmental Protection Agency, Washington, D.C. (June 1973).
832. M. E. Hinkle, *U.S. Geol. Surv. Profess. Paper 750B*, B171 (1971).
833. M. E. Hinkle and R. E. Learned, *U.S. Geol. Surv. Profess. Paper 650D*, D251 (1970).
834. M. Abu-Elgneit, 169th Amer. Chem. Soc. Natl. Meeting, *Amer. Chem. Soc. Div. Petrol. Chem. Preprints 20*, 35 (February 1975).
835. W. K. T. Gleim, J. T. Gatsis, and C. J. Perry, in *The Role of Trace Elements in Petroleum*, (T. F. Yen, ed.), Ann Arbor Science Publishers, Inc., Ann Arbor, Mich., (1975), Chap. 9.
836. R. A. Hofstader, O. I. Milner, and J. H. Runnels, eds., *Analysis of Petroleum for Trace Metals*, Advan. Chem. Ser. 156, American Chemical Society, Washington, D.C. (1976).
837. J. H. Runnels, R. Merryfield, and H. B. Fisher, *Anal. Chem. 47*, 1258 (1975).
838. H. E. Knauer and G. E. Milliman, *Anal. Chem. 47*, 1263 (1975).
839. W. K. Robbins and H. H. Walker, *Anal. Chem. 47*, 1269 (1975).
840. H. H. Walker, J. H. Runnels and R. Merryfield, *Anal. Chem. 48*, 2065 (1976).
841. W. K. Robbins, *Anal. Chem. 46*, 2177 (1974).
842. G. F. Kirkbright, M. Marshall, and T. S. West, *Anal. Chem. 44*, 2379 (1972).
843. M. Sarkovic, *Nafta (Zagreb) 20*, 479 (1969).
844. O. I. Milner, *Analysis of Petroleum for Trace Elements*, Pergamon Press, Inc., Elmsford, N.Y. 1963.
845. W. Schuknecht and H. Schinkel, *Brennstoff-Chem. 42*, 292 (1961).
846. ASTM Method D1318; Deutsche Normen, DIN 51797, *Erdoel Kohle 14*, 942 (1961).
847. O. Popescu, E. Florescu, and W. Popescu, *Petrol. Gaze (Bucharest) 16*, 468 (1965).
848. K. Augsten, *Freiberger Forschungsh. A380*, 9 (1966).

849. K. Augsten, *5th Schnierstoffe U. Schmierungstech. Symp., Chemnitz* (1963).
850. R. J. Heemstra and N. G. Foster, *Anal. Chem. 38*, 492 (1966).
851. B. E. Buell, Flame atomic spectrometric determination of Ni and V directly in shale oils, crude oils, crude oil distillates and residuals; submitted for publication.
852. L. G. Mashireva, S. B. Sorakina, and V. A. Korovin, *Chem. Technol. Fuels Oils 11*, 68 (1975); translated from *Khim. i Tekhnol. Topliv i Masel 11*, 55 (1975).
853. V. A. Korovin, E. F. Kotenko, L. G. Mashireva, and Z. T. Yunusov, Zavodsk. Lab. *41*, 1093 (1975).
854. J. A. Burrows, J. C. Heerdt, and J. B. Willis, *Anal. Chem. 37*, 148 (1965).
855. J. F. P. Lush, in: *Recent Analytical Developments in the Petroleum Industry*, (D.R. Hodges, ed.), Halsted Press, New York (1974), Chap. 11.
856. W. L. Nelson, *Oil Gas J. 56*(51), 75 (1958).
857. W. L. Nelson, *Oil Gas J. 57*, 102 (1959).
858. W. L. Nelson, *Oil Gas J. 64*, 127 (1966).
859. W. L. Nelson, *Oil Gas J. 69*, 111 (1971).
860. W. L. Nelson, *Oil Gas J. 70*, 49 (1972).
861. W. L. Nelson, *Oil Gas J. 74*: series of eight articles on "Guide to World Crudes" from nos. 13, 98 (March 29) through 27, 98 (July 5).
862. G. W. Hodgson and B. L. Baker, *Amer. Assoc. Petrol. Geol. Bull. 43*, 311 (1959).
863. G. W. Hodgson, *Amer. Assoc. Petrol. Geol. Bull. 38*, 2537 (1954).
864. B. L. Baker and G. W. Hodgson, *Amer. Assoc. Petrol. Geol. Bull. 43*, 472 (1959).
865. E. W. Baker, *J. Chem. Eng. Data 9*, 307 (1964).
866. C. T. Brown, *Petrol. Eng. 28*(1), 250 (1956).
867. J. V. Brunnock, D. F. Duckworth, and G. G. Stephens, *J. Inst. Petrol. 54*, 310 (1968).
868. E. R. Adlard, *J. Inst. Petrol. 58*, 63 (1972).
869. D. E. Bryan, V. P. Guinn, R. P. Hackelman, and H. R. Lukens, *U.S. At. Energy Comm. Rept GA-9889* (Jan. 21, 1970).
870. D. J. Veal, *Anal. Chem. 38*, 1080 (1966).
871. R. H. Filby and K. R. Shah, in *The Role of Trace Elements in Petroleum*, (T. F. Yen, ed.), Ann Arbor Science Publishers, Inc., Ann Arbor, Mich. (1975), p. 615.
872. R. F. Mast and R. R. Ruck, and W. F. Meents, *Illinois State Geol. Surv. Environ. Geol. Notes 65*, 1 (November 1973).
873. R. F. Mast, R. R. Ruch, and W. F. Meents, *Illinois State Geol. Surv. Circ. 483* (1973).
874. H. Al-Shahristani and M. J. Al-Atyia, *Geochim. Cosmochim. Acta 36*, 929 (1972).
875. G. Arich and G. Constantinides, *Riv. Combust. 14*, 695 (1960).
876. I. Lakatos, *Banyasz. Kohasz. Lapok Koolaj Foldgaz 5*, 97 (1972).
877. Z. Biernat and M. Solecki, *Biul. Inst. Naftawego 8*, 7 (1958).
878. F. M. Farhan and H. Pazandeh, *Analusis 3*, 201 (1975).
879. M. Cornu, *XX Congr. Group. Avan. Methodes Spectrog*. (1957), pp. 185–194.
880. E. Faulhaber and L. Liebetrau, *Erdoel Kohle 18*, 270 (1965).
881. E. Hohmann, K. Gottlieb, and U. Mueller, *Erdoel Kohle Erdgas Petrochem. Bunnst. Chem. 26*, 647 (1973).
882. Z. Gregorowicz and P. Orzechowski, *Nafta (Katowice) 14*, 106 (1958).
883. A. Serbanescu, *Petrol. Gaze (Bucharest) 16*, 219 (1965).
884. A. Serbanescu, *Petrol. Gaze (Bucharest) 11*, 408 (1960).
885. A. Serbanescu, *Bull. Inst. Petrol. Gaz Geol. 17*, 107 (1971); *Chem. Abstr. 79*, 106572j (1973).
886. A. N. Aleksandrov, M. I. Dement'eva, and Ya. E. Shmulyakovskii, eds., *Metody Issled. Produktov Neftepereabotki i Neftekhim. Sinteza*, Gostoptekhizdat, Leningrad (1962).
887. L. A. Gulyaeva and I. F. Lositskaya, *Geokhimiya 1959*, 152.

888. E. M. Poplavko, V. V. Ivanov, T. G. Karasik, A. D. Miller, V. S. Orekhov, S. D. Taliev, Yu. A. Tarkhov, and V. A. Fadeeva, *Geokhimiya 1974*(9), 1399 *Chem. Abstr. 82*, 61491a (1975).
889. P. Ya. Demenkova, L. N. Zakharenkova, A. P. Kurbatskaya, and M. M. Pastova, *Tr. VNIGRI 174*, 68, (1961).
890. P. Ya. Demenkova, L. N. Zakharenkova, and A. P. Kurbatskaya, *Tr. VNIGRI 117*, 186 (1958).
891. P. Ya. Demenkova and L. N. Zakharenkova, *Tr. VNIGRI 82*, 182 (1955).
892. S. M. Katchenkov, *Low Level Chemical Elements in Sedimentary Rocks and Petroleum*, Gostoptekhizdat, Leningrad (1959). A monograph containing a survey of metals in Russian crudes.
893. S. M. Katchenkov, *Dokl. Akad. Nauk SSSR 67*, 503 (1949).

References to Section 2.9.4

894. R. McBrian and L. E. Atchinson, *Soc. Automotive Engrs. J. 58*(7), 19 (1950).
895. R. McBrian, *J. Pacific Ry. Club, 32*(2), 6 (1948); see discussion of history in Ref. 22.
896. W. A. Cassidy, *Mech. Eng. 72*(10), 854 (1950).
897. F. M. Burt, *Diesel Progr. 1950*, 68 (April).
898. J. P. Pagliasotti and F. W. Porsche, *Anal. Chem. 23*, 198 (1951).
899. J. P. Pagliasotti and F. W. Porsche, *Anal. Chem. 23*, 1820 (1951).
900. L. E. Calkins and M. M. White, *Natl. Petrol. News 38*, R519 (1946).
901. R. G. Russell, *Anal. Chem. 20*, 296 (1948).
902. A. G. Gassman and W. R. O'Neill, *Anal. Chem. 21*, 417 (1949).
903. A. G. Gassman and W. R. O'Neill, *Proc. Amer. Petrol. Inst. Sect. III, 29M*, 79 (1949).
904. M. J. Murray and H. A. Plagge, *Proc. Amer. Petrol. Inst. Sect. III, 29M*, 84 (1949).
905. H. K. Hughes, J. W. Anderson, R. W. Murphy, and J. B. Rather, *Proc. Amer. Petrol. Inst. Sect. III, 29M*, 89 (1949).
906. D. R. Jackson, *Lubrication Eng. 28*, 76 (1972).
907. Symposium on Spectrographic Analysis of Diesel Engine Lubricating Oil, American Locomotive Co., Schenectady, N.Y., May 5–6, 1952.
908. Symposium on Diesel Locomotive Engine Maintenance–Including Spectrographic Analysis of Lubricating Oils, American Locomotive Co., Schenectady, N.Y., May 11–12, 1953.
909. J. Rigaux, *Le Contrôle des moteurs diesel par la spectrographie des huiles de graissage*, Dunod, Paris (1961).
910. Anon., *Modern Railroads 1953*, 233 (April).
911. J. T. Rozsa and L. E. Zeeb, *Petrol. Process. 8*, 1708 (1953).
912. J. M. Gillette, B. R. Boyd, and A. A. Shurkus, *Appl. Spectrosc. 8*, 162 (1954).
913. E. T. Myers, *Modern Railroads 1955* (November).
914. A. J. Mitteldorf, *Spex Speaker 3*(2) (1958). A review.
915. A. J, Mitteldorf, *Spex Speaker 13*(1) (1958). A review.
916. D. C. Kittinger and J. L. Ellis, *U.S. Clearinghouse Fed. Sci. Tech. Inform. AD-671115* (1968).
917. *U.S. Natl. Bur. Std. Tech. News Bull. 57*(6), 135 (1973).
918. A. Beerbower, *Lubrication Eng., 32*, 285 (1976).
919. J. V. Jolliff, *Naval Engrs. J. 1966*, 1073 (Dec.).
920. D. R. Barr and H. J. Larson, Jr., *Appl. Spectrosc. 26*, 51 (1972).
921. Anon., *Arcs and Sparks 4*(1) (1958). Labara-Story of the Month about the Pacific Intermountain Express Co. maintenance program using ESD oil analyses.

922. R. E. Linnard, C. B. Threlkeld, and R. T. Blades, *Trans. Amer. Soc. Mech. Engrs. 79*, 709 (1957).
923. Anon., *Petrol. Week* (July 3, 1959).
924. A. Esenwein and H. Preis, *Schweiz. Arch. Angew. Wiss. Tech. 38*, 359 (1972); *Chem. Abstr. 78*, 149480z (1973).
925. V. C. Barth, Direct-Reading Spectrographic Evaluation of Used Railroad Oils, paper 9, 2nd Pacific Area ASTM National Meeting, September 1956. See *Amer. Soc. Testing Mater. Spec. Tech. Publ. 214* (1957) for Symposium on Railroad Materials Lubricating Oils.
926. W. D. Perkins, J. R. Miller, and J. H. Moser, Direct Determination of Metals in Used Lubricating Oils by Emission Spectrography, paper 13, 2nd Pacific Area ASTM National Meeting, September 1956.
927. W. A. Rappold and R. E. Ramsay, "Direct-Reading Spectrometric Control of Lubricating Oil Additive Manufacture," paper 14, 2nd Pacific Area ASTM National Meeting, September 1956.
928. ASTM, Suggested Methods for Spectrochemical Analysis of Used Diesel Lubricating Oils, 2nd Pacific Area ASTM National Meeting, September 1956.
929. F. R. Bryan, Report of ASTM Committee D-2, Methods of Spectrographic Analysis of Lubricating Oils, 2nd Pacific Area ASTM National Meeting, September 1956.
930. ASTM Committee D-2, Sampling Diesel Locomotive Lubricating Oil for Spectrographic Analysis, *Amer. Soc. Testing Mater. Bull. 208*, 24 (1955).
931. S. K. Kyuregyan, *Emission Spectral Analysis of Petroleum Products*, Izdatelstvo Khimiya, Moscow (1969); *Chem. Abstr. 74*, 5285j (1971).
932. S. K. Kyuregyan, *Determination of Internal Combusion Engine Wear by Spectral Analysis,* Izdatelstvo Mashinostroenie, Moscow (1966); *Chem. Abstr. 68*, 35621j (1968).
933. S. K. Kyuregyan and M. M. Marenova, *Khim. i Tekhnol. Topliv. i Masel 12*, 57 (1967); *Chem. Abstr. 67*, 101630U (1967).
934. S. K. Kyuregyan, K. N. Golovanov, A. I. Turkevich, and M. M. Marenova, *16th Prikl. Spektrosk. Mater. Soveshch.*, 1965 (publ. 1969), Vol. 1, pp. 239–242; *Chem. Abstr. 72*, 68944c (1970).
935. A. N. Aleksandrov, M. I. Dement'eva, and Ya. E. Shmulyakovskii, eds., *Metody Issled, Produktov Neftepereabotki i Neftekhim. Sinteza*, Gostoptekhizdat, Leningrad (1962).
936. E. V. Il'ina, *Zavodsk. Lab. 29*, 1317 (1963).
937. J. W. McCoy, *The Inorganic Analysis of Petroleum*, Chemical Publishing Company, New York (1962).
938. O. I. Milner, *Analysis of Petroleum for Trace Elements*, Pergamon Press, Inc., Elmsford, N.Y. (1963).
939. M. Pinta, *Detection and Determination of Trace Elements*, translated from French, Ann Arbor Science Publishers, Ann Arbor, Mich. (1971); originally publ. 1962).
940. J. Noar, *Inst. Petrol. Rev. 9*, 187, 209 (1955).
941. E. L. Gunn, *Anal. Chem. 32*, 1449 (1960).
942. R. A. Cummins, and P. R. Mason, *J. Inst. Petrol. 48*, 237 (1962).
943. P. J. Coulter, N. L. Bottone, and H. W. Leggon, *Develop. Appl. Spectrosc. 10*, 293 (1972).
944. D. L. Fry, *Appl. Spectrosc. 10*, 65 (1956).
945. G. S. Golden, *Appl. Spectrosc. 25*, 668 (1971).
946. H. Hauptmann and G. Jager, *Schmierungstechnik 4*, 235 (1973).
947. H. P. Woods, Jr., *Appl. Spectrosc. 27*, 490 (1973).
948. E. Jantzen, *Deut. Luft. Reumfabn. Forschugsber. 73*, (1973); *Chem. Abstr. 78*, 149483c (1973).

949. I. Lakatos, *Magy. Kem. Folyoirat 78*, 624 (1972); *Chem. Abstr. 78*, 74389r (1973).
950. V. C. Westcott and W. W. Seifert, *Wear 21*, 27 (1972).
951. B. D. Pickles and C. C. Washbrook, *Proc. Soc. Anal. Chem. 7*, 13 (1970).
952. T. H. Zink, *Appl. Spectrosc. 13*, 94 (1959).
953. R. J. McGowan, *Appl. Spectrosc. 15*, 179 (1961).
954. C. Feldman, *Anal. Chem. 21*, 1041 (1949).
955. A. G. Gassman and W. R. O'Neill, *Anal. Chem. 23*, 1365 (1951).
956. E. L. Gunn, *Anal. Chem. 26*, 1895 (1954).
957. M. T. Carlson and E. L. Gunn, *Anal. Chem. 22*, 1118 (1950).
958. A. J. Hamm, J. Noar, and J. G. Reynolds, *Analyst 77*, 766 (1952).
959. R. O. Clark, E. L. Baldeschwiller, C. M. Gambrill, C. E. Headington, H. Levin, J. M. Powers, and J. B. Rather, *Anal. Chem. 23*, 1348 (1951).
960. T. G. Biktimirova and L. G. Mashireva, *Zavodsk. Lab. 39* 1086 (1973).
961. T. G. Biktimirova and A. A. Baibazarov, *Khim. i Tekhnol. Topliv. i Masel 14*(7), 50 (1969); *Chem. Abstr. 71*, 10372e (1969).
962. D. Hoggan, C. E. Marquart, and W. R. Battles, *Khim. i Tekhnol. Topliv. i Masel 36* 1955 (1964).
963. H. D. Veldhuis, S. Cohen, and G. A. Nohstoll, *Lab. Pract. 7*, 1311 (1952).
964. E. L. Gunn, *Anal. Chem. 26*, 1895 (1954).
965. J. Hansen, P. Skiba, and C. R. Hodgkins, *Anal. Chem. 23*, 1362 (1951).
966. B. K. Barney, *Anal. Chem. 26*, 567 (1954).
967. B. K. Barney II and W. A. Kimball, *Anal. Chem. 24*, 1548 (1952).
968. R. F. Meeker and R. C. Pomatti, *Anal. Chem. 25*, 151 (1953).
969. P. L. Work and A. L. Juliard, *Anal. Chem. 28*, 1261 (1956).
970. J. E. McEvoy, T. H. Millikan, and A. L. Juliard, *Anal. Chem. 27*, 1869 (1955).
971. V. A. Soroka, A. P. Lizogub, V. V. Simashko, and A. G. Gulaeva, *Neftepererab. Neftekhim. 7*, 17 (1972); *Chem. Abstr. 79*, 94331q (1973).
972. G. V. Dyroff, J. Hansen, and C. R. Hodkins, *25*, 1898 (1953).
973. L. W. Gamble and C. E. Kling, *Spectrochim. Acta 4*, 439 (1952).
974. L. L. Gent, C. P. Miller, and R. C. Pomatti, *Anal. Chem. 27*, 15 (1955).
975. J. H. Karchmer and E. L. Gunn, *Anal. Chem. 24*, 1733 (1952).
976. O. I. Milner, J. R. Glass, J. P. Kirchner, and A. N. Yurick, *Anal. Chem. 24*, 1728 (1952).
977. H. R. Sennstrom, *Ry. Mech. Elec. Eng. 126*, 65 (1952).
978. E. B. Childs and J. A. Kanehann, *Anal. Chem. 27*, 222 (1955).
979. J. A. Kanehann, *Anal. Chem. 27*, 1873 (1955).
980. I. P. Polkanov and A. P. Sotnikov, *Khim. i Tekhnol. Topliv. i Masel 17*(11), 60 (1972).
981. J. Hansen and C. R. Hodgkins, *Anal. Chem. 30*, 368 (1958).
982. B. B. Agrawal and S. F. Fish, *India J. Technol. 10*, 117 (1972).
983. S. S. Karacki and F. L. Corcoran, Jr., *Appl. Spectrosc. 27*, 41 (1973).
984. R. J. Heemstra and N. C. Foster, *Anal. Chem. 38*, 492 (1966).
985. P. M. McElfresh and M. L. Parsons, *Anal. Chem. 48*, 1021 (1974).
986. S. Greenfield and P. B. Smith, *Anal. Chim. Acta 59*, 341 (1972).
987. A. W. Varnes, M. S. Vigler, and A. Eskaman, Trace Metal Determination in Petroleum and Petroleum Products by AA and Induction Coupled Plasma-ES, Meeting of the Dutch Atomic Spectroscopy Working Group, Amsterdam, The Netherlands, May 23, 1975.
988. G. Pforr and O. Aribot, *Z. Chem. 10*, 78 (1970).
989. V. A. Fassel, C. A. Peterson, F. N. Abercrombie, and R. N. Kniseley, *Anal. Chem. 48*, 516 (1976).
990. S. Sprague and W. Slavin, *At. Absorption Newsletter* 12 (4) (1963).
991. J. A. Burrows, J. B. Willis, and J. C. Keerdt, *Anal. Chem. 37*, 579 (1965).

992. S. Kaegler, *Instr. News 17*, 1 (1966).
933. E. A. Means and D. Ratcliff, *At. Absorption Newsletter 4*, 174 (1965).
994. E. J. Moore, O. I. Milner, and J. R. Glass, *Microchem J. 10*, 148 (1966).
995. R. A. Mostyn and A. F. Cunningham, *J. Inst. Petrol. 53*, 101 (1967).
996. S. Slavin and W. Slavin, *At. Absorption Newsletter 5*, 106 (1966).
997. P. Prazak, *Ropa Uhlie 15*, 330 (1973); *Amer. Petrol. Inst. Abstr. 21* 2371 (1974).
998. G. K. Billings and P. C. Ragland, *Can. J. Spectrosc. 14*, 8 (1969).
999. H. L. Kahn, G. E. Peterson, and D. C. Manning, *At. Absorption Newsletter 9*, 79 (1970).
1000. D. R. Jackson, *Lubrication Eng. 28*, 76 (1972).
1001. D. R. Jackson, C. Salama, and R. Dunn, *Can. J. Spectrosc. 15* 17 (1970).
1002. M. Koshiki and S. Oshima, *Anal. Chim. Acta 55*, 436 (1971).
1003. A. M. Manjarrez and C. Pereda, *Rev. Soc. Quim. Mex. 13*, 53A (1969).
1004. V. Masek, *Ropa Uhlie 9*, 319 (1967).
1005. G. E. Peterson and H. L. Kahn, *At. Absorption Newsletter 9*, 71 (1970).
1006. C. Salama, *At. Absorption Newsletter 10*, 72 (1971).
1007. J. E. Shallis and H. L. Kahn, *At. Absorption Newsletter 7*, 84 (1968).
1008. G. R. Supp, *At. Absorption Newsletter 11*, 122 (1972).
1009. G. F. Kirkbright, M. Marshall, and T. S. West., *Anal. Chem. 44*, 2379 (1972).
1010. J. B. Sanders, Applications of A. A. Spectroscopy to the Analysis of Petroleum Products, Varian-Techtron brochure, Springvale, Australia (1971).
1011. S. Skujins, The Analysis of Lubricating Oil Additives, Varian-Techtron Application Notes No. 4/70.
1012. W. B. Barnett, H. L. Kahn, and G. E. Peterson, *At. Absorption Newsletter 10*, 106 (1971).
1013. J. F. Lush, in: *Recent Analytical Developments in the Petroleum Industry* (D. R. Hodges, ed.), Halsted Press, New York (1974), Chap. 11.
1014. B. Welz, *Chim Ind. (Milan) 58*, (8) 562 (1976).
1015. M. S. Vigler and V. F. Gaylor, Jr., *App. Spectrosc. 28*. 342 (1974).
1016. S. H. Kaegler, 3rd DGMK Colloq., Hannover, October 6–8, 1975, *Deut. Ges. Mineralolwiss. Kohlechem, eV. Compend.*, 503 (1975–76); *Amer. Petrol. Inst. 33*, 8018 (1976).
1017. M. Tuzar, M. Novak, and V. Stepina, *Ropa Uhlie 15*, 389 (1973).
1018. K. W. Jackson, K. M. Aldous, and D. G. Mitchell, Jr., *Appl. Spectrosc. 28*, 569 (1974).
1019. K. G. Brodie and J. P. Matousek, *Anal. Chem. 43*, 1557 (1971).
1020. R. D. Reeves, C. J. Molnar, M. T. Glenn, J. R. Ahlstrom, and J. D. Winefordner, *Anal. Chem. 44*, 2205 (1972).
1021. R. D. Reeves, C. J. Molnar, and J. D. Winefordner, *Anal. Chem. 44*, 1913 (1972).
1022. J. F. Alder and T. S. West, *Anal. Chim. Acta 58*, 331 (1972).
1023. M. P. Bratzel, Jr., and C. L. Chakrabarti, *Anal. Chim. Acta 61* 25 (1972).
1024. S. H. Omang, *Anal. Chim. Acta 56*, 466 (1971).
1025. G. Hall, M. P. Bratzel, Jr., and C. L. Chakrabarti, *Talanta 20*, 755 (1973).
1026. B. M. Patel and J. D. Winefordner, *Anal. Chim. Acta 64*, 135 (1973).
1027. F. S. Chuang, B. M. Patel, R. D. Reeves, M. T. Glenn, and J. D. Winefordner, *Can. J. Spectrosc. 18*, 6 (1973).
1028. F. S. Chuang and J. D. Winefordner, *Appl. Spectrosc. 28*, 215 (1974).
1029. A. Prevot and M. Gente, *Rev. Franc. Corps Gras 20*, 95 (1973); *Chem. Abstr. 78*, 131628g (1973).
1030. R. J. Lukasiewicz and B. E. Buell, *Anal. Chem. 47*, 3673 (1975).
1031. M. Kashiki, S. Yamazoe, and S. Oshima, *Anal. Chim. Acta 54*, 533 (1971).
1032. S. T. Holding and J. W. Noar, *Analyst 95*, 1041 (1970).
1033. S. T. Holding and J. J. Rowson, *Analyst 100*, 465 (1975).

1034. J. Guttenberger and M. Marold, *Fresenius' Z. Anal. Chem. 262*, 102 (1972).
1035. R. L. Miller, L. M. Fraser, and J. D. Winefordner, *Appl. Spectrosc. 25*, 477 (1971).
1036. D. Demers, M. Gardels, and D. Mitchell, Pittsburgh Conference on Analytical Chemistry and Applied Spectroscopy, paper 45, 1971.
1037. W. O. Davis, *U.S. Clearinghouse Fed. Sci. Tech. Inform., AD-670568* (1968).
1038. R. Smith, C. M. Stafford, and J. D. Winefordner, *Can. J. Spectrosc. 14*, 2 (1969).
1039. D. J. Johnson, F. W. Plankey, and J. D. Winefordner, *Anal. Chem. 47*, 1739 (1975).
1040. D. H. Cotton and D. R. Jenkins, *Spectrochim Acta 25B*, 283 (1970).
1041. A. L. Conrad and W. C. Johnson, *Anal. Chem. 22*, 1530 (1950).
1042. B. E. Buell, in *Encyclopedia of Spectroscopy* (G. L. Clark, ed.), Reinhold Publishing Corporation, New York (1960), p. 365.
1043. K. Augsten, *5th Schnierstoffe U. Schmierungstech. Symp., Chemnitz* (1963).
1044. K. Augsten, *Freiberger Forshungsh.* A, 380, 9 (1966).
1045. Beckman Instruments, Inc., Beckman Technical Data Sheet DU-47-M (T) (September 1955).
1046. R. E. Curtis and R. W. Scott, *Anal. Chem. 26*, 1851 (1954).
1047. H. D. Dubois and R. Barieau, Improved Flame Photometry by Direct Recording Study of Selected Flame Area Emission, Amer. Petrol. Inst. Meeting, Los Angeles, Calif., May 12–15, 1958.
1048. M. L. Moberg, B. Waithman, W. H. Ellis, and D. D. DuBois, *Amer. Soc. Testing Mater. Spec. Tech. Publ. 116*, 92 (1951).
1049. A. M. Kuliev, F. G. Suleimanova, V. E. Brashaev, and A. G. Vladimirov, *Prisadki Smaz. Maslam (1969)*, 202.
1050. O. Popescu, E. Florescu, and W. Popescu, *Petrol. Gaze (Bucharest) 16*, 468 (1965).
1051. I. Muntean and S. Badea, *Petrol Gaze (Bucharest) 22*, 115 (1971).
1052. M. Sarkovic, *Nafta (Zagreb) 20*, 479 (1969).
1053. S. I. Gleason and G. Hold, *Beckman Bull. 16* (1955).
1054. M. Whisman and B. H. Eccleston, *Anal. Chem. 27*, 1861 (1955).
1055. B. E. Buell, *Anal. Chem. 30*, 1514 (1958).
1056. B. E. Buell, unpublished data, Union Oil memos (1975–1976).
1057. B. E. Buell, unpublished data, Union Oil memos (1958).
1058. J. H. Taylor, T. T. Bartels, and N. L. Crump, *Anal. Chem. 43*, 1780 (1971).
1059. T. T. Bartels and M. P. Slater, *At. Absorption Newsletter 9*, 75 (1970).
1060. M. Freegarde and W. J. Barnes, *Analyst, 97*, 406 (1972).
1061. R. H. Kriss and T. T. Bartels, *At. Absorption Newsletter 9*, 78 (1970).
1062. J. A. Bowman and J. B. Willis, *Anal. Chem. 39*, 1210 (1967).
1063. L. Capacho-Delgado and D. C. Manning, *At. Absorption Newsletter 5*, 1 (1966).
1064. W. B. Barnett, *Anal. Chem. 44*, 695 (1972).
1065. I. Maruta, M. Suzuki, and T. Takeuchi, *Anal. Chim Acta. 51*, 381 (1970).
1066. J. Y. Marks and G. G. Welcher, *Anal. Chem. 42*, 1033 (1970).
1067. R. M. Dagnall, G. F. Kirkbright, T. S. West, and R. Wood, *Anal. Chem. 42*, 1029 (1970).
1068. A. P. Ferris, W. B. Jepson, and R. C. Shapland, *Analyst 95* 574, 578 (1970).
1069. M. J. Martin, *Analyst 95*, 567 (1970).
1070. J. M. Ottoway, D. T. Coker, W. B. Rowston, and D. R. Bhattarai, *Analyst 95*, 567 (1970).
1071. M. Yanagisawa, H. Kihara, M. Suzuki, and T. Takeuchi, *Talanta 17*, 888 (1970).
1072. M. Yanagisawa, M. Suzuki, and T. Takeuchi, *Anal. Chim. Acta. 53*, 386 (1970).
1073. J. A. Hurlburt and C. D. Chriswell, *Anal. Chem. 43*, 465 (1971).
1074. S. S. Krishnon, K. A. Gillespie, and D. R. Crapper, *Anal. Chem. 44*, 1469 (1972).
1075. T. Maruta, T. Takeuchi, and M. Suzuki, *Anal. Chim. Acta. 58*, 452 (1972).
1076. M. Marimkovac, P. J. Slevin, and T. J. Vickers, *Applied Spectrosc. 25*, 322 (1971).

1077. T. Nakahara, M. Munemori, and S. Musha, *Applied Spectrosc. 62*, 267 (1972).
1078. S. L. Sachdev, J. W. Robinson, and P. W. West, *Anal. Chim. Acta. 37*, 12 (1967).
1079. M. D. Amos, P. A. Bennett, W. K. Brodie, P. W. Y. Lung, and J. P. Matousek *Anal. Chem. 43*, 211 (1971).
1080. K. W. Jackson and T. S. West, *Anal. Chem. 45*, 249 (1973).
1081. T. Maruta and T. Takeuchi, *Anal. Chim. Acta 62*, 253 (1972).

References to Section 2.9.5

1082. L. Lykken, *Anal. Chem. 17*, 353 (1945).
1083. G. L. Clark and H. A. Smith, *J. Phys. Chem. 33*, 659 (1929).
1084. A. G. Gassman and W. R. O'Neill, *Proc. Amer. Petrol. Inst. Sect. III, 29M*, 79 (1949).
1085. P. T. Gilbert, Jr., *Amer. Soc. Testing Mater. Spec. Tech. Publ. 116*, 67 (1951).
1086. G. T. Buress and J. A. Grant, Amer. Chem. Soc. Meeting, September 15, 1952.
1087. J. H. Jordan, *Petrol. Refiner 32*, 139 (1953).
1088. W. Meine, *Erdoel Kohle 8*, 711 (1955).
1089. T. Okada, T. Ueda, and T. Kohzuma, *Bunko Kenkyu 4*, 30 (1956).
1090. G. W. Smith and A. K. Palmby, *Anal. Chem. 31*, 1798 (1959).
1091. B. E. Buell, *Amer. Soc. Testing Mater. Spec. Tech. Publ. 269*, 157 (1960).
1092. B. E. Buell, *Anal. Chem. 34*, 635 (1962).
1093. H. W. Wilson, *Anal. Chim. Acta 38*, 921 (1966).
1094. R. M. Dagnall and T. S. West, *Talanta 11*, 1553 (1964).
1095. M. Kashiki, S. Yamazoe, and S. Oshima, *Anal. Chim. Acta. 53*, 95 (1971).
1096. Du Pont Method M113-17 (1971).
1097. T. Nishishita, S. Yamazoe, W. R. Mallett, M. Kashiki, and S. Oshima, *Anal. Letters 8*, 849 (1975).
1098. R. J. Lukasiewicz, P. H. Berens, and B. E. Buell, *Anal. Chem. 47*, 1045 (1975).
1099. L. L. McCorriston and R. K. Ritchie, *Anal. Chem. 47*, 1137 (1975).
1100. J. W. Robinson, *Anal. Chem. 24*, 451 (1961).
1101. C. L. Chakrabarti, *Appl. Spectrosc. 21*, 160 (1967).
1102. D. J. Trent, *At. Absorption Newsletter 4*, 348 (1965).
1103. R. A. Mostyn and A. F. Cunningham, *J. Inst. Petrol. 53* 101 (1967).
1104. G. Nagypataki and Z. Tamasi, *Magy. Kem. Lapja 17*, 140 (1962).
1105. S. L. Sachdev and P. W. West, *Environ. Sci. Technol. 4*, 749 (1970).
1106. E. J. Moore, O. I. Milner, and J. R. Glass, *Microchem. J. 10*, 148 (1966).
1107. M. P. Bratzel and C. L. Chakrabarti, *Anal. Chim. Acta 61*, 25 (1972).
1108. K. H. Nelson and M. D. Grimes, *Anal. Chem. 32*, 594 (1960).
1109. U. Kapitaniak. *Gospodarka Paliwami Energ. 2*, 78 (1963).
1110. S. Kaegler, *Instr. News 17*, 1 (1966).
1111. J. H. Jordan, *Petrol. Refiner 33*, 158 (1954).
1112. R. Goux, *Methodes Phys. Analyse* 6, 118 (1970).
1113. A. M. Manjarrez and C. Pereda, *Rev. Soc. Quim. Mex. 13*, 53A (1969).
1114. B. E. Buell, in: *Flame Emission and Atomic Absorption Spectrometry*, Vol. 1 (J. A. Dean and T. C. Rains, eds.), Marcel Dekker, New York (1969), Fig. 9, p. 286.
1115. B. E. Buell, *Anal. Chem. 30*, 1514 (1958).
1116. H. D. Dubois and R. Barieau, Improved Flame Photometry by Direct Recording Study of Selected Flame Area Emission, Amer. Petrol. Inst. Meeting, Los Angeles, Calif., May 12–15, 1958.
1117. M. S. Vigler and J. K. Failoni, *Appl. Spectrosc. 19*, 57 (1965).
1118. T. T. Bartels and C. E. Wilson, *At. Absorption Newsletter 8*, 3 (1969).

1119. W. F. Schoer and R. L. Pontious, *Petro/Chem. Engr. 36* 22 (1964).
1120. D. H. Cotton and D. R. Jenkins, *Spectrochim. Acta. 25B* 283 (1970).
1121. V. Sychra, J. Matousek, and S. Marek, *Chem. Listy 63*, 177 (1969).
1122. V. Sychra, J. Matousek, and S. Marek, *Anal. Chim. Acta 52*, 376 (1970).
1123. J. Matousek and V. Sychra, *Anal. Chem. 41*, 518 (1969).
1124. M. E. Griffing, C. T. Leacock, W. R. O'Neill, A. L. Rozek, and G. W. Smith, *Anal. Chem. 32*, 374 (1960).
1125. D. J. von Lehmden, R. H. Jungers, and R. E. Lee, Jr., *Anal. Chem. 46*, 239 (1974).

References to Section 2.9.6

1126. D. H. Stormont, *Oil Gas J. 64*, 51 (October 24, 1966).
1127. B. Peralta and J. Sosnowski, The Unicracking-JHC Family of Processes, Union Oil Company of California Technical Sales Division, Brea, Calif.
1128. B. J. Duffy and H. M. Hart, *Chem. Eng. Progr. 48*, 344 (1952).
1129. G. A. Mills, *Ind. Eng. Chem. 42*, 182 (1950).
1130. M. J. Murray and H. A. Plagge, *Proc. Amer. Petrol. Inst. Sect. III, 29M*, 79 (1949).
1131. G. V. Dyroff, J. Hansen, and C. R. Hodgkins, *Anal. Chem. 25*, 1898 (1953).
1132. E. L. Gunn and J. M. Powers, *Anal. Chem. 24*, 742 (1952).
1133. J. R. Weaver and R. R. Brattain, *Anal. Chem. 21*, 1038 (1949).
1134. J. H. Karchmer and E. L. Gunn, *Anal. Chem. 24*, 1733 (1952).
1135. P. L. Work and A. L. Juliard, *Anal. Chem. 28*, 1261 (1956).
1136. J. E. McEvoy, T. H. Milliken and A. L. Juliard, *Anal. Chem. 27* 1869 (1955).
1137. L. W. Gamble and W. H. Jones, *Anal. Chem. 27* 1456 (1955).
1138. R. A. Dean and E. V. Whitehead, *6th World Petrol. Congr. 9*, 1 (1963).
1139. O. I. Milner, J. R. Glass, J. P. Kirchner, and A. N. Yurick, *Anal. Chem. 24*, 1728 (1952).
1140. J. E. Barney, II, and G. P. Haight, Jr., *Anal. Chem. 27*, 1285 (1955).
1141. J. E. Barney, II, *Anal. Chem. 27* 1283 (1955).
1142. J. Hansen and C. R. Hodgkins, *Anal. Chem. 30*, 368 (1958).
1143. D. Hoggan, C. E. Marquart, and W. R. Battles, *Anal. Chem. 36*, 1955 (1964).
1144. B. E. Buell, U.S. Patent 3,583,811 (June 8, 1971).
1145. J. Agazzi, D. C. Burtner, D. J. Crittenden, and D. R. Patterson, *Anal. Chem. 35*, 332 (1963).
1146. W. A. Rowe and K. P. Yates, *Anal. Chem. 35* 368 (1963).
1147. J. E. Shott, Jr., T. J. Garland, and R. V. Clark, *Anal. Chem. 33*, 506 (1961).
1148. J. T. Horeczy, B. N. Hill, A. E. Walters, H. G. Schutze, and W. H. Bonner, *Anal. Chem. 27*, 1899 (1955).
1149. Ya. E. Shmulyakovskii, A. A. Baibazarov, F. P. Khapaeva, T. G. Zamilove, and L. M. Zamilova, *Khim i Tekhnol. Topliva i Masel 18*(4), 55 (1973); *Anal. Chem. 47*, 200R (1975).
1150. R. C. Barras, *Jarrell-Ash Newsletter*, No. *13* (June 1962).
1151. R. C. Barras and J. D. Helwig, *Amer. Petrol. Inst. Midyear Meeting Div. Ref. Preprint 20-63* (May 14, 1963).
1152. R. C. Barras and H. W. Smith, *7th World Petrol. Congr. 9*, 65 (1967).
1153. Private communications involving exchange data from Union Oil and other oil companies.
1154. D. Trent and W. Slavin, *At. Absorption Newsletter 3*, 131 (1964).
1155. J. D. Kerber, *Appl. Spectrosc. 20*, 212 (1966).
1156. L. Capacho-Delgado and D. C. Manning, *At. Absorption Newsletter 5*, 1 (1966).

1157. J. A. Bowman and J. B. Willis, *Anal. Chem. 39*, 1210 (1967).
1158. E. S. Obidinski and K. W. Johnson, Gas Turbine Conference and Products Show, Brussels, Belgium, May 24–28, 1970; *Amer. Soc. Mech. Engrs. Paper 70-GT-8.*
1159. A. J. Smith, J. O. Rice, W. C. Shaner, Jr., and C. C. Cerato, 166th Natl. Amer. Chem. Soc. Meeting, Chicago, Ill., August 1973.
1160. Manual on Requirements, Handling and Quality Control of Gas Turbine Fuel, 75th Amer. Soc. Testing Mater. Meeting, Los Angeles, Calif., June 1972; *Amer. Soc. Testing Mater. Spec. Tech. Publ. 531* (1973).
1161. V. Sychra and J. Matousek, *Anal. Chim. Acta 52*, 376 (1970).

References to Section 2.9.7

1162. G. A. Mills and H. A. Shabaker, *Petrol. Refiner 30*, 97 (September 1951).
1163. G. A. Mills, *Ind. Eng. Chem. 42*, 182 (1950).
1164. B. J. Duffy and H. M. Hart, *Chem. Eng. Progr. 48*, 344 (1952).
1165. J. B. Marling, *Anal. Chem. 20*, 299 (1948).
1166. E. L. Gunn, *Anal. Chem. 21*, 599 (1949).
1167. E. L. Gunn, *Anal. Chem. 23*, 1354 (1951).
1168. L. W. Gamble, *Anal. Chem. 23*, 1817 (1951).
1169. C. W. Key and G. D. Hoggan, *Anal. Chem. 24*, 1921 (1952).
1170. J. P. Pagliassotti and F. W. Porsche, *Anal. Chem. 24*, 1403 (1952).
1171. T. Okada, S. Nakai, and T. Kozuma, *J. Chem. Soc. Japan Ind. Chem. Sect. 59*, 495 (1956).
1172. A. N. Aleksandrov, M. I. Dement'eva, and Ya. E. Shmulyakovskii, eds., *Metody Issled. Produktov Neftepereabotki i Neftekhim. Sinteza*, Gostoptekhizdat, Leningrad (1962).
1173. K. A. Rayburn, *Appl. Spectrosc. 22*, 726 (1968).
1174. Union Carbide Corp., Linde Div., Analytical Method LAM-109.
1175. N. Zaidman and D. Orechkin, *Novosti Neft. Tekhn. Geol. 5* 25 (1955); *Chem. Abstr. 51*, 5625d (1957).
1176. M. Sato, T. Kwan, Y. Shimizv, K. Inoue, Y. Koenuma, H. Nishikata, Y. Takenuma, R. Aizawa, S. Kobayashi, K. Egi, and K. Matsumoto, *Pollut. Control* (*Kogai*) *5*, 61 (1970); in Japanese.
1177. M. A. Coudert and J. M. Vergnaud, *Anal. Chem. 42*, 1303 (1970).
1178. M. Kashiki and S. Oshima, *Anal. Chim. Acta 51*, 387 (1970).
1179. J. J. Labrecque, *Appl. Spectrosc. 30*, 625 (1976).
1180. N. M. Potter, *Anal. Chem. 48*, 531 (1976).

References to Section 2.9.8

1181. C. J. Perry and D. A. Keyworth, *Can. J. Spectrosc. 12*, 47 (1967).
1182. J. W. Anderson and H. K. Hughes, *Anal. Chem. 23*, 1358 (1951).
1183. G. V. Dryoff, J. Hansen, and C. R. Hodgkins, *Anal. Chem. 25*, 1898 (1953).
1184. P. L. Work and A. L. Juliard, *Anal. Chem. 28*, 1261 (1956).
1185. E. B. Childs and J. A. Kanehann, *Anal. Chem. 27*, 222 (1955).
1186. L. L. Gent, C. P. Miller, and R. C. Pomatti, *Anal. Chem. 27*, 15 (1955).
1187. C. W. Key and G. D. Hoggan, *Anal. Chem. 25*, 1673 (1953).
1188. C. W. Key and G. D. Hoggan, *Anal. Chem. 26*, 1900 (1954).
1189. I. Lakatos, *Banyasz. Kohasz. Lapok Koolaj Foldgas*, *6*, 115 (1973); *Chem. Abstr. 79*, 94236n (1973).

1190. D. J. von Lehmden, R. H. Jungers, and R. E. Lee, Jr., *Anal. Chem. 46*, 239 (1974).
1191. D. C. Burrell, *At. Absorption Newsletter 4*, 328 (1965).
1192. R. D. Thomas, *Anal. Chem. 38M*, 785 (1966).
1193. R. A. Mostyn and A. F. Cunningham, *J. Inst. Petrol. 53*, 101 (1967).
1194. *Institute of Petroleum Standards for Petroleum and Its Products, Part I: Methods of Analysis and Testing*, 34th ed., Applied Science Publishers Ltd., Essex, England (1975).
1195. R. Goux, *Methodes Phys. Analyse 6*, 118 (1970).
1196. A. Serbanescu, G. Banateanu, T. Fedin, and S. Badea, *Rev. Chim. (Bucharest) 26*, 863 (1975).
1197. M. Kashiki, S. Amazoe, and S. Oshima, *Maruzen Sekiyu Giho #16*, 88 (1971); *Chem. Abstr. 80*, 50126t (1974).
1198. J. Ropars, *Analysis 1973*, March 2(3), 199; *At. Absorption and Emission Spectrometry Abstr. 5*, #753, 300 (1973).
1199. R. M. Yutkevich and V. M. Minut, *Khim. i Tekhnol. Topliva i Masel 17*, 54 (1972); *Anal. Chem. 47*, 200R (1975).
1200. J. F. Alder and T. S. West, *Anal. Chim. Acta 61*, 132 (1972).
1201. L. A. May and B. J. Presley, *Spectrosc. Letters 8*, 201 (1975).
1202. L. A. May and B. J. Presley, *Microchem. J. 21*, 119 (1976).
1203. G. Norwitz and H. Gordon, *Talanta 20*, 905 (1973).
1204. F. Herman, *Metallwertschaft 15*, 1124 (1936).
1205. R. C. Wells, *U.S. Geol. Surv. Bull. 950*, 3 (1946).
1206. M. S. Vigler and A. L. Conrad, *Appl. Spectrosc. 11*, 122 (1959).
1207. B. E. Buell, U.S. Patent 3,583,811 (June 8, 1971).

References to Section 2.9.9

1208. H. Lundegardh, *Die Quantitative Spektralanalyse der Elemente*, Gustav Fisher, Jena (Part I, 1929; Part II, 1934).
1209. A. P. Vanselow and B. M. Laurance, *Ind. Eng. Chem. Anal. Ed. 8*, 240 (1936).
1210. R. L. Mitchell, *J. Soc. Chem. Ind. 55*, 267T (1936).
1211. A. P. Vanselow, and G. R. Bradford, in: *Analysis for Soils, Plants, and Waters* (H. D. Chapman and P. F. Pratt, eds.). Univ. Calif. Div. Agri. Science (1961).
1212. R. L. Mitchell, *Biol. Rev. Cambridge Phil. Soc. 22*, 1 (1947). A review with 260 references.
1213. R. L. Mitchell, in: *Trace Analysis*, (J. H. Yoe and H. J. Koch, Jr., ed.), John Wiley & Sons, Inc., New York (1955), Chap. 14.
1214. R. L. Mitchell and R. O. Scott, *Appl. Spectrosc. 11*, 6 (1957).
1215. R. O. Scott, in symposium on Advances in the Chemical Analysis of Soils, Fertilizers, and Plants, *Soc. Chem. Ind. (London) 1960*; also appears in *J. Sci. Food Agr. 10*, 584 (1960).
1216. R. L. Mitchell, *Spectrochim. Acta 4*, 62 (1960).
1217. G. E. Heggan and L. W. Strock, *Anal. Chem. 25*, 859 (1953).
1218. R. O. Scott, *J. Sci. Food Agr. 11*, 584 (1960).
1219. W. G. Schrenk, *Amer. Soc. Testing Mater. Spec. Tech. Publ. 221*, 58 (1957).
1220. F. Smith, W. G. Schrenk, and H. King, *Anal. Chem. 20*, 941 (1948).
1221. A. Berneking and W. G. Schrenk, *J. Agr. Food Chem. 5*, 742 (1957).
1222. S. J. Toth, A. L. Prince, A. Wallace, and D. S. Mikkelsen, *Soil Science 66*, 459 (1948).
1223. C. L. Grant, *Spex Speaker 6* (September 1961).
1224. W. H. Allaway, in *Trace Analysis: Physical Methods* (G. H. Morrison, ed.), Wiley-Interscience, New York (1965), Chap. 3.

1225. J. E. Allan, The Preparation of Agricultural Samples for Analysis by A. A. Spectroscopy, Varian-Techtron brochure, Springvale, Australia (1970).
1226. D. J. David, *At. Absorption Newsletter*, No. 9 (December 1962).
1227. W. Slavin, *At. Absorption Newsletter*, No. *4* (June 1962).
1228. A. Strasheim and D. J. Eve, *Appl. Spectrosc. 14*, 97 (1960).
1229. K. E. Knutson, *Analyst 82*, 241 (1957).
1230. F. Breck, *J. Assoc. Offic. Agr. Chemists 51*, 132 (1968).
1231. M. H. Chaplin and A. R. Dixon, *Appl. Spectrosc. 28*, 5 (1974).
1232. J. B. Jones, Jr., *J. Assoc. Offic. Agr. Chemists 58*, 764 (1975).
1233. W. D. Basson and R. G. Bohmer, *Analyst 97*, 482 (1972).
1234. E. E. Cary and O. E. Olson, *J. Assoc. Offic. Agr. Chemists 58*, 433 (1975).
1235. D. R. Boline and W. G. Schrenk, *Appl. Spectrosc. 30*, 607 (1976).
1236. J. C. M. Pau, E. E. Pickett, and S. R. Koirtyohann, *Analyst 97*, 860 (1972).
1237. S. Pawluk, *At. Absorption Newsletter 6*, 53 (1967).
1238. R. Linville, *Perkin-Elmer Instr. News 18*, 10 (1968).
1239. J. L. Seeley, D. Dick, J. H. Arvik, R. L. Zimdahl, and R. K. Skogerboe, *Appl. Spectrosc. 26*, 456 (1972).
1240. H. L. Kahn, F. J. Fernandez, and S. Slavin, *Appl. Spectrosc. 11*, 42 (1972).
1241. C. H. McBride, *At. Absorption Newsletter 3*, 144 (1964).
1242. S. J. Weger, Jr., L. R. Hassner, and L. W. Ferrara, *J. Agr. Food Chem. 17*, 1276 (1969).
1243. J. A. Brabson and W. D. Wilhide, *Anal. Chem. 26*, 1060 (1954).
1244. M. M. Ferraris and G. Proksch, *Anal. Chim. Acta 59*, 177 (1972).
1245. B. Gutsche, K. Rudiger, and R. Herrmann, *Deut. Lebensm.-Rundschau 68*, 13 (1972).

References to Section 2.10

1246. A. Ye. Fersman, ed., *U.S. Geol. Surv. Cir. 127* (1952); translated from Russian.
1247. H. E. Hawkes and J. S. Webb, *Geochemistry in Mineral Exploration*, Harper & Row, Publishers, New York (1962).
1248. R. E. Wainerdi and E. A. Uken, eds., *Modern Methods of Geochemical Analysis*, Plenum Press, New York (1971).
1249. B. Mason, *Principles of Geochemistry*, 3rd ed., John Wiley & Sons, Inc., New York (1966).
1250. R. L. Parker, *U.S. Geol. Surv. Prof. Paper 440-D* (1967).
1251. R. W. Fairbridge, ed., *Encyclopedia of Geochemistry and Environmental Sciences*, Van Nostrand Reinhold, New York (1972). See p. 367 for FE applications by B. E. Buell.
1252. L. H. Ahrens, *Spectrochemical Analysis*, 2nd ed., Addison-Wesley Publishing Company, Inc., Reading, Mass. (1961).
1253. K. J. Murata, *Amer. Soc. Testing Mater. Spec. Tech. Publ. 221*, 67 (1957).
1254. E. F. Cruft and D. L. Giles, *Economic Geol. 62*, 406 (1967).
1255. H. Schwander, *Schweiz. Mineral. Petrog. Mitt. 40*, 2 (1960).
1256. A. W. Helz and B. F. Scribner, *J. Res. Natl. Bur. Std 38*, 439 (1947).
1257. H. A. Camacho-Calderon and C. A. Vallecilla-Risscos, *Rev. Univ. Ind. Santander, 5*, 447 (1963).
1258. R. Mavrodineanu, *Bibliography on Flame Spectroscopy: Analytical Applications 1800–1966*, Natl. Bur. Std. Misc. Publ. 281, Washington D. C. (1967).
1259. P. H. Lundegardh, *Geol. Foren. Stockholm Forh. 72*, 151 (1950).
1260. G. H. Osborn and H. Johns, *Analyst 76*, 410 (1951).
1261. N. S. Poluektov, M. P. Nikonova, and R. A. Vitkun, *Zh. Analit. Khim. 13*, 48 (1958).

1262. A. Willgallis, *Z. Anal. Chem. 157*, 239 (1957).
1263. R. J. Brumbaugh and W. E. Fanus, *Anal. Chem. 26*, 463 (1954).
1264. R. B. Ellestad and E. Horstmann, *Anal. Chem. 27*, 1229 (1955).
1265. C. E. White, M. H. Fletcher, and J. Parks, *Anal. Chem. 23*, 478 (1951).
1266. R. Pouget, *Comm. Energie At. (France) Rappt.* (1963); 2176 (1962); see *Chem. Abstr. 60*, 1107 (1964) and *58*, 5024 (1963).
1267. E. L. Horstman, *Anal. Chem. 28*, 1417 (1956).
1268. L. Biagi, R. Pironi, and G. Simboli, Parts I and II, *Mineral. Petrog. Acta 9*, 111, 163 (1963); *Chem. Abstr. 63*, 1211 (1965).
1269. T. F. Borovik-Romanova, *Spectral Analytical Determinations of Alkali and Alkaline Earth Elements in Water, Plants, Soils, and Rocks*, Izdatelstvo. Akad. Nauk SSSR, Moscow (1956); *Chem. Abstr. 52*, 7034 (1958).
1270. R. E. Mosher, J. Bird, and A. J. Boyle, *Anal. Chem. 22*, 715 (1950).
1271. H. Kramer, *Anal. Chim. Acta 17*, 521 (1957).
1272. W. A. Dippel, C. E. Bricker, and N. H. Furman, *Anal. Chem. 26*, 553 (1954).
1273. R. Ratner and D. Scheiner, *Analyst 89*, 136 (1964).
1274. G. K. Billings, *At. Absorption Newsletter 3*, 1 (1964).
1275. G. K. Billings, *At. Absorption Newsletter 4*, 312 (1965).
1276. G. K. Billings, The Preparation of Geological Samples for Analysis by Atomic Absorption, Varian-Techtron, T-1001, Springvale, Australia (1970).
1277. E. E. Angino and G. K. Billings, *Atomic Absorption Spectrometry in Geology*, American Elsevier Publishing Company, Inc. (1967).
1278. W. Slavin, *At. Absorption Newsletter 4*, 243 (1965).
1279. F. N. Ward, H. M. Nakagawa, T. F. Harms, and G. H. Van Sickle, *U.S. Geol. Surv. Bull 1289* (1969).
1280. B. Bernas, *Anal. Chem. 40*, 1682 (1968); *At. Absorption Newsletter 9*, 52 (1970).
1281. J. C. Van Loon, *At. Absorption Newsletter 11*, 60 (1972).
1282. F. J. Flanagan, *Geochim Cosmochim. Acta 37*, 1189 (1973).
1283. S. Abbey, Studies in Standard Samples of Silicate Rocks and Minerals, Part 3: 1973, Extension and Revision of Usable Values, Geol. Survey of Canada, papers 73-76 (1973).
1284. J. I. Dinnen, *Anal. Chem. 47*, 97R (1975).
1285. M. H. Hey, *Mineral Mag. 39*, 4 (1973).

Analytical Emission Spectroscopy in Biomedical Research 3

William Niedermeier

3.1 INTRODUCTION

The role of trace metals in the etiology and pathogenesis of disease has been a subject of interest for many years. Early studies were principally concerned with acute toxicities caused by metals or with the effect that severe dietary deficiencies of the metals produced on growth, development, and reproduction of laboratory animals. More recently, interest has developed in the possibility that chronic dietary excesses or deficiencies of certain trace metals might play a role in the etiology and pathogenesis of chronic diseases.

Of equal importance is the possibility that certain chronic diseases might result from failure of the organism to metabolize trace metals by normal processes rather than from dietary excess or deficiency of the metal per se. It is well known that trace metals rarely occur in tissues in the ionic state, but are usually found complexed with proteins or other organic ligands. Many of these ligands perform important biological functions, and probably represent the active forms of the trace metals. Elucidation of the role of trace metals in human disease is therefore contingent on an understanding of the metabolism of the trace metals both in normal and in pathologic tissues.

Identification of abnormal concentrations of a trace metal in blood serum or other tissues of patients with a chronic disease is only the first step in establishing the relationship between the trace metal and the disease process. Any study

William Niedermeier • Division of Clinical Immunology and Rheumatology, Department of Medicine, University of Alabama in Birmingham, Birmingham, Alabama

to establish the relationship between a trace metal and the etiology and pathogenesis of a disease should include the following phases:

1. Identification of the presence of abnormal quantities of a trace metal in tissues of patients who have the disease
2. Identification of the ligands with which the trace metal is associated in tissues
3. Distribution of the ligand in tissues
4. Elucidation of the chemical, physical, and biological properties of the ligand, and correlation of these properties with signs and symptoms of the disease

Studies of this kind may lead to complete understanding of the role of trace metals in disease processes and to the ultimate development of a means for treatment or prevention of the disease.

The ambitious studies of Tipton *et al.*[1-3] have generated considerable interest in the possible role of trace metals in diseases. These investigators performed spectrographic analyses to determine the concentrations of 24 trace metals present in 10 different types of tissue taken from several hundred cadavers of individuals that were apparently normal but died suddenly from accidental causes, who lived in widely scattered parts of the world. The results of this study were discussed in a series of publications by Schroeder *et al.*,[4-22] and studies to evaluate the effects of chronic dietary deficiencies and excesses of some of the metals on laboratory animals have been reported.

Research relating trace metals to human disease has suffered from the lack of suitable methods of analysis. The problem of accurately determining the concentrations of trace metals in the range of a few micrograms per 100 g of tissue is complicated by the fact that the tissues in which these metals are present are composed of a variety of organic materials which must first be removed. The resultant inorganic matrix is composed predominantly of alkali metals which present further problems for the analyst.[23] Although many analytical methods have been used, emission spectroscopy has found favor with a number of investigators for the analysis of biological materials.[24-32] It is considered an especially useful procedure because it is (a) highly sensitive, (b) requires only small quantities of sample, and (c) allows simultaneous analysis for many metals.

Development of the direct-reading emission spectrometer and computerized methods of data reduction and analysis have added new dimensions to studies on the relationships between trace metal metabolism and health. It is anticipated that, through the use of modern methods of emission spectrometric analysis, questions relating trace metal metabolism to diseases of unknown etiology may be answered.

3.2 SPECTROMETRIC METHODS

The method of analysis described here is applicable to blood serum, synovial fluid, saliva, and perhaps other body fluids which have relatively constant macroelement compositions. Two milliliters of the biological fluid to be analyzed provides sufficient material to permit performance of the analysis in quadruplicate.[33] Analyses may be performed simultaneously for as many as 22 trace metals, although in practice the number is generally limited to 14. Samples and standards were wet-ashed and aliquots of the solution were transferred to graphite electrodes. The emission produced in a dc arc is analyzed with a direct-reading emission spectrometer. Conversion of photomultiplier response to concentration and statistical evaluation of the data was accomplished with a digital computer.

3.2.1 Equipment and Materials

The operating parameters used with the direct-reading emission spectrometer are listed in Table 3.1. A Simeon-Twyman lens system[34] was used to prevent the continuum emitted by the incandescent electrode tips from entering the spectrometer. Table 3.2 lists the 14 elements that were analyzed, the analytical lines used, and the respective exit (secondary) slit widths.

Standards were prepared from spectroscopically pure reagents. Nitric, perchloric, and hydrochloric acids used for digestion, and ammonium chloride used as a spectroscopic buffer, were special high-purity reagents. Distilled water was prepared from deionized water in a borosilicate glass distillation apparatus. All glassware was cleaned with reagent-grade concentrated nitric acid at 100°C. This was followed by rinsing in distilled water, submerging in 0.01% aqueous ethylenediaminetetraacetic acid solution, and finally rinsing with doubly distilled water.

Blood specimens were collected with disposable plastic syringes fitted with stainless steel needles that had plastic hubs. In a preliminary study, the use of

Table 3.1 Apparatus and Spectrometric Operating Conditions[a]

Electrodes:	Sample: AGKSP graphite, $\frac{1}{4}$-inch-diameter, $\frac{3}{16}$-inch cup with center post (ASTM P-1)
	Counter: AGKSP graphite, $\frac{1}{4}$-inch-diameter (ASTM C-5)
Analytical gap:	4 mm
Excitation unit:	Jarrell-Ash Spectro varisource, continuous dc arc
	Voltage, dc: 230
	Current (A), dc: 10.0
Spectrometer:	Jarrell-Ash 1.5-m direct-reading compact Atomcounter, 30,000-grooves-per-inch grating
	Dispersion: 5.6 Å/mm
	Wavelength range (Å): 2000–8000
	Slit width (μm):
	Entrance: 25
	Exit: 10–75
Preburn (s):	0
Exposure (s):	50

[a]From Niedermeier *et al.*,[33] p. 54; by courtesy of the Society of Applied Spectroscopy.

Table 3.2 Analytical Lines Used for Elements Studied[a]

Element	Wavelength (Å)	Slit width (μm)
Zn	2138.5	10
Cd	2288.0	75
Au	2427.9	75
Pb	2833.0	75
Fe	3020.6	75
Mo	3193.9	75
Cu	3247.5[b]	75
Sn	3262.3	75
Ni	3414.7	10
Al	3961.5	75
Mn	4030.7	75
Rb	4201.8	10
Cr	4254.3	75
Cs	4555.3	10
Sr	4607.3	75
Ba	5535.5	10
V	5698.5	75

[a]From Niedermeier *et al.*,[33] p. 54, by courtesy of the Society of Applied Spectroscopy.
[b]Second order.

platinum needles and acid-washed, glass syringes demonstrated that this disposable equipment contributed no detectable contamination to the specimens.

3.2.2 Methods

3.2.2.1 Sample Preparation

After coagulation of the blood specimen at room temperature, the serum was separated from the clot by centrifugation at 1500 rpm. Exactly 2.00 ml of the blood serum or other biological fluid was pipetted into 20- X 150-mm ignition tubes. To each tube, the high-purity digestion acids were added: 1.00 ml of nitric acid and 0.25 ml of perchloric acid. The tubes containing the samples were placed in a constant-temperature block and slowly heated to and then maintained at a temperature of 130°C for 5-6 hours. When digestion of the specimens was complete, the colorless solutions were heated at 180°C until dry (5-6 hours). To convert the metals to their chlorides, 1 ml of high-purity hydrochloric acid was added to each of the cooled samples and the contents of the tubes were again evaporated at 130°C until dry. The dried residues were dissolved in 1.00 ml of 0.9% NH_4Cl spectroscopic buffer solution that was prepared from high-purity materials by dissolving the salt in 3 N hydrochloric acid. By dissolving the ash from 2 ml of biological fluid in 1 ml of buffer solution, a twofold concentration of the specimens was achieved.

The equivalent of 0.4 ml of the original sample (i.e., 0.2 ml of reconstituted ash) was transferred to the crater of the sample electrode with an acid-washed disposable micropipette and evaporated to dryness at reduced pressure in a vacuum dessiccator at room temperature. All specimens were analyzed in quadruplicate. Results were expressed as the mean value of the four determinations.

3.2.2.2 Preparation of Standards

Standards were prepared in matrix solution that contained macroelements in concentrations closely approximating those of normal human blood serum. The equivalent of 0.4 ml of biological fluid, or 0.1 ml of standard solution, was placed in the electrodes. The matrix solution used to prepare the standard contained 33 g of sodium chloride, 1.93 g of potassium dihydrogen phosphate, and 1.08 g of calcium chloride per liter of solution. These concentrations of the macroelements were four times higher than those found in normal blood serum. Thus, the total mass of matrix elements present in 0.1 ml of standard solution was equal to that in the 0.4-ml aliquot of ashed sample, which was subjected to analysis. Spectroscopically pure reagents were used.

The standard solutions of the trace elements were prepared by diluting a 0.03% stock solution of the elements to concentrations of 1000, 400, 200, 40, 20, and 4.0 μg% with matrix solution. Thus, 0.1 ml of each of the six standard solutions was equivalent to 0.4 ml of biological fluid that contained 250, 100, 50, 10, 5, and 1.0 μg%, respectively, of each element. Each of these standard solutions contained all 14 of the trace elements in the same concentration.

To minimize errors that might arise from the possible presence of minute impurities in the reagents used for ashing, the standards were treated by the same procedure described for the specimens of blood serum. The dried residue was dissolved in a 1.8% NH_4Cl spectroscopic buffer solution and made to 1.00 ml, the original volume of unashed standard. Electrodes were prepared by transferring 0.1 ml of the ashed standard solutions (equivalent to 0.4 ml of blood serum on the basis of total ash) into the crater of the sample electrodes and evaporating to dryness at reduced pressure in a vacuum desiccator at room temperature.

3.2.2.3 Spectrometric Procedure

The instrument was calibrated each day on the basis of quadruplicate determinations performed on each of the six standard solutions. The equations of the curves that described the concentration response of the instrument for each of the elements was calculated with the use of digital computer by the method described below. For each element a mathematical expression was developed relating concentration to spectrometer response. Using the calibration curve derived from standards, the computer then calculated the concentration of the elements in the unknown specimens. Concentrations were expressed in terms of micrograms of the element per 100 ml of sample ($\mu g\%$). One microgram percent is the equivalent of 0.01 ppm.

3.2.3 Discussion

To evaluate the effect of the blood serum matrix on the spectrometer response, a set of standards was prepared with water to which none of the matrix salts were added. As shown in Fig. 3.1, the presence of matrix considerably improved the response obtained for a number of the trace metals, particularly in the lower concentration ranges. As indicated by these curves, some elements could be determined at much lower concentrations than others. The threshold values for each of the metals calculated according to Johnson *et al.*[35] are shown in Table 3.3. Although the threshold value is a mathematical concept, it can be

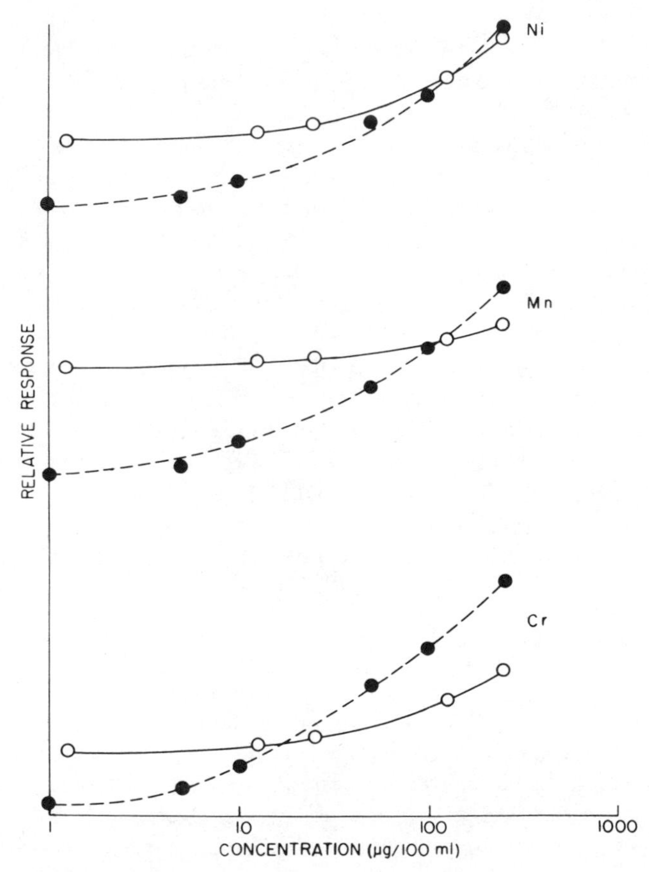

Fig. 3.1 Comparison of spectrometer response in the presence of (●—●) and in the absence of (○—○) blood serum matrix.

Table 3.3 Standard Deviations Obtained by Emission Spectrometric Method of Analysis of Trace Metals[a]

Element	Threshold value	Concentration of solution tested ($\mu g\%$)					
		1	5	10	50	100	250
Cu	0.5	0.4	0.8	1.3	5.1	10.4	48.9
Fe	1.6		2.4	2.9	4.9	8.2	16.0
Al	5.7			4.1	11.6	13.5	34.5
Ba	3.2		2.5	3.0	6.7	10.6	31.8
Mn	1.1	0.8	1.1	1.4	4.4	8.2	22.5
Ni	1.1	1.1	1.4	1.7	4.3	7.6	18.8
Cs	4.3		3.0	3.4	10.2	18.4	62.2
Sn	4.2		3.8	4.2	6.6	10.1	35.1
Sr	0.5	0.4	0.8	1.4	5.4	10.6	29.5
Cr	1.0	0.8	1.0	1.3	3.9	7.4	19.4
Zn	13.5				12.4	23.5	243.2
Pb	2.3		1.7	1.9	3.6	6.4	24.6
Mo	2.4		1.9	2.2	4.1	7.7	24.6
Cd	1.5		1.3	1.7	3.2	11.7	53.4

[a]From Niedermeier *et al.*,[33] p. 55, by courtesy of the Society of Applied Spectroscopy.

looked upon as a realistic definition of the minimum detectable quantity of each element. All the elements, except zinc, had threshold values considerably less than 10 $\mu g\%$.

In a study to determine the precision of the method, 10 determinations were made on each of the six standard solutions. After the usual computer computations on the standards, these data were treated as unknowns. The standard deviations of the computed concentrations were then calculated and are summarized in Table 3.3. For the elements that had threshold values of about 1 $\mu g\%$ or less, the standard deviations were approximately equal to the threshold values. Standard deviations of all elements were higher for the more concentrated solutions, but when calculated on the basis of the percentage of the element in the standard (coefficient of variation), the best results were obtained with the 50 or 100 $\mu g\%$ standards. At concentrations of 50 $\mu g\%$ or greater, the coefficients of variation were generally less than 10%.

Table 3.4 shows the results of analyses performed by this method on 105 specimens of normal human blood serum. The mean values were calculated for all 105 specimens. A value of zero was assumed for those specimens in which the concentration was below the threshold value. For most of the elements, the mean concentration in the normal blood serum was at least double that of the respective threshold value (Table 3.3). For comparison, the results reported by other investigators are also listed in this table. In general, the results are in close agreement with those previously reported. The mean lead concentration ap-

Table 3.4 Trace Element Content of Normal Human Blood Serum[a]

Element	Average concentration (μg%)	
	This study	Literature values
Cu	83.2	70–140[b]
Fe	86.9	39–170[c]
Al	20.6	46[c]
Ba	15.0	1–19[d]
Mn	6.6	1–7[d]
Ni	10.1	1–6[d]
Cs	9.8	—
Li	30.5	<1[d]
Sr	3.3	<1[d]
Cr	9.8	1–38[d]
Zn	103.7	121[e]
Pb	8.5	2[f]
Mo	10.1	3[g]
Cd	6.3	38[g]

[a] From Johnson *et al.*,[35] p. 556, courtesy of the Society of Applied Spectroscopy.
[b] H. C. Damm and J. W. King, *Handbook of Clinical Laboratory Data* (CRC Press, Cleveland, 1965), p. 140.
[c] P. L. Altman, *Blood and Other Body Fluids* (Federation of American Societies for Experimental Biology, Washington, D.C., 1961), p. 22.
[d] Niedermeier, (1962), p. 443.
[e] J. G. Ferrante, Jr., *Boston Med. Quart. 12*, 95 (1961).
[f] R. A. Kehoe, J. Cholak, and R. V. Story, *J. Nutr. 19*, 579 (1940).
[g] E. M. Butt *et al.*, *Arch. Environ. Health 8*, 52 (1964).

peared high. Subjects for this study lived in an area where illicit whiskey was readily available. Excessive consumption of this product may have accounted for the high lead content, or it may have been attributable to an increase in the levels of environmental pollution over levels existing when results from the earlier studies were reported. Iron is the only metal present in blood serum that produces a fairly strong spectral line in the region of Pb 2833. The addition of 500 μg% of iron to the 10 μg% standard, however, had no influence on the spectrometer readout for lead, which strongly implies that interference from iron does not explain the presence of high lead values.

In contrast to the relatively simple method used in preparing samples of metallurgical or geological materials for analysis by emission spectroscopy, the preparation of biological samples is tedious and time consuming. Of all methods attempted, the most satisfactory results were obtained by the wet digestion procedure, followed by evaporation in a hollow cup graphite electrode that contained a center post. Reproducible results were not realized when we used a previously described method[29] in which the dried sample mixed with graphite and dry spectroscopic buffers was packed into a hollow cup electrode. No significant improvement in reproducibility or sensitivity was noted in the results from several tests conducted with inert atmospheres created by modified Stallwood jets. Recent reports[30,31] indicated that the use of a controlled dc arc in a static argon atmosphere resulted in markedly improved reproducibility and sensitivity. Potential advantages in achieving lower limits of detection and greater precision may also be realized from the use of plasma sources.[36,37]

Dry-ashing techniques were found to be unsatisfactory for these studies because of volatilization of certain metals and to problems of quantitatively recovering insoluble oxides produced by these methods. Wet digestion with sulfuric acid, which is sometimes used for sample preparation, involves high temperatures for elimination of excess acid from the digested sample, and, unlike perchloric and nitric acids, sulfuric acid is difficult to obtain in a state that is free of trace metals. The low-temperature, wet-digestion procedure minimizes possible losses of certain elements by vaporization that sometimes accompanies higher-temperature techniques.

The adjustable slit[34] placed between the analytical gap and the entrance slit of the spectrograph effectively prevented the continuum emitted by the incandescent electrode tips from entering the spectrograph. Thus, the background was appreciably reduced and the limit of determination of most of the elements was extended. Several types of excitation were tried. Although an ac spark consistently gave excellent reproducibility, the sensitivity from it was insufficient for this application. An ignited ac arc, which compromises reproducibility and sensitivity, also afforded insufficient sensitivity. Therefore, the dc arc was deemed the only acceptable method of excitation. To minimize errors caused by the instability of the dc arc, each sample (and standard) was analyzed four times and the results were expressed as the mean of the four determinations.

3.3 COMPUTERIZED EVALUATION OF RESULTS

Because of the wide variation in trace metal composition of biological materials, a large number of samples must be analyzed to obtain meaningful data. Since every sample is analyzed in our laboratory for as many as 17 trace metals, conversion of spectrometer response to concentration by classical methods becomes a monumental task. Therefore, methods of computerized computation were investigated with the use of a digital computer programmed in Fortran IV.

The graphic procedure is the most rudimentary method for the calculation of concentration from spectrometric data. For each element the appropriate emission data are plotted as a function of concentration of standards. The graphic solution for the concentration of unknown samples is then obtainable. Plotting boards and other mechanical computational devices have been described.[38] The use of electronic analog computers with photoelectric instrumentation has also been discussed.[39] Another method that seems practical is the use of digital computers for analyzing the data and printing the corresponding concentrations.[35]

3.3.1 Mathematical Relationships

Several different functional relationships were used in an effort to obtain an empirical equation relating mean counts (x) with concentration (y) for the 17 different elements. These equations included polynomials through the third order as well as logarithmic functions. The best general empirical relationship appears to be the four-constant transcendental equation

$$y = A' + Bx + 10^{C+Dx} \tag{3.1}$$

For this purpose, Eq. (3.1) has many desirable mathematical attributes. It is single-valued, monotonic, and analytic. The equation is fitted to the standardization data by a three-step successive approximation. With the four constants evaluated it is a simple matter to insert the mean values of x from unknowns and to solve for the corresponding y values.

For each element, the standards with concentrations (y's) of 1, 5, 10, and 50 μg% and their corresponding mean counts (x's) are fitted[40] by the method of averages, with two groups of two each to the equation

$$y_a = A + Bx \tag{3.2}$$

Thus, each point has equal weight. The procedure that appeared to give the best results, as shown by the lowest sum of the residuals squared (the F test), was the grouping of $y_1 + y_2$ and $y_3 + y_4$, wherein the subscripts refer to the succession of points starting with the 1-μg% level. The sum of the residuals squared was about equal to that obtained by the least-squares method.

Having found the values of A and B for each element, it was possible to calculate concentrations based on the number of counts for the two highest points, x_5 and x_6, which corresponded to the known concentrations of y_5 and y_6 for the levels of 100 and 250 μg%, respectively. These differences (z's) were then $z_5 = y_5 - (A + Bx_5)$ and $z_6 = y_6 - (A + Bx_6)$. When the data for all 17 elements were considered, three types of cases resulted.

Case 1 is the situation that prevails when both z_5 and z_6 are greater than zero. In this event, the two z values are solves simultaneously for the constants, C and D, in the exponential term of Eq. (3.1)

$$\begin{aligned} \log z_5 &= C + Dx_5 \\ \log z_6 &= C + Dx_6 \end{aligned} \tag{3.3}$$

Although this makes an exact fit of the equation to the data at the two highest points, the exponential term does not equal zero for low values of x.

The third step in the approximation is to evaluate 10^{C+Dx} and to subtract this constant from all computed y values. In the generalized transcendental relationship, Eq. (3.1), $A' = A - 10^{C+Dx}$. There is then no longer an exact fit to the data for the points 5 and 6, but the error incurred can be overlooked. This procedure ensures that the exponential term does not contribute to the computed y_1 value. Because of the size of C, D, and x, the increase in the computer y values for the points corresponding to 5, 10, and 50 μg% is very small, for at these levels 10^{C+Dx} is essentially negligible. In general, this three-step approximation is applicable to 14 of the 17 elements.

Case 2 is the situation that prevails when either z_5 or z_6 is equal to, or less than, zero. In this event, the linear equation, Eq. (3.2), gives a satisfactory fit to the data, and the second and third steps of the successive approximation are omitted. This can also be considered the trivial case of Eq. (3.1), wherein the constants C and D are both zero. Then, $A' = A - 1$, since any real number raised to the zero power is unity. Generally, two or three of the 17 elements conform to case 2.

Case 3 is the situation that prevails when both z_5 and z_6 are less than, or equal to zero; therefore, a good fit can be obtained with the linear relationship, Eq. (3.2). The constants A and B can be obtained by the method of averages, with the data in two groups of three each; thus, $y_1 + y_2 + y_3$ and $y_4 + y_5 + y_6$. The method of least squares can also be used. The resultant final equation makes case 3 appear to be a repetition of case 2, but in actuality the difference lies in the grouping of the points. However, case 3 is described by Eq. (3.1), with C and D both equal to zero. Chromium frequently occurs as case 3.

3.3.2 Fit of the Equation to the Data

The procedure for fitting Eq. (3.1) to the data will be illustrated using the results obtained for copper shown in Table 3.5. These data are based on the means of four measurements. In the standardization, the y values are accurately known and the variation in x for a given value of y is attributable solely to random errors. The upper and lower limits of the counts were computed from the standard deviation and the appropriate factors from t tables, in the usual manner.

This is a case 1 example, and the mathematical computations are summarized in Table 3.5. The four constants in the complete equation were obtained by the three-step approximation procedure described previously. The y_a values were computed from the equation $y_a = A + Bx$ by the substitution of the mean values of x. The constants A and B were obtained from the first four points, by the method of averages, through the simultaneous solution of the two equations $6 = 2A + 332.8B$ and $60 = 2A + 1749.8B$, which lead to $y_a = -3.3 + 0.0381x$. The

Table 3.5 Typical Standardization Data for the Element Copper[a]

Point No.	Concentration y_{obs} (μg%)	Mean counts, x_{mean}	Upper and lower limits of: 80% level	Upper and lower limits of: 95% level	y_a	Difference $z = y_{obs} - y_a$	y_b	Calculated values: y	Calculated values: x	Calculated values: δ_y
1	1	106.0	117.5; 94.5	124.7; 87.3	0.7	+0.3	3.1	0.7	113	0.4
2	5	226.8	243.2; 210.4	258.4; 175.3	5.3	−0.3	7.9	5.5	212	0.9
3	10	370.0	406.0; 334.0	440.0; 300.0	10.8	−0.8	13.8	1.4	337	1.5
4	50	1379.8	1535.2; 1224.2	1681.9; 1077.6	49.3	+0.7	54.1	53.9	1293	6.1
5	100	2308.0	2462.9; 2154.7	2608.3; 2009.2	84.7	+15.3	100.0	97.6	2356	11.5
6	250	4359.3	4923.3; 3795.3	5290.2; 3428.4	162.8	+87.2	250.0	247.6	4380	39.0

[a] From Johnson *et al.*,[35] p. 553, by courtesy of the Society of Applied Spectroscopy.

z_5 and z_6 values were then used to solve for C and D, as shown by Eq. (3.3). The calculated concentrations are tabulated in the column headed y_b. Here $y_b = A + Bx + 10^{C+Dx}$. The exponential term 10^{C+Dx} was evaluated, with $x_1 = 106.0$. The resultant quantity, 2.4, was then subtracted from each y_b value. The results are tabulated in the column headed as calculated y.

The complete equation is thus

$$y = -5.7 + 0.381x + 10^{0.3327+0.00369x} \tag{3.4}$$

The calculated y values compare favorably with the observed y's, as shown in Table 3.5 and Fig. 3.2. The latter is a log–log plot in the manner customarily used for graphical presentation of spectrometric data. To make a point-by-point comparison, Eq. (3.4) was solved for the x values corresponding to y equal to 1, 5, 10, 50, 100, and 250 μg%. The method of *regula falsi* gave sufficiently accurate results with this transcendental equation. The x values are shown in Table 3.5 under the heading "Calculated x."

The departure of the upper and lower limits from the mean counts were calculated as a percent of the mean observed counts at the 80 and 95% confidence levels, as shown in Fig. 3.3. The average values of ±9.8% are represented by the horizontal lines marked 80%. A logarithmic scale was used on the horizontal axis in this figure because of the several decades involved in the concentration range.

From the solutions of Eq. (3.4), the difference in the mean observed value of x minus its calculated value was expressed as a percent of the mean observed counts for the six known values of y used in the standardization. The resultant deviations are plotted in Fig. 3.3, where the individual points are connected by the line segments. The 80% confidence level represents a reasonable considera-

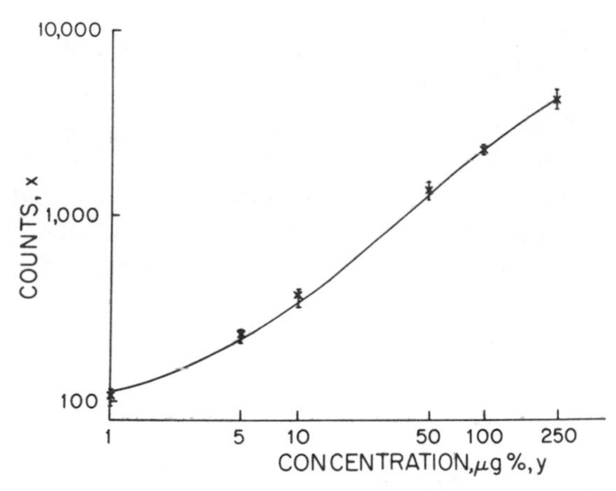

Fig. 3.2 Typical log–log standardization curve for the element copper. The mean values are given by the X's and the range of the counts at the 80% confidence level are shown by the line segments. The curve represents the equation $y = -5.7 + 0.0381x + 10^{0.3327+0.000309}$ [From Johnson *et al.*,[35] p. 554, by courtesy of the Society of Applied Spectroscopy.]

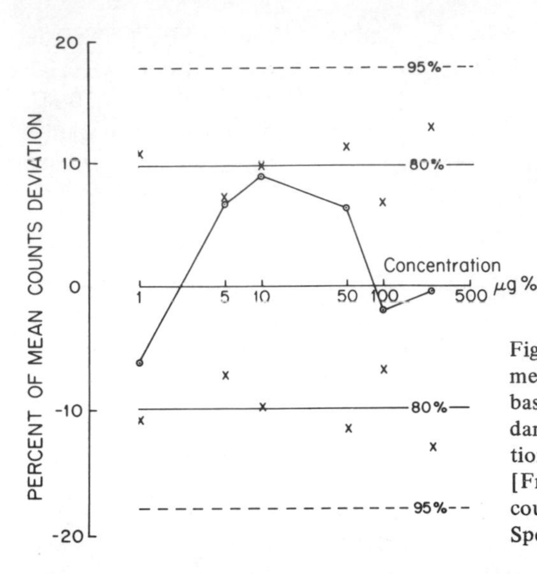

Fig. 3.3 Percentage deviation of experimental (X) and calculated (o) results, based on the mean values for one standardization of copper. The concentrations are shown on a logarithmic scale. [From Johnson *et al.*,[35] p. 552, by courtesy of the Society of Applied Spectroscopy.]

tion of the evaluation of these spectroscopic data. Based on this consideration and the overall appearance of Fig. 3.3, one can see that the generalized functional relationship, Eq. (3.1), fits the data within the range of the experimental error.

3.3.3 Confidence Limits

In the replicate determinations for a given element, the standard deviation, and hence the reproducibility of the counts, is essentially a constant percentage of the observed counts. This is illustrated in Fig. 3.3, where the mean value of the standard deviation of four determinations was ±9.8% of the observed counts at the 80% confidence level. The values are 4.7% and 17.8%, respectively, at the 50% and 95% confidence levels. Even when the number of replicate determinations was 10 or 30, the standard deviation was about 10% of the mean value, at the 80% confidence level, for all six concentrations of this element.

When the four constants in the transcendental function, Eq. (3.1), are known, the reliability of the determination of the concentration of unknowns can be calculated. The first derivative of the function, with deltas substituted for d's, shows how small changes in x influence the y values. Thus,

$$\delta y = B\,\delta x + 10^{C+Dx}(2.303)D\,\delta x \tag{3.5}$$

The δy values for the case of copper are shown in Table 3.5 along with the calculated x and y values. These figures are based on the constants of Eq. (3.4) as applied to Eq. (3.5) with a δx value of $0.1x$. Hence, they are computed at the 80% confidence level. In the particular case of copper, with the mean number of counts of 370.0, which occurred at 10 μg%, the corresponding computed concentration is correctly stated as $y = 11.4 \pm 1.5\ \mu g\%$. This computed value is within the range of the experimental error.

3.3.4 Background

The generalized transcendental relationship, Eq. (3.11), fits the data for any element. Independent of the case type, the approximate functional relationship for small values of y is given by Eq. (3.2). Extrapolating and solving this equation for x when y equals zero, i.e., obtaining the root of the equation, leads to $x' = -A/B$. The term x' is thus the value of the background. Conversely, the background values of the first four points of standardization are equally weighted in the evaluation of the constants A and B. In the example of copper previously cited, the value is $x' = 88$.

3.3.5 Threshold Values

Considering the background value and the errors in the analytical procedure, it is possible to calculate the lowest concentration (threshold concentration) of each element that can be determined. A reasonable definition of threshold concentration, y_0, is $y_0 = \delta y_0$. For small values of y, Eq. (3.2) is approximately correct. The first derivative of this equation, with deltas replacing d's, then leads to $y = B\,\delta x$. In the example given for copper, $\delta x = 0.1x$. This leads to $A + Bx = 0.1Bx$, or $x_0 = 1.1x'$, where x' is the background with a value of 88. Thus, $x_0 = 97$. The corresponding y_0 value, Eq. (3.2), is 0.4 $\mu g\%$. The use of the threshold value is as follows: if the calculated value of an unknown were 0.2 $\mu g\%$, then instead of stating that the copper concentration was $y = 0.2 \pm 0.4\ \mu g\%$, the result would be reported as $y < 0.4\ \mu g\%$. This circumvents the difficulty of obtaining negative concentrations and is a realistic approach to the subject of minimum detectable quantities.

3.3.6 Discussion

A rapid method of checking standardization data is to compute corresponding y values (concentrations) from mean values of observed counts (spectrometer response) obtained for standards. Table 3.6 shows a summary of such a check on the data from one set of standards that was determined in duplicate. For each element the case number is shown and the threshold value is given.

An inspection of the calculated y values in Table 3.6 shows that the poorest results were obtained with aluminum and vanadium. Such results are consistent and can be attributed to the fact that these elements, throughout the entire concentration range, display not only a high background, but also a low slope in the curve relating x with y. Poor results were also obtained for rubidium and zinc, particularly at lower concentrations. It is possible that better results could be ob-

Table 3.6 Summary of Typical Standardization Data for All 17 Elements[a]

Element	Case No.	Computed values (μg%)						Threshold
		1	5	10	50	100	250	
Cu	1	1.0	5.0	10.4	50.2	99.8	249.8	0.5
Fe	2	0.8	5.2	13.5	46.5	103.2	235.0	1.6
Al	1	1.0	5.5	−0.5	69.1	68.2	218.3	5.7
Rb	1	1.0	5.0	1.7	61.6	99.7	249.6	15.5
Au	1	2.0	4.2	8.9	61.0	79.6	229.6	3.3
Ba	2	1.9	4.1	9.4	50.6	93.5	326.6	3.2
V	1	−5.1	15.9	30.1	52.3	45.2	195.2	56.2
Mn	1	1.0	5.1	9.7	52.3	97.6	247.6	1.1
Ni	1	1.1	5.0	10.7	51.2	96.2	246.2	1.1
Cs	1	−0.3	6.4	9.5	51.7	98.2	248.2	4.3
Li	1	0.5	5.5	10.9	49.4	99.9	249.9	4.2
Sr	1	1.0	5.1	11.1	51.2	96.6	246.6	0.5
Cr	3	2.1	5.0	8.9	44.8	94.3	260.8	1.0
Zn	1	5.6	0.4	8.3	51.7	100.0	250.0	13.5
Pb	1	1.7	4.4	9.8	52.7	99.0	249.0	2.3
Mo	1	0.4	6.1	12.5	55.5	94.5	244.5	2.4
Cd	1	−0.7	7.0	14.3	55.0	99.3	249.3	1.5

[a]From Johnson *et al.*,[35] p. 555, by courtesy of the Society of Applied Spectroscopy.

tained by the selection of other prominent lines. However, the choices must be made with due regard for the elements present in the matrix.

An inspection of Table 3.6 and other standardization data makes possible the generalization that in the range of the three lower concentrations, the calculated y values are within 1 or 2 μg% of the observed values. With the higher concentrations, the computed values are within approximately 10% of the observed values. Based on error analyses identical to that for copper, the computed values for concentration are, in general, within the limits of the experimental errors when the values of counts are considered at the 80% confidence level.

With a sample of unknown composition, it is unlikely that the concentration of an element will coincide with that of a standard. Thus, the major use of the transcendental equation is that of interpolation. To investigate this point, samples were made with some of the elements at a concentration of 150 μg%. This appeared to be a good choice, for on a linear scale the 100–250 μg% standards represent a large concentration difference. Moreover, with case 1 elements, the xy curve is rapidly increasing in slope in this third decade of concentration. For standards that contained 150 μg to nickel and zinc, the computer values were 157.8 and 142.1 μg%, respectively. These were case 1 elements. Iron and copper, which are examples of case 2 and case 3 elements, yielded 160.2 and 165.8 μg%, respectively. These results indicate that the generalized equation describes a curve that has the correct shape.

In addition to its use for computational purposes, the method used to fit the data to the generalized equation, wherein y is expressed as a function of x, results in a smoothing action for the data. This is sometimes necessary and totally valid, since emission spectroscopy data never display structured curves. The standardization data of the elements shown in Table 3.6 were all fitted to the generalized equation by use of the method and rules previously described. In this table, two elements, iron and barium, were considered as case 2. Had they been treated as case 3 elements, high errors in the lower concentration ranges would have resulted. Thus, the criteria by which the case numbers are defined were not arbitrarily fixed, but were chosen to give repeated best results for all 17 elements. Moreover, these rules can be readily expressed in computer language.

The reason that some elements behave in the manner of cases 2 and 3 can be explained in terms of background counts, linear-phototube response, and proportionality of line intensity with concentration. At low concentrations, the rationale for case 1 elements is the same, but at higher concentrations, self-absorption apparently occurs, resulting in line reversal. Although the exponential term in the transcendental equation is empirical, it adequately adjusts mathematically for the nonlinearity of x and y within the ranges of the variables encountered in this study. Line reversal is not predicted by the complete equation.

A comparison among numerous standardizations shows that for each element, the constants A', B, C, and D vary from one standardization to the next, and both C and D exhibit marked changes. If only A and B were to change, the shifts might be attributable to a change in the background counts. The observation that all four constants may vary and the case numbers change for such elements as barium and iron suggests that something more is involved. Several variable transformations were tried in an effort to establish a coincidence in the curves that represented two different standardizations of the same element. In this operation, vertical and horizontal movements, as well as rotation, of the curves were tried. None of these transformations, either singly or in combination, was successful. This appears to rule out the possibility of obtaining one master curve to describe the relationship between x and y for a given element and then shifting the curve in some manner to describe a second standardization, wherein the latter is based upon a single concentration and the corresponding mean counts. These variations are no doubt explained by the slight movement of the exit slits, which is attributable to minor vibrations or shifts of the building and to small changes of ambient temperature in the laboratory. The problem is largely circumvented by making daily standardizations.

Table 3.4 illustrates the application of the generalized transcendental equation coupled with computer handling of the data for the analysis of trace elements in the blood serum of normal adults. The concentrations of all these metals in blood serum are at least twice as great as the threshold values. It can be seen that this method gives highly acceptable values.

We and others[41-43] have found that the saving of time and the reduction of calculating errors are inherent advantages of computerized computation of emission spectrometric data. The results of quadruplicate determination on 50 unknown samples can be calculated in less than 10 min of computer time. This can be contrasted to the 40 man-hours of hand calculations, where the spectrometer readings are averaged on a desk calculator and corresponding concentrations are obtained by graphic interpolation.

The method described here was originally used with a manual keypunch machine located on the console of the spectrometer. The spectrometer operator directly transcribed the spectrometer readings to punched cards. We have more recently interfaced the spectrometer directly to a teletype machine, which provides a record of the spectrometer responses both in typewritten form and in the form of punched tape. At the end of the day, the punched tape is carried to the computer laboratory for data processing. Alternatively, the teletype machine could be interfaced directly to a time-shared computer to provide immediate data processing and retrieval of the results. The subject of interfacing instruments to computers to facilitate the handling of analytical data is currently a topic of interest to many investigators.[44-46] Spectrographic instruments and analog computers with automatic digital-data transmission have been installed in some steel mills to achieve automation in the routine analysis of metal production samples.[47]

Although the present work was done with biological specimens, the mathematical procedure would no doubt be applicable to other systems. With digital computers, once the mathematical model, i.e., the generalized functional relationship between x and y, is known and the program written, the computation of element concentration is straightforward.[48]

3.4 STUDIES ON TRACE METALS IN PATIENTS WITH RHEUMATOID ARTHRITIS

Although many trace metals have been identified in mammalian tissues, only seven (Cu, Fe, Mn, Co, Cr, Zn, Mo) are known to possess essential functions in metabolism. Rarely have the etiology or the pathogenesis of chronic human diseases been related to deficiency, toxicity, or abnormal metabolism of either the essential or the nonessential trace metals.[13] Any chronic disease, for which the cause has not been established, might be related to an abnormality in trace metal metabolism. In this respect rheumatoid arthritis is a disease of particular interest. Many of the therapeutic agents used in the treatment of this disease have the capacity to form complexes with one or more of the trace metal ions.[49,50] The administration of gold[51] in treating arthritis has stimulated interest because it may compete with other trace metals as prosthetic groups, as cofactors, or for active sites in enzymes.

Several trace metals have been shown to be active in connective tissue metabolism. Zinc has a marked influence on growth and skeletal development.[52] Disturbances in mucopolysaccharide synthesis have been noted in experimental animals which have a manganese deficiency.[53–55] Manganese turnover is low in patients with rheumatoid arthritis; treatment with corticosteroids restores the turnover rate to normal.[56] Copper plays an essential role in the biosynthesis of elastin.[57]

It was shown[58, 59] that the low viscosity of the synovial fluid in patients with rheumatoid arthritis was at least partly due to the low molecular weight of the hyaluronic acid which it contained. Investigations[60] directed at elucidation of factors that depolymerize purified hyaluronic acid *in vitro* revealed that cuprous and ferrous ions in trace amounts actively evoked depolymerization of the polysaccharide. Subsequent studies[61] showed that, in patients with rheumatoid arthritis, the concentration of copper in synovial fluid was approximately three times greater than that in normal subjects. The presence of ceruloplasmin accounted for most of the excess of copper in both blood serum and synovial fluid of these patients. It was more recently demonstrated[62] that ceruloplasmin fails to depolymerize hyaluronic acid and, in the presence of ascorbic acid, may even inhibit depolymerization of high-molecular-weight hyaluronic acid.[63]

In early spectrographic studies[64] synovial fluid from patients with rheumatoid arthritis was shown to contain elevated concentrations of iron, aluminum, and copper. Abnormal concentrations of several other metals were suggested by these studies, but the semiquantitative methods used led to inconclusive results.

Continued interest in the possible role of trace elements in this disease prompted investigations in which emission spectrometric analyses were used to determine the concentrations of 14 trace elements in blood serum and synovial fluid of patients with rheumatoid arthritis. It was anticipated that the precision of this analytical method would resolve some of the questions and extend the observations of previous investigations.

3.4.1 Collection of Specimens

Fifty specimens of synovial fluid were aspirated from the knee joints of 44 patients who had rheumatoid arthritis with effusion. In a similar manner, 50 specimens were collected from 50 patients seen at postmortem. The deaths were attributable to various causes but there was no evidence of connective tissue disease. Specimens in which blood could be visually detected were discarded. The mean age of the patients with rheumatoid arthritis and of the subjects seen at postmortem was 45 and 51 years, respectively.

One hundred and five blood specimens were collected from 68 patients with rheumatoid arthritis and a like number from normal volunteers who included

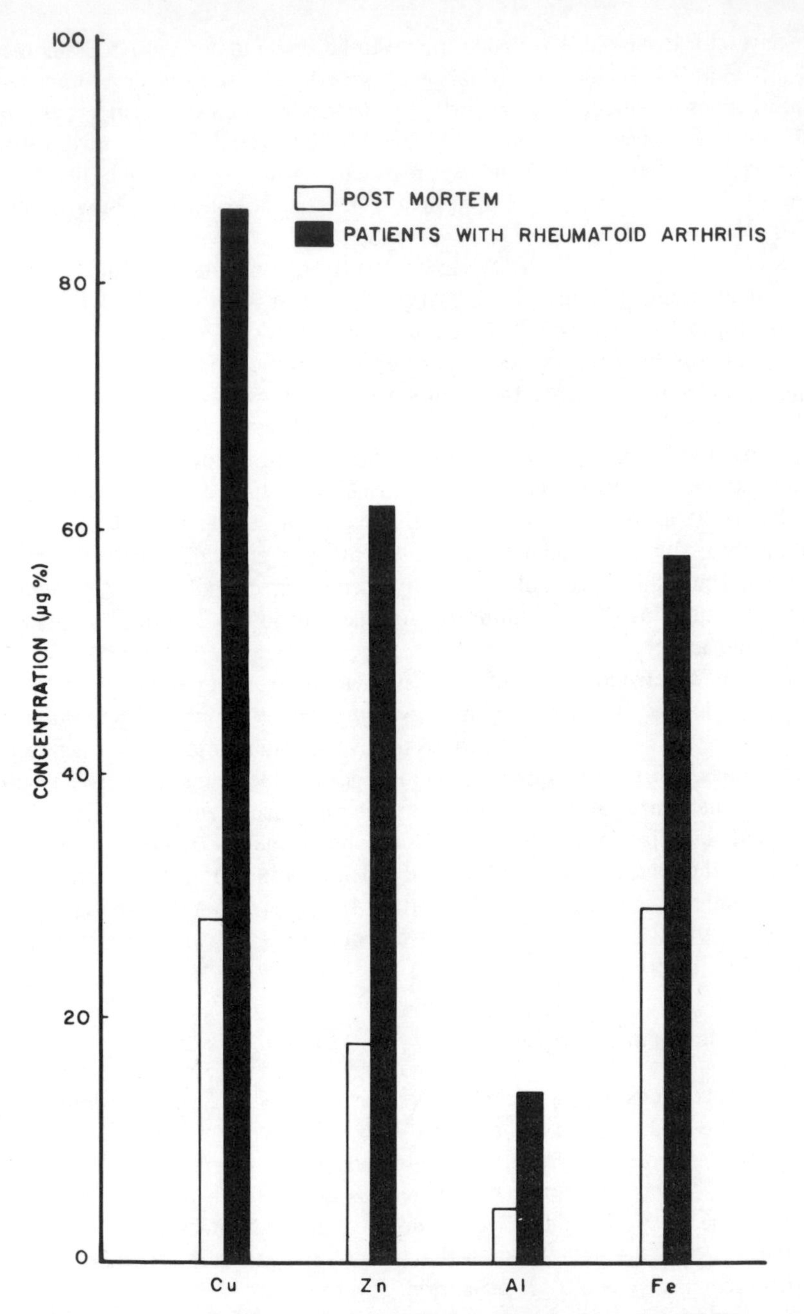

Fig. 3.4 Concentrations of copper, zinc, aluminum, and iron in synovial fluid of patients with rheumatoid arthritis. Mean concentrations were calculated from results obtained for 50 patients with rheumatoid arthritis and 50 patients seen at postmortem who were without evidence of connective-tissue disease.

laboratory, office, and maintenance personnel. The mean ages of the patients with rheumatoid arthritis and of the normal volunteers were 51 and 31, respectively. All specimens were ashed by the wet digestion procedure and analyzed by emission spectroscopy as described in Section 3.2.

3.4.2 Results

The mean values, calculated from the results obtained on 50 specimens of synovial fluid from patients with rheumatoid arthritis and 50 postmortem specimens, are compared in Figs. 3.4 and 3.5. Certain trace elements were either absent or appeared in concentrations below the limits of detection of our method. Copper and iron were detected in all specimens analyzed. Strontium was detected in all postmortem specimens but in only 78% of the specimens from patients with rheumatoid arthritis. Zinc was detected in only 48% of the postmortem specimens but in 88% of the specimens from patients with rheumatoid arthritis. The incidence of detection of most other elements was greater in postmortem specimens than in specimens from the patients with rheumatoid arthritis.

In synovial fluid from patients with rheumatoid arthritis, the mean concentrations of copper, iron, aluminum, and zinc were significantly elevated ($p <$

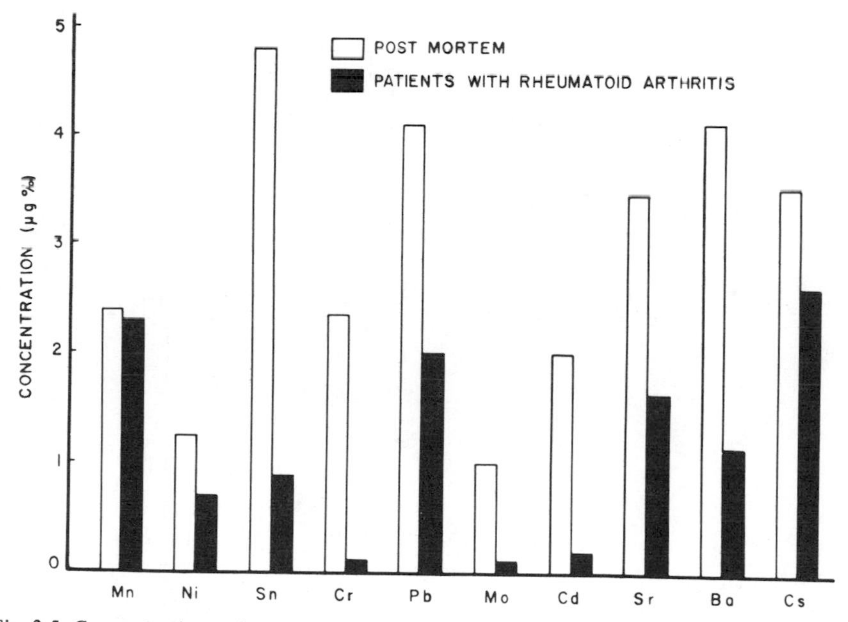

Fig. 3.5 Concentrations of manganese, nickel, tin, chromium, lead, molybdenum, cadmium, strontium, barium, and cesium in synovial fluid of patients with rheumatoid arthritis. Mean concentrations were calculated from results obtained for 50 patients with rheumatoid arthritis and 50 patients seen at postmortem who were without evidence of connective-tissue disease.

0.001) whereas the mean concentrations of barium, tin, strontium, and cadmium were significantly lower ($p > 0.001$). Although the mean values for some of the other metals were markedly different in synovial fluid of patients with rheumatoid arthritis when compared with the controls, statistical analysis of the data revealed that the apparent differences noted for manganese, nickel, cesium, chromium, lead, and molybdenum were not significant ($p > 0.001$).

Figures 3.6 and 3.7 compare the mean values calculated from the results of analysis of 105 blood serum specimens from patients with rheumatoid arthritis with the same number from normal volunteers who had no signs of connective tissue disease. Copper, strontium, and zinc were detected in all specimens of blood serum analyzed. Iron was detected in all the normal specimens and in all except three of the rheumatoid specimens analyzed. With the exception of lead, all other elements were detected in more of the rheumatoid than in normal specimens. The mean concentrations of copper, barium, cesium, tin, manganese, and molybdenum were significantly higher ($p > 0.001$) in blood serum from patients with rheumatoid arthritis than in that of normal patients. The

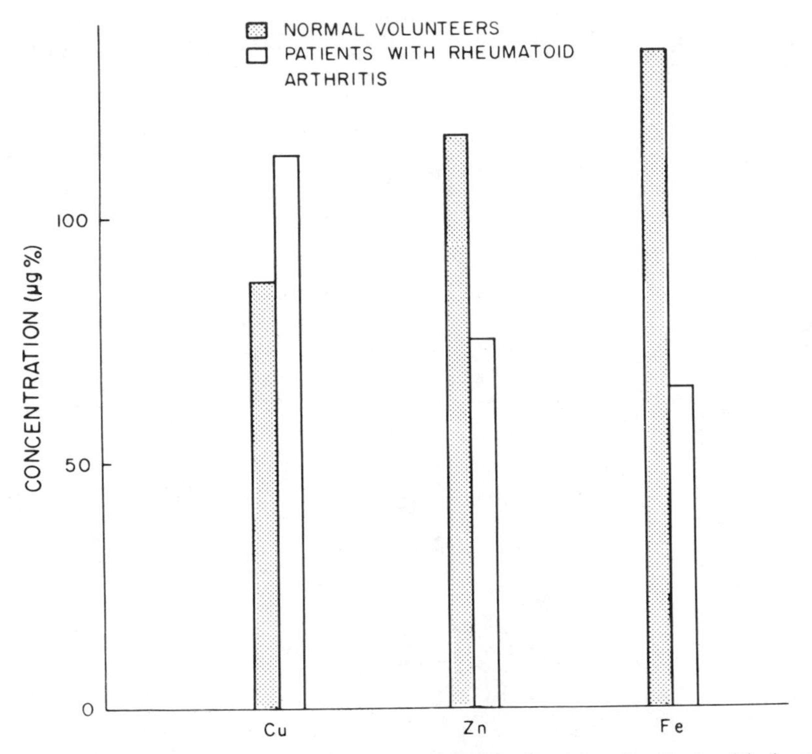

Fig. 3.6 Concentrations of copper, zinc, and iron in blood serum of patients with rheumatoid arthritis. Mean concentrations were calculated from results obtained for 105 patients with rheumatoid arthritis and 105 normal volunteers.

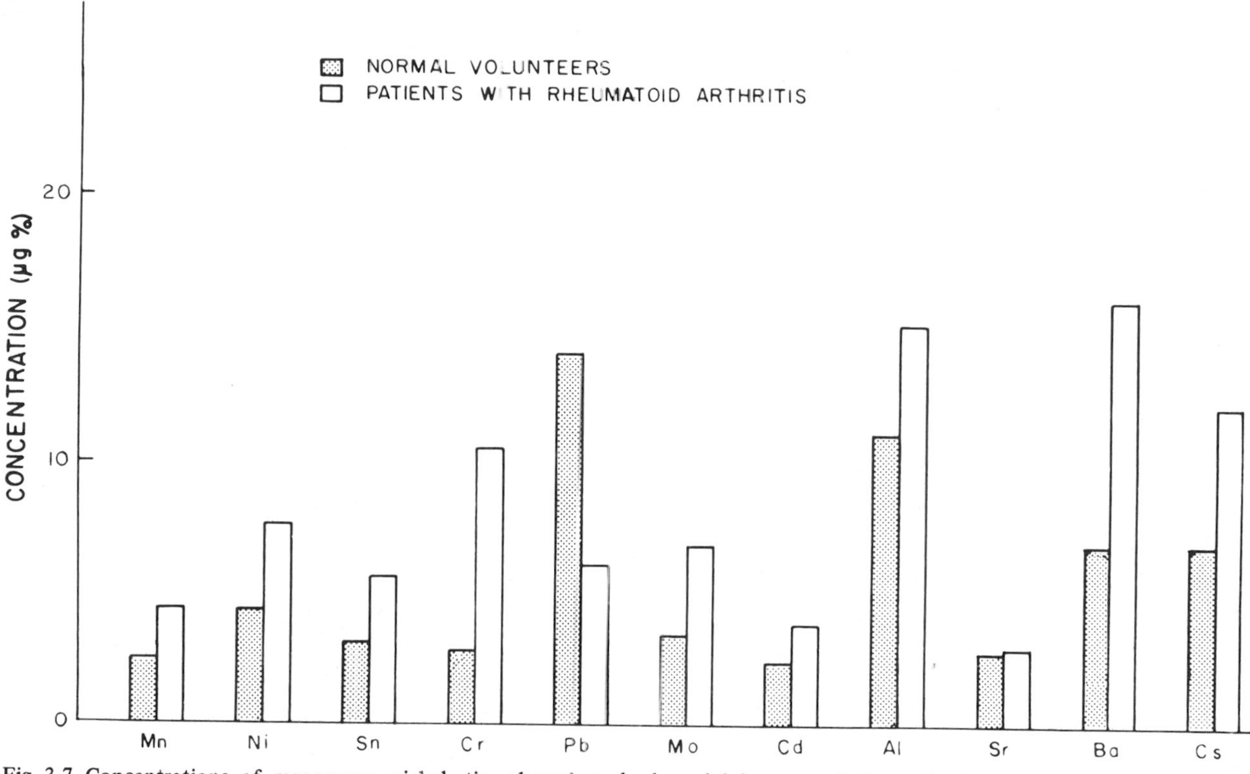

Fig. 3.7 Concentrations of manganese, nickel, tin, chromium, lead, molybdenum, cadmium, aluminum, strontium, barium, and cesium in blood serum of patients with rheumatoid arthritis. Mean concentrations were calculated from results obtained for 105 patients with rheumatoid arthritis and 105 normal volunteers.

mean concentrations of iron, zinc, and lead were significantly lower ($p < 0.001$) in blood serum from patients with rheumatoid arthritis than in normal blood serum. No significant differences ($p > 0.001$) were noted for aluminum, nickel, strontium, chromium, or cadmium.

3.4.3 Discussion

Copper, iron, manganese, chromium, zinc, and molybdenum are probably essential elements for one or more mammalian species.[13] Aluminum, barium, nickel, cesium, tin, lead, strontium, and cadmium appear to serve no physiological function in mammalian metabolism and their appearance in tissue may be adventitious and associated only with their wide distribution in nature. Either an excess or a deficiency of an essential trace element might have pathological consequences, but only an excess of a nonessential trace element would be expected to evince biological significance. The possibility remains, however, that a physiological function may be discovered in the future for some of the metals that are presently considered nonessential.

Statistical evaluation of the data obtained for blood serum indicated that, of the essential trace elements for which analyses were performed, the mean concentrations were higher for copper, molybdenum, and manganese and lower for iron and zinc in rheumatoid than in normal blood serum. The concentrations of strontium and chromium were essentially the same in rheumatoid and normal blood serum. Evaluation of the data obtained for synovial fluid indicated that the mean concentrations were higher for copper, iron, and zinc in rheumatoid than in postmortem synovial fluid. The concentrations of manganese, molybdenum, and chromium were essentially the same in both rheumatoid and postmortem synovial fluid.

Among the nonessential trace elements[13] for which analyses were performed, only barium, cesium, and tin appeared in higher concentrations in rheumatoid than in normal blood serum. The concentrations of all other nonessential trace elements in rheumatoid and normal blood serum and synovial fluid were either the same or were lower in the rheumatoid patients. Therefore, if trace metal metabolism is related to the etiology or pathogenesis of rheumatoid arthritis, the metals most likely to be involved are copper, iron, barium, cesium, tin, zinc, molybdenum, manganese, and strontium. These metals were the subject of further investigations, which are described in Section 3.5.

The effects of postmortem changes on the composition of synovial fluid are not well understood. To minimize these changes, the specimens used in the present studies were collected as soon as possible after the death of the patient. Early studies showed that postmortem synovial fluid had a lower hyaluronic acid concentration than that from normal living volunteers,[59] although no differences in intrinsic viscosity of the polysaccharide could be detected. The concentrations of total copper and ceruloplasmin were essentially the same in both postmortem and normal synovial fluid.[61]

Copper was the only trace metal found to occur in higher concentrations in both blood serum and synovial fluid of patients with rheumatoid arthritis. The high levels of serum copper are not specific for rheumatoid arthritis but have been observed in patients with various other chronic diseases.[16] Ceruloplasmin accounts for the excess of copper in both serum and synovial fluid from patients with rheumatoid arthritis.[61] The concentration of nonceruloplasmin copper was essentially the same in rheumatoid synovial fluid and blood serum as it was in the respective fluids obtained from normal subjects. In contrast to this observation is the report[65] that nonceruloplasmin copper was markedly elevated in blood serum from patients with rheumatoid arthritis. A third group of investigators[66] more recently reported that nonceruloplasmin copper levels are nearly the same in rheumatoid and normal patients. These discrepancies, which are apparently attributable to differences in analytical methods,[67] require further study for resolution.

The increased concentration of copper in synovial fluid may be a reflection of the elevation in serum copper, or it may result from an increased permeability of the synovial membrane. Attempts to relate high copper concentration with the low viscosity of synovial fluid from patients with rheumatoid arthritis were unsuccessful.[58]

The anemia associated with rheumatoid arthritis apparently accounts for the low serum levels of iron. The high concentration of iron in synovial fluid of patients with this disease may be related to the recent observation that the iron content of the synovial membrane of these patients was markedly increased over that of normal subjects. Histological demonstration of iron deposits in synovial membrane is a constant feature of the pathology of rheumatoid arthritis.[68]

The concentration of zinc in whole blood of patients with rheumatoid arthritis has been reported to be normal.[69] Recent reports indicate that dietary supplements of this element given to man facilitates normal healing[70] and reverses the atherosclerotic process.[71] The relationship of these observations to the results of the present studies is not apparent.

A recent study[72] indicated that the rate of total body turnover of manganese was low in patients with active rheumatoid arthritis. The mean concentrations of serum manganese, which were obtained by activation analysis, were markedly lower than the same values in the present study. Also, in contrast to the results of the present study, it was reported that the mean serum manganese level of patients with rheumatoid arthritis was the same as that of the control subjects. Red blood cells from patients with rheumatoid arthritis, however, contained significantly higher concentrations of manganese than those from normal subjects.[73]

Changes in the composition of synovial fluid that accompany rheumatic diseases were considered by earlier investigators to be due to changes in the permeability of the synovial membrane,[58] which was regarded as a filtering device. It was held that the membrane pores were enlarged by rheumatic diseases, and thus allowed more rapid passage of blood plasma constituents into the synovial fluid. This hypothesis accounted for the higher concentrations of proteins in synovial fluid from patients with rheumatoid arthritis. More recent studies[61, 74, 75] on the distribution of specific plasma proteins suggest a more complex action, in which specific mechanisms in the synovial membrane regulate transport from blood serum to synovial fluid. The mean concentration of aluminum, barium, tin, strontium, molybdenum, and cadmium in synovial fluid were lower in rheumatoid arthritics than in the postmortem controls. In blood serum, however, the mean concentrations of these metals were higher in patients with rheumatoid arthritis than in the normal subjects. This observation suggested that transport across the synovial membrane involved processes other than passive diffusion.

The elevated concentrations of trace elements in synovial fluid might also be attributed to an increase in their release from cartilage and synovial cells or to increased lysis of white blood cells because of increased rates in turnover of these tissues during periods of inflammation. The mean concentrations of zinc and iron for rheumatoid patients were higher in the synovial fluid and lower in blood serum than that in the normal postmortem controls. This finding suggests that these two metals may appear in synovial fluid of patients with rheumatoid arthritis from a source other than blood serum.

The significance of these observations in respect to the pathogenesis of rheumatoid arthritis is not known. It would appear that abnormal trace metal metabolism may be the result of a chronic inflammatory condition. Changes in trace metal metabolism may reflect nonspecific physiologic responses of the organism to pathologic stimuli. These changes may be analogous to the nonspecific elevation in erythrocyte sedimentation rate, or to the increased concentration of acute phase reactants in blood serum of patients with chronic inflammatory diseases. To determine whether or not the observations reported here are specific for rheumatoid arthritis, further study will be needed on trace metal metabolism of patients with other rheumatic diseases, and of patients with chronic inflammatory diseases who have no joint involvement.

3.5 EFFECT OF CHRYSOTHERAPY ON TRACE METALS

Gold salts, which were effective in treating several chronic infectious diseases were used in the treatment of rheumatoid arthritis on the assumption that the disease was of infectious origin.[76] Studies to elucidate pathways of gold metabolism[77, 78] failed to reveal the mechanism by which this heavy metal produces its beneficial effects in rheumatoid arthritis. It was shown by isotope studies that gold enters synovial fluid, can be phagocytized, and is found in the proliferating inner cell lining of the synovial membrane.[79, 80] It inhibits the activity of such enzymes as glucosaminidase,[81] β-glucuronidase,[82] acid phosphatase,[82] and cytofibrokinase[83]; can uncouple oxidative phosphorylation[84]; increases the cross-linkages in collagen by forming complexes with organically bound sulfur[85]; and can block the sulfhydryl groups of glutathione and cystine.[76] The activity of many enzymes, which are required for biosynthesis of mucopolysaccharides, are contingent on the presence of certain trace metals. The observation[86] that abnormal concentrations of these trace metals occur in blood serum and synovial fluid of patients with rheumatoid arthritis suggested that the therapeutic action of gold might be mediated by metabolic action on a trace metal. It was postulated that gold might displace certain essential trace metals from their combination with biologically active compounds that normally served as prosthetic groups or cofactors, thereby modifying the biologic activity of the compound in such a way as to alter a series of disease-producing events.

The presence of higher than normal concentrations of certain nonessential trace elements in blood serum and synovial fluid of patients with rheumatoid arthritis suggested that alternatively the activity of some biologically active compounds might be modified if they complexed with these nonessential, adventitious metals, so as to create a metabolic lesion. It was further postulated that gold might displace this trace metal (which is normally not present) from its combination with the biologically active compound, and thus unblock the metabolic lesion. It, therefore, seemed expedient to determine whether chrysotherapy influenced the concentrations of certain trace metals reported[86] to be abnormal in blood serum of patients with rheumatoid arthritis.

3.5.1 Methods

Seventeen patients with rheumatoid arthritis who were on long-term gold therapy were included in this study. The average age of the subjects was 54.5 years, and the average duration of their diseased condition was 9.4 years with a range of 2 and 25 years. All had been on myochrysine therapy for an average of 4.3 years with a range of 18 months to 12 years. Sixty-seven patients with rheumatoid arthritis who had never received chrysotherapy were arbitrarily selected from the outpatient clinic, without regard for age, medication, or for the duration or stage of the disease. Their average age was 51 years. A second group of control subjects consisted of 105 normal volunteers, arbitrarily selected from laboratory, office, and maintenance personnel. The average age of this group was 31 years. Specimens from the patients were ashed by a wet-digestion procedure and analyzed by the spectrochemical method already described.

3.5.2 Results

The trace metals studied and a summary of the analytical results are listed in Fig. 3.8. Copper, iron, manganese, zinc, and molybdenum are considered essential for one or more mammalian species.[16] Statistical evaluation[87] of the data obtained for these essential metals indicated that the mean concentrations of

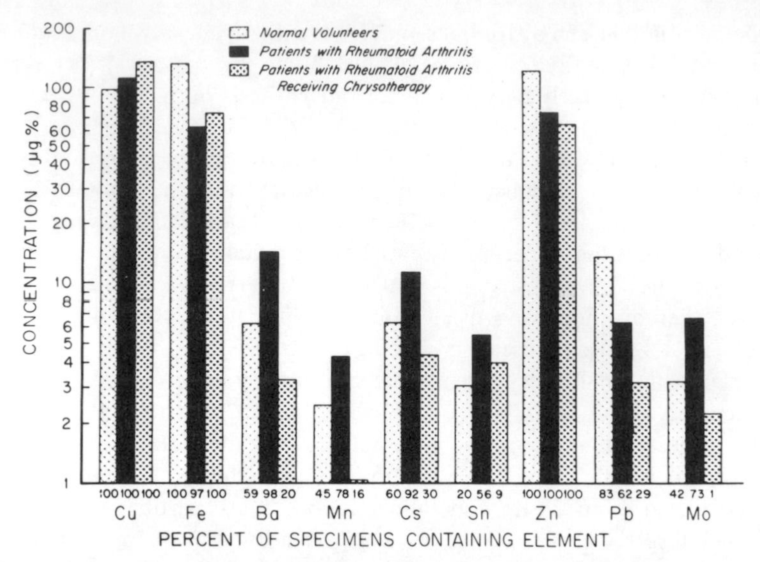

Fig. 3.8 Effect of chrysotherapy on trace metal composition of blood serum of patients with rheumatoid arthritis. Mean concentrations were calculated from results obtained for 105 patients with rheumatoid arthritis, 105 normal volunteers, and 67 patients with rheumatoid arthritis who were receiving chrysotherapy.

copper, manganese, and molybdenum were higher in rheumatoid than in normal blood serum. Even higher blood serum copper concentrations were noted in patients who received chrysotherapy than in those who did not. The elevation of molybdenum and manganese levels in blood serum of patients with rheumatoid arthritis were lower in the patients who received chrysotherapy. The mean concentrations for these metals appeared to be even lower than those observed for the normal subjects, but statistical evaluation of the data showed that these differences were not significant.

The mean concentrations of iron and zinc, the other two essential trace metals studied, were lower in blood serum of rheumatoid arthritics than in blood serum from the normal control group. Chrysotherapy appeared to have no influence on the concentration of iron or zinc in blood serum of patients with rheumatoid arthritis.

The mean concentration of the nonessential trace metals, tin, barium, and cesium, were significantly higher in blood serum of patients with rheumatoid arthritis than in the normal controls. Chrysotherapy lowered the concentrations of these metals below the mean concentrations of the normal subjects. These results are consistent with the hypothesis that gold may displace some other metal ion from a biologically active compound and thereby modify a series of disease-producing events. It is further suggested that of the metals studied: molybdenum, manganese, tin, barium, and cesium are the ones most likely to be involved in such disease-producing events. The observation that the concentrations of these metals in blood serum responded to chrysotherapy by shifting toward the concentrations observed in normal blood serum is consistent with the earlier postulation on the mechanism of action for gold.

3.6 TRACE METALS IN DENTAL CARIES

At least 44 trace elements have been identified as being either constituents or adventitious contaminants of human enamel.[88] The approximate concentrations ranged from below 1 ppm to between 100 and 1000 ppm. It has been reported[89] that some trace metals have a tendency to accumulate in surface enamel in amounts that vary according to the external environment of the tooth. Wah Leung[90] has reviewed the evidence supporting the finding that exposure to saliva may produce chemical alterations in enamel through ion exchange and/or ion capture.

Despite considerable interest in the microelements as possible caries determinants,[91] comparatively few investigations have been made on the trace mineral content of human saliva[92–95] and these have been limited in scope for experimental or technical reasons. The present study was undertaken to gain further insight into the microelemental composition of primate saliva by comparing the levels of 14 different trace metals in caries-prone (man) and caries-resistant (marmoset) species when the two were subjected to identical and to different salivary stimulants.[96]

3.6.1 Collection of Samples

Thirty pilocarpine- and 30 paraffin-stimulated samples containing 2–3 ml of fasting human whole saliva were collected, and immediately frozen, in sterile polystyrene 15- × 100-mm test tubes fitted with plastic caps. All 30 donors had historical evidence of caries experience. Each contributed one pilocarpine- and one paraffin-elicited specimen on separate occasions. Pilocarpine was administered subcutaneously at a dose level of 6 mg of the hydrochloride salt.

Marmoset saliva was provided by 30 healthy, caries-free, adult, cotton-top animals (*Saguinus oedipus*). They were housed in all-plastic cages with wooden perches for 48–72 hr and fasted overnight. Prior to sampling, each marmoset was sedated with 0.25 mg of sernyl given intramuscularly. They were then injected subcutaneously with 0.5 mg of pilocarpine hydrochloride/100 g of body weight and a 2- to 3-ml saliva sample was obtained by holding the sedated animal over a plastic funnel inserted into the collection tube. The entire procedure was repeated on a subsequent day using 0.12 mg of mecholyl (acetyl-β-methylcholine chloride) per 100 g of body weight as the salivary stimulant. Each specimen was immediately capped and stored in a deep freeze until the entire series was ready for analysis by the emission spectrometric method already described (Section 3.2).

3.6.2 Results

The distribution of the number of trace elements in the human and marmoset saliva samples arranged according to type of stimulant is depicted in Table 3.7. The collection times in minutes for the 2- to 3-ml volumes used in the analyses were 0.7-3.0 (human-paraffin), 1.8-3.2 (human-pilocarpine), 12.3-29.2 (marmoset-pilocarpine), and 14.2-30.3 (marmoset-mecholyl). There were no significant intragroup correlations between the number or quantity of trace minerals and the flow rate.

The results of the qualitative and quantitative analyses for the trace metals in human whole saliva are summarized in Table 3.8. Mean concentrations in the pilocarpine-stimulated salivas exceeded those in the paraffin-induced specimens for all trace metals tested except copper, tin, aluminum, and strontium. Statistically only, the mean difference in the tin and chromium content were significant ($p = <0.001$).

The frequency of detection and concentration patterns of the test trace metals in pilocarpine- and mecholyl-stimulated marmoset whole saliva are shown

Table 3.7 Distribution of Detectable Trace Metals in Saliva Samples from Man and Marmoset[a]

Number of trace metals in sample	Human saliva (paraffin)		Human saliva (pilocarpine)		Marmoset saliva (pilocarpine)		Marmoset saliva (mecholyl)	
	No.	%	No.	%	No.	%	No.	%
1								
2								
3								
4	3	10.0						
5			1	3.5				
6	8	26.7	2	6.9				
7	5	16.7	4	13.8				
8	6	20.0	4	13.8	2	6.9		
9	2	6.7	4	13.8	1	3.5		
10	3	10.0	6	20.8	1	3.5		
11	2	6.7	4	13.8	2	6.9		
12	1	3.3	1	3.5	9	31.1		
13			2	6.9	8	27.7	11	36.7
14			1	3.5	6	20.8	19	63.3
	30		29		29		30	

[a]From reference 96, p. 183, by courtesy of Pergamon Press.

in Table 3.9. Quantitatively, the mean levels contained in the pilocarpine-stimulated salivas were higher than the mecholyl-induced salivas for iron, molybdenum, copper, tin, aluminum, strontium, and barium and lower for zinc, cadmium, lead, nickel, manganese, chronium, and cesium. The mean differences for zinc, nickel, aluminum, manganese, chromium, strontium, and barium were statistically significant ($p = <0.05$ to $p = <0.001$).

The mean concentrations of the 14 trace metals in human and marmoset saliva obtained in response to pilocarpine are contrasted in Fig 3.9. Mean values for marmoset saliva exceeded those of human saliva for all the trace metals tested except tin and chromium. All differences between means, exclusive of tin and cadmium, were statistically impressive ($p = <0.05$ to $p = <0.001$).

3.6.3 Discussion

Chemically, marmoset saliva is considerably more alkaline and has a substantially greater buffer (sodium, calcium, and bicarbonate) content than human saliva. The present findings demonstrate that marmoset saliva is also more richly endowed with trace metals per unit volume than the human secretion when the same salivary stimulant is used to induce flow. Thus, it would appear that the secretory cells of the marmoset salivary glands may be more permeable to most of the metallic ions present in blood and/or that the salivary secretory process in the marmoset glands is less discriminating than that in man.

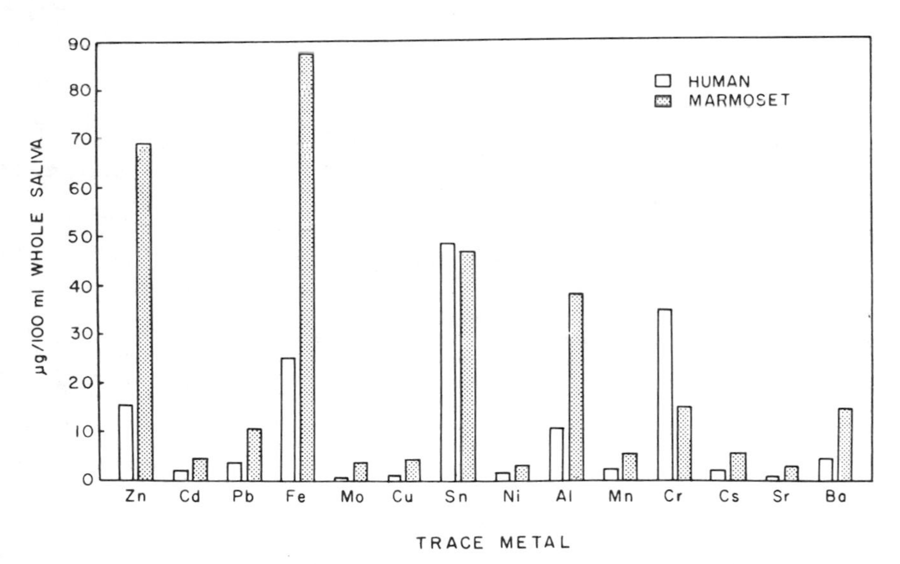

Fig. 3.9 Trace metal concentrations in pilocarpine-stimulated human and marmoset saliva. [From Dreizen *et al.*,[96] p. 186, by courtesy of Pergamon Press.]

Table 3.8 Trace Metal Content of Stimulated Human Whole Saliva[a]

Metal	Stimulant	Positive samples No.	Positive samples (%)	Range (μg%)	Mean (μg%)	Standard deviation	Mean difference (μg%)	p
Zinc	Pilocarpine[b]	10	34.6	<13.5–137.1	16.5	30.0	>3.0	>0.05
	Paraffin[c]	28	93.3	<13.5–90.5	<13.5	17.8		
Cadmium	Pilocarpine	12	41.5	<1.5–11.0	2.2	3.2	>0.7	>0.05
	Paraffin	4	13.3	<1.5–10.4	<1.5	2.2		
Lead	Pilocarpine	22	76.1	<2.3–16.4	3.7	3.7	0.2	>0.05
	Paraffin	28	93.3	<2.3–9.9	3.5	2.2		
Iron	Pilocarpine	29	100.0	<1.0–92.1	25.0	29.3	7.9	>0.05
	Paraffin	30	100.0	<6.5–46.2	17.1	8.2		
Molybdenum	Pilocarpine	8	27.7	<2.4–3.3	<2.4	1.0		
	Paraffin	4	13.3	<2.4–3.7	<2.4	0.8		
Copper	Pilocarpine	23	79.6	<0.5–8.9	1.1	1.7	0.8	>0.05
	Paraffin	28	93.3	<0.5–6.9	1.9	1.9		

Tin	Pilocarpine	24	83.0	<0.5–75.0	48.3	24.3	16.5	>0.005
	Paraffin	30	100.0	<25.0–119.3	64.8	20.7		
Nickel	Pilocarpine	18	62.3	<1.1–12.1	1.4	2.5	>0.3	>0.05
	Paraffin	6	20.0	<1.1–2.7	<1.1	0.6		
Aluminum	Pilocarpine	18	62.3	<5.7–43.1	10.4	12.2	0.1	>0.05
	Paraffin	14	46.7	<5.7–46.0	10.5	14.3		
Manganese	Pilocarpine	19	65.7	<1.1–9.3	2.1	2.7	0.4	>0.05
	Paraffin	12	40.0	<1.1–8.8	1.7	2.7		
Chromium	Pilocarpine	21	72.7	<1.0–110.0	34.7	20.8	26.1	>0.001
	Paraffin	27	90.0	<1.0–32.0	8.6	7.3		
Cesium	Pilocarpine	23	79.6	<4.3–6.8	<4.3	1.6		
	Paraffin	1	3.3	<4.3–1.0	<4.3	0.2		
Strontium	Pilocarpine	15	51.9	<0.5–2.4	0.5	0.6	0.8	>0.05
	Paraffin	11	36.7	<0.5–8.9	1.3	2.3		
Barium	Pilocarpine	26	90.0	<3.2–11.5	4.1	2.7	0.9	>0.05
	Paraffin	2	6.7	<3.2–1.3	<3.2	0.3		

[a] From Dreizen *et al.*,[96] p. 184, by courtesy of Pergamon Press.
[b] 29 samples.
[c] 30 samples.

Table 3.9 Trace Metal Content of Stimulated Marmoset Whole Saliva[a]

Metal	Stimulant	Positive samples No.	Positive samples (%)	Range (μg%)	Mean (μg%)	Standard deviation	Mean difference (μg%)	p
Zinc	Pilocarpine[b]	28	96.9	<13.5–320.0	69.0	64.5	33.1	<0.05
	Mecholyl[c]	30	100.0	30–220.0	102.1	45.3		
Cadmium	Pilocarpine	13	45.0	<1.5–34.4	4.4	7.6	3.4	>0.05
	Mecholyl	30	100.0	2.0–22.0	7.8	5.6		
Lead	Pilocarpine	20	69.2	<2.3–46.0	10.7	10.8	3.0	>0.05
	Mecholyl	30	100.0	5.5–34.6	13.7	7.3		
Iron	Pilocarpine	29	100.0	15.0–378.5	87.2	92.4	21.6	>0.05
	Mecholyl	30	100.0	11.0–376.2	65.6	80.0		
Molybdenum	Pilocarpine	26	90.0	<2.4–19.4	3.7	4.0	1.0	>0.05
	Mecholyl	21	70.0	<2.4–10.0	2.7	3.1		
Copper	Pilocarpine	29	100.0	0.8–10.0	4.5	2.2	1.1	>0.05
	Mecholyl	30	100.0	0.9–12.6	3.4	2.4		

Tin	Pilocarpine	29	100.0	1.0–110.0	46.7	25.0	5.8	>0.05
	Mecholyl	30	100.0	1.0–71.1	40.9	18.4		
Nickel	Pilocarpine	17	58.8	<1.1–15.3	3.0	4.0	8.2	<0.001
	Mecholyl	30	100.0	5.0–19.0	11.2	4.5		
Aluminum	Pilocarpine	26	90.0	<5.7–85.0	38.1	26.6	20.5	<0.05
	Mecholyl	29	96.7	<5.7–52.0	17.6	13.2		
Manganese	Pilocarpine	21	72.7	<1.1–18.8	5.3	5.2	4.1	<0.01
	Mecholyl	30	100.0	2.3–22.8	9.4	6.2		
Chromium	Pilocarpine	27	93.4	<1.0–34.0	14.9	9.5	9.3	<0.005
	Mecholyl	30	100.0	4.0–62.0	24.2	14.1		
Cesium	Pilocarpine	29	100.0	0.6–23.0	5.4	4.6	1.5	>0.05
	Mecholyl	30	100.0	0.8–59.0	6.9	10.3		
Strontium	Pilocarpine	29	100.0	1.0–9.5	2.8	1.8	0.8	<0.05
	Mecholyl	29	96.7	<0.5–4.5	2.0	1.2		
Barium	Pilocarpine	29	100.0	1.5–33.8	14.4	7.4	6.5	<0.01
	Mecholyl	30	100.0	1.0–72.0	7.9	12.5		

[a] From Dreizen *et al.*,[96] p. 185, by courtesy of Pergamon Press.
[b] 29 samples.
[c] 30 samples.

The data clearly indicate that, in response to various stimuli and to varying intensities of the same stimulus, the salivary glands produce secretions that differ in inorganic composition. The number and amount of test trace metals was almost always lower in paraffin-stimulated saliva than in pilocarpine-elicited saliva from the same subject. Similarly, the frequency and magnitude of the measurable trace metals in marmoset saliva induced by mecholyl (a pure parasympathomimetic sialogogue)[97] was almost invariably greater than when the pilocarpine (a combined parasympathetic and sympathetic salivary stimulant)[98] was used in the same animal. The marked qualitative and quantitative differences in the trace metal partition between pilocarpine-stimulated human and marmoset saliva may be attributable, in part, to the comparatively large dose of drug required to produce the volume of marmoset saliva needed for analysis.

The nonessential trace elements were found in marmoset saliva more frequently[59, 99] than in human saliva; this may reflect the compositional uniformity of the marmoset diet as opposed to the intersubject variations in the human diets. Analysis of the marmoset diet by emission spectrometry disclosed the presence of each of the 14 trace metals in amounts ranging from 0.08 of cesium to 10.8 mg of zinc per 100 g (wet weight). The moisture content of the primate diet was 47.3%.

Whether the high trace metal content of marmoset saliva may be a contributing factor to the caries immunity of this primate remains to be established. Further studies, with special emphasis on those elements that have been provisionally equated with caries resistance are certainly indicated.

3.7 CARDIOVASCULAR DISEASE

Many diseases of the cardiovascular system appear to involve chronic, insidious, and progressive injury to the blood vessel wall that becomes obvious only in later years. These eventual clinical indicators of disease may be secondary to, or even further removed from, the fundamental defect(s) responsible for the disease. The ultimate prevention and cure of these diseases, many of which have a polyetiology, will require the identification of those fundamental early defects and the elucidation of the initial insults to the tissues.

With the present limited understanding of the etiology of many cardiovascular diseases and of the metabolism of some trace metals, it is difficult to discern molecular mechanisms that might explain the results of the many studies which inferentially relate trace metal metabolism to cardiovascular disease. Thus far, nickel (Section 3.7.1) is the only example in which the observation of an unusual concentration of a trace metal in blood has stimulated the isolation of a new biological compound. We envision this as a first step in relating altered levels of a trace metal to the etiology of a disease. An even more cogent example of the important influence of a trace metal in cardiovascular disease may be that of chromium (Section 3.7.2), which has been suggested as a factor in maintaining the structural integrity of the blood vessel wall.

3.7.1 Myocardial Infarction

Early studies of the concentrations of trace metals in blood serum of patients with myocardial infarction showed[100, 101] a striking increase in copper levels and a significant decrease in zinc levels. The copper levels paralleled an increase in the serum levels of ceruloplasmin, the copper-containing protein of human serum. The decrease in zinc levels was accompanied by an increase in the activity of lactate dehydrogenase, a zinc-containing enzyme. This apparent contradiction could be explained by the *in vitro* observation that binding of zinc ions at positions other than the active site markedly inhibits the enzyme.

Two recent studies have reported the appearance of increased concentrations of serum nickel within 24 h of the onset of acute myocardial infarction.[102, 103] The normal serum level of 2.6 (S.D. ± 0.8) micrograms per liter is doubled in this period but returns to normal after about 48 h. Since the mean concentration of nickel in cardiac muscle of healthy individuals who died by accident or homicide has been reported to be about 6 μg/kg of wet weight, it appears that this twofold increase is too great to be attributed simply to the release of nickel that is normally present in cardiac muscle.[104]

The report of this abnormal concentration of serum nickel prompted research that brought about the isolation of a specific nickel-binding protein from human and rabbit serums. This protein, called nickeloplasmin, has been characterized as an α_2-macroglobulin that binds 1 gram-atom of nickel per mole[105] and is similar in other properties to the earlier reported zinc α_2-macroglobulin.[106] The identification of a specific nickel binding protein strongly suggests that this metal has an essential physiological role.[107–112] However, the molecular basis for the hypernickelemia observed in cases of myocardial infarction remains to be determined.

In studies attempting to relate trace metal metabolism to myocardial infarction, it was recognized that abnormal serum levels of copper, zinc, and nickel also occurred in varying amounts in other pathological states. It, therefore, appeared that the trace metal concentrations could be of only limited value as aids in the diagnosis of this cardiovascular disease. In most of these studies the analytical determinations were limited to a specific metal. Atomic absorption spectroscopy was generally the analytical method used. Because of the well-known competitive and synergistic interactions among trace metals, it is highly desirable in studies of this kind to analyze simultaneously for many trace metals. Accordingly, a study was initiated to evaluate changes in 14 trace metals as a function of time following an acute myocardial infarction, using emission spectrometric methods of analysis.[113] The object was to identify changes in trace metal patterns that might be specific indicators of an acute myocardial infarction. Time-related changes in certain serum enzymes as a function of time following a myocardial infarction have been used diagnostically for a number of years. Among the enzymes most commonly used for this purpose are creatine phosphokinase (CPK), serum glutamic oxaloacetic transaminase (SGOT), and lactic dehydrogenase (LDH). Although many conditions other than a myocardial infarction are known to increase serum levels of these enzymes, the extent and duration of the increases correlate well with the occurrence of an acute myocardial infarction.

3.7.1.1 Methods

Venous serum for trace metal analysis was obtained on admission and in the fasting state in the morning of days 1, 2, 3, and 7 from 42 patients admitted to the Myocardial Infarction Research Unit with a tentative diagnosis of acute myocardial infarction. CPK, SGOT, and LDH levels were determined routinely in the hospital clinical laboratories. The emission spectrometric procedures employed were the same as those described in Section 3.2.

Criteria employed for the diagnosis of an acute myocardial infarction were (1) compatible history and (2) serial ECG ST-T changes with development of new Q waves. As shown in Table 3.10, 27 patients were found to have sustained an acute myocardial infarction on the basis of these criteria. The remaining 15 patients that were included in the study as a control group were judged not to have sustained an infarction. These patients were found to have arrhythmias or chest pain of unknown origin. Complete enzyme data were obtained for 23 of the patients that had sustained an acute myocardial infarction and 11 of the control group.

Statistical analysis of the data was carried out on a computer using the Statistical Analysis System.[114] As has been observed in other investigations[2, 115] the distributions of the trace metal concentrations were generally skewed toward smaller values. Similar distributions were observed for the serum enzyme levels. Consequently, all statistical manipulations were carried out after the data had been subjected to a natural logarithmic transformation that provided distributions approximating the normal distribution. Hence, the geometric mean rather than the arithmetic mean was used as the central measure of the metal concentration and the enzyme levels.

A complete multivariate analysis of variance was applied to the data. Significance was determined according to the values of the Wilks lambda statistic,[116] or likelihood ratio criterion. In addition, the data for the 14 trace metals and 3 enzymes for each patient were fitted using a generalized quadratic form, i.e., $b_0 + b_1x + b_2x^2$, where x is the time, in days, after admission. The values of b_0, b_1, and b_2 accounted for the observed behavior quite well (R-square > 0.80) and were used in the subsequent classification analysis.

Classification was achieved by the pattern recognition technique of discriminant analysis (see also section 3.9). In this procedure the b_0, b_1, and b_2 values

Table 3.10 Composition of Patient Groups[a]

		Trace metal study						Enzyme study					
		MI[b]: Age			No MI[c]: Age			MI: Age			No MI: Age		
Sex	Race	No.	Mean	Range	No.	Mean	Range	No.	Mean	Range	No.	Mean	Range
Male	Caucasian	18	57.5	43–71	7	65.0	54–79	17	56.7	43–70	7	61.9	45–79
Female	Caucasian	2	46.0	39–53	7	54.4	39–78	1	30		4	47.8	39–66
Male	Negro	2	61.8	56–69	1	65.0		5	61.8	56–69	0		
Female	Negro	2	69.5	67–72	0			1	67		0		

[a]From Webb *et al.*,[113] by courtesy of Academic Press, Inc.
[b]Patients that sustained an acute myocardial infarction.
[c]Patients that did not sustain an acute myocardial infarction.

of, for example, CPK, SGOT, and LDH for each patient were used to represent the patient in the multidimensional space created by these b values. In this example the multidimensional space is constructed from b_0, b_1, and b_2 for each of the three enzymes, i.e., the space is nine-dimensional. Using prior information on the clinical diagnosis, the centroid for the two types of patient (with and without an acute myocardial infarction) was calculated from the "points" in the multidimensional array. After these centroids were defined, the location of each patient was considered in turn. The probability (from 0 to 1) of identifying each patient as having or not having sustained an infarction was then determined on the basis of its location in the multidimensional space relative to the previously calculated centroids. The individual was then classified as one of the two types of patients according to the larger calculated probability. This discriminant analysis procedure was carried out for a variety of combinations of enzyme and trace metal parameters.

3.7.1.2 Results

The mean concentrations of the 14 trace metals in the blood serum of patients sustaining an acute myocardial infarction and in the control group are listed in Table 3.11. Analysis for tin was not done on all samples. The statistical significance of the differences between these two patient groups was tested by a multivariate analysis of variance. For every metal the variations from patient to

Table 3.11 Serum Levels of Trace Metals in Patients Sustaining a Myocardial Infarction and in the Control Group[a]

	MI patient group		No MI patient group	
Trace metal	Number of analyses	Mean concentration (μg%)	Number of analyses	Mean concentration (μg%)
Copper	139	123	71	129
Iron	139	78.0	71	83.5
Aluminum	139	18.8	71	7.6
Nickel	139	0.54	71	1.3
Strontium	139	0.57	71	1.2
Barium	139	10.7	71	15.0
Manganese	139	4.0	71	6.3
Cesium	139	8.3	71	12.4
Tin	46	3.7	30	18.6
Chromium	139	0.72	71	0.78
Zinc	99	113	59	163
Lead	139	3.0	71	3.1
Molybdenum	139	0.75	71	1.5
Cadmium	139	1.3	71	2.1

[a] From Webb *et al.*,[113] by courtesy of Academic Press, Inc.

patient were large and significant ($p < 0.001$), and no significant differences between the mean values listed in Table 3.11 for the two groups of patients were noted. The results for molybdenum (Table 3.11) were borderline at the 0.05 level.

The statistical significance of the effect of time after hospital admission on the serum concentrations of trace metals was investigated by factoring this variable into linear, quadratic, and residual components. These were then tested, both individually and in combination with other variables, by a multivariate analysis of variance. The analysis indicated that statistically significant effects were limited to several trace metals: copper, iron, and aluminum and, to a less extent, nickel, strontium, and zinc. The results of this analysis are shown in Table 3.12. No statistically significant effects were observed for any of the time-factor variables with the trace metals barium, manganese, cesium, tin, chromium lead, molybdenum, or cadmium.

Mean levels of the three serum enzymes in the study and control groups are shown in Table 3.13. The results of the analysis of variance are listed in Table 3.14. For each enzyme the mean value in the group of patients that had sustained an acute myocardial infarction is significantly higher (Table 3.13) than in the control group. This relatively higher mean serum level was maintained throughout the entire time period of study.

The results of the multivariate analysis of variance indicated that serum levels of iron, copper, aluminum, and nickel, and of the three enzymes SGOT, CPK, and LDH, might be useful in the classification of patients. Consequently, a discriminant analysis was performed on the enzyme data and on various combinations of trace metal data. Sufficient data on serum enzyme levels were available for 34 patients. The results of the discriminate analysis are shown in Fig. 3.10 as a 2 X 2 matrix of actual and inferred identities of the patients.

Table 3.12 Statistical Significance (p) of the Variation of Serum Trace Metal Concentrations with Time after Hospital Admission (t) and with the Combined Variable, Time and the Occurrence of an Acute Myocardial Infarction (MI)[a]

	t			$t \times$ MI		
	Linear	Quadratic	Residual	Linear	Quadratic	Residual
Copper	0.001	0.49	NS[b]	0.001	NS	NS
Iron	NS[c]	NS	0.0001	0.044	NS	NS
Aluminum	NS	0.14	0.024	NS	NS	NS
Nickel[d]	0.01	NS	NS	NS	NS	NS
Zinc	0.009	NS	NS	NS	NS	NS
Strontium	NS	NS	NS	0.029	NS	NS

[a]From Webb *et al.*,[113] by courtesy of Academic Press, Inc.
[b]Not significant at 0.05 level; i.e., $p > 0.05$.
[c]Significance level between 0.05 and 0.10.
[d]Significant effect ($p = 0.017$) of race/sex on time course of trace metals.

Table 3.13 Serum Enzyme Levels in Patients Sustaining a Myocardial Infarction and Control Group

	MI patient group			No MI patient group		
Enzyme	Mean age (yr)	Number of analyses	Mean concentration	Mean age (yr)	Number of analyses	Mean concentration
SGOT	58	118	79.2	58	59	32.9
CPK	58	116	12.3	57	57	2.44
LDH	58	117	320	57	59	182

Since 23 patients had sustained an acute myocardial infarction and 11 had not, the sum of the entries in the first row of the matrix must be 23, and the sum of second row entries must be 11. Diagonal entries correspond to correct inferences. Off-diagonal entries represent misclassifications, or errors. The figure illustrates the complete success of the classification based on the three serum enzymes. In fact, not one of the 34 cases had a greater than 0.01 probability of being misclassified.

More extensive data were available on the serum levels of the trace metals, allowing 42 patients to be classified by the discriminant analysis procedure. Of these patients, 27 had sustained an acute myocardial infarction, while 15 had not. The results of this classification are shown in Fig. 3.11. All three combinations lead to errors, i.e., off-diagonal entries in the matrix. The off-diagonal entries with the superscript (solid dot) are those for which the probability of correct classification is less than 0.05; i.e., they constitute major errors in classification. All other off-diagonal entries have a greater than 5% probability of being correctly classified; i.e., they are less serious errors. For example, using the

Table 3.14 Statistical Significance (p) of the Variation of Serum Enzyme Levels with the Occurrence of an Acute Myocardial Infarction (MI), with Time after Hospital Admission (t) and with the Combined Variable, Time after Admission and the Occurrence of an Infarction[a]

		t			$t \times$ MI		
Enzyme	MI	Linear	Quadratic	Residual	Linear	Quadratic	Residual
SGOT	0.001	0.003	0.035	0.17	NS[b]	NS	0.031
CPK	0.001	0.0001	0.001	0.001	NS	NS[c]	NS
LDH	0.001	NS	0.0001	NS[c]	NS	0.038	NS

[a] From Webb *et al.*,[113] by courtesy of Academic Press, Inc.
[b] Not significant at 0.05 level; i.e., $p > 0.05$.
[c] Significance level between 0.05 and 0.10.

CPK,SGOT,LDH		INFERRED MI	INFERRED not MI
ACTUAL	MI	23	0
	not MI	0	11

Fig. 3.10 Classification by discriminant analysis of patients on the basis of sequential determinations of the serum enzymes creatine phosphokinase (CPK), glutamic oxaloacetic transminase (SGOT), and lactic dehydrogenase (LDH) [From Webb *et al.*,(113) by courtesy of Academic Press, Inc.]

copper, iron, and aluminum data (Cu, Fe, Al), one patient with an infarction was classified, on the basis of a probability of 0.96, as a patient without an infarction. However, the three patients who had not sustained an infarction but who were classified as having sustained an infarction, i.e., classification errors, had probabilities of 0.36, 0.38, and 0.19 of being correctly classified. The data of Fig. 3.11 indicate that the Cu, Fe, Al data lead to a slightly greater success in classification than the Cu, Fe, Ni data. The combined data, Cu, Fe, Al, and Ni, show additional improvement, particularly in the absence of the superscripts on the off-diagonal elements indicative of serious errors.

Extending the trace metal data to include separately each of the covariates age, race, and sex leads to greater diagonalization in the classification matrix. These results are shown in Fig. 3.12. In all three cases, the few off-diagonal entries have a probability in excess of 0.20 of being diagonal entries. A further extension, combining four trace metals (Cu, Fe, Al, Ni) with pairs of the covariates age, race, and sex, leads to complete success in classification. The matrices shown in Fig. 3.13 contain no off-diagonal entries, reproducing the success

Cu,Fe,Al		INFERRED MI	INFERRED not MI
ACTUAL	MI	25	2•
	not MI	3	12

Cu,Fe,Ni		INFERRED MI	INFERRED not MI
ACTUAL	MI	25	2
	not MI	5••	10

Cu,Fe,Al,Ni		INFERRED MI	INFERRED not MI
ACTUAL	MI	26	1
	not MI	3	12

Fig. 3.11 Classification by discriminant analysis of patients on the basis of sequential determinations of up to four serum trace metals: copper (Cu), iron (Fe), aluminum (Al), and nickel (Ni). [From Webb *et al.*,(113) by courtesy of Academic Press, Inc.]

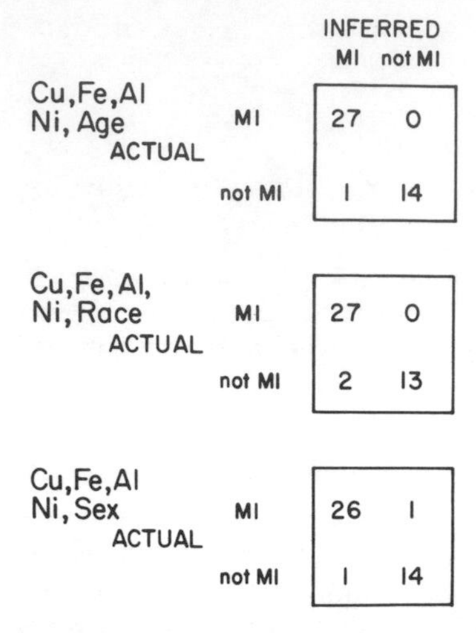

Fig. 3.12 Classification by discriminant analysis of patients on the basis of sequential determinations of the four serum trace metals, copper (Cu), iron (Fe), aluminum (Al), and nickel (Ni), with data of age, race, and sex, taken separately. [From Webb *et al.*,[113] by courtesy of Academic Press, Inc.]

achieved by classifying on the basis of the three serum enzymes (Fig. 3.10). It should be pointed out that this analysis (Fig. 3.13) required 14 of the 15 available statistical degrees of freedom (b_0, b_1, b_2, for four trace metals, with one each for age, race, or sex), while that for the serum enzymes (Fig. 3.10) required only 9 (b_0, b_1, and b_2 for three enzymes).

3.7.1.3. Discussion

The major objective of this study was to compare, using objective statistical criteria, the relative ability of serum levels of trace metals and of appropriate enzymes to discriminate between patients who had sustained an acute myocardial

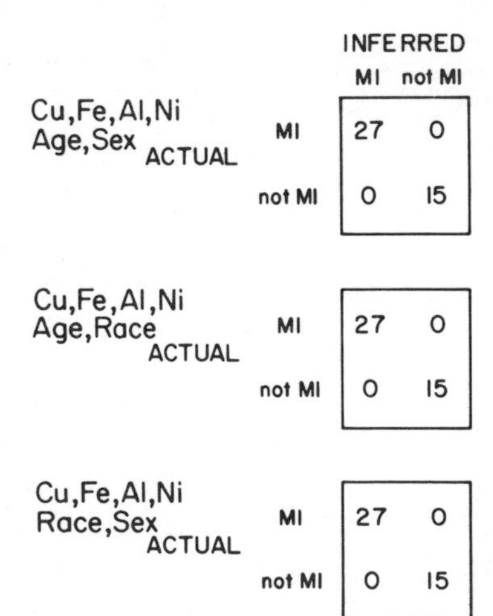

Fig. 3.13 Classification by discriminant analysis of patients on the basis of sequential determinations of the four serum trace metals, copper (Cu), iron (Fe), aluminum (Al), and nickel (Ni), with data of age, race, and sex, taken in pairs. [From Webb *et al.*,[113] by courtesy of Academic Press, Inc.]

infarction and those who had not. A profile of the statistically significant trace metal parameters, i.e., serum levels of copper, iron, aluminum, and nickel, was used in this part of the study. Although the actual serum concentrations are of interest, it is the overall pattern created for each patient by these four trace metals over time that is of prime concern. Using such a profile, created either by serum enzyme or serum trace metal data, the discriminant analysis technique provides a quantitative means of recognizing patterns within the clinical data. It is apparent that serum levels of the enzymes CPK, SGOT, and LDH provide a strong pattern that unambiguously and completely discriminates between the two patient groups (Fig. 3.10). These changes are quite striking[117] and are widely used by clinicians with considerable success without exploiting curve-fitting and statistical procedures. When data on serum levels of the four trace metals of interest were used to classify patients, almost complete (91%) success was obtained in identifying patients that had sustained an acute myocardial infarction (Fig. 3.11). Not unexpectedly the inclusion of additional patient information, e.g., sex, race, age, further improved the discrimination between the two groups (Figs. 3.12 and 3.13). The results of this study clearly indicate that a limited amount of trace metal information coupled with appropriate statistical analysis can achieve success in the diagnosis of a myocardial infarction comparable to that obtained using serum enzyme data.

The success achieved in the present study raises the possibility that trace metal analysis of blood serum may find clinical application in the diagnosis of acute myocardial infarction. In patients with left bundle branch block, chest pain, and tissue injury, either from surgery or other trauma, it is at present virtually impossible to diagnose an acute myocardial infarction with certainty. The changes in serum levels of trace metals reported herein may well lead to more definitive methods of diagnosis, particularly in difficult or equivocal cases.

3.7.2 Atherosclerosis

A number of metal ions affect molecular parameters that are generally recognized as important factors in the development of atherosclerosis. Vanadium influences the biosynthesis of cholesterol, phospholipids, and triglycerides.[118-120] Manganese interferes with the inhibition of cholesterol that is induced by vanadium.[121] In a study of the effect of 13 trace elements on serum cholesterol levels in rats, it was noted that chromium, nickel, and niobium exerted anti-cholesterogenic properties, whereas tellurium was possibly cholesterogenic.[122] It was more recently suggested[123] that lithium depletion might be a cause of atherosclerotic heart disease.

A study of 72 patients with atherosclerosis revealed a significant decrease in the zinc content and an increase in the manganese content of plasma. The erythrocyte content of zinc and cobalt was considerably increased. In the aorta, the content both of zinc and manganese was greatly reduced and that of cobalt was slightly elevated. No molecular basis for these observations was proposed.[124] Serum copper levels have been found to increase linearly from 124 mg/100 ml at age 20, to 145 at age 60.[125] It was suggested that this increase may be paralleled by an increase in the rate of lipid peroxidation, which, in turn, might result in an acceleration of the process of atherosclerosis.

Chromium is the trace metal most likely to be of importance in atherosclerosis.[126-131] As will be described later, the extremely well characterized chemistry of this element provides a ready hypothesis for its role in the etiology and pathogenesis of atherosclerotic heart disease.[126,127,132-135] Epidemiological data provided the first evidence to suggest that atherosclerosis and chromium might be related. These studies showed that human tissues from subjects in the United States (a country where atherosclerosis is endemic) were deficient in chromium. Chromium deficiency was rarely found in tissues of subjects from the Far East, the Near East, or Africa, where atherosclerosis is a relatively rare disease. Among normal individuals (those whose death was accidental) the mean concentration of chromium found in aortas of U.S. subjects was approximately one third that found in aortas of subjects from other countries. It was further noted that the chromium concentrations in aortas from the latter subjects remained relatively constant during their lifetimes, but that, in U.S. subjects, the concentration decreased markedly with age. These differences were thought to stem from the low chromium content of the highly refined foods which constitute the American diet.[121,128,129]

One study of atherosclerosis included 15 patients who died of the disease, 3 with mild or moderate atherosclerosis who died accidentally, and 11 without evidence of atherosclerosis who died accidentally. Detectable (1 ppm of Cr in the ashed tissue) concentrations of chromium were found in only two aortas from patients with atherosclerosis, but it was found in aortas of nine persons without evidence of atherosclerosis who died accidentally.[126] The mean concentration of chromium in the nonatherosclerotic aortas was five times greater than that found in the atherosclerotic aortas. The concentrations of chromium found in aortas of patients who died from myocardial infarction were likewise markedly lower than those of patients who died accidentally.

Chromium is a requisite for the metabolism of glucose and of lipids, both of which are intimately involved in the development of atherosclerosis.[130,131,136-138] Experimental lifetime chromium deficiency leads to elevated serum cholesterol levels and the deposition of aortic plaques.[139-141] Unfortunately, the studies referred to were conducted with rats, whose atherosclerosis is a comparatively poor model for the human disease. Nevertheless, the biochemical and pathological changes that are associated with human atherosclerosis do occur in this experimental animal, and prevention of such changes by the dietary intake of chromium suggests that it has some role in the causation of this cardiovascular disease.

Chromium affects calcification, at least *in vitro*.[142] When added to metastable solutions of calcium phosphate, Cr(III) destabilizes the solution, thereby initiating precipitation of hydroxyapatite. Precipitation occurred when chromium chloride was added to concentrations of 1 μM or less. Similar results were obtained with lead(II), iron(II), iron(III), bismuth(III), and aluminum(III). It is possible that, *in vivo*, the chromium that is released from the aorta, as atherosclerosis develops, serves to initiate calcification on and in the blood vessel walls.

The wide clinical use of radioactive hexavalent chromium led to the demonstration that chromium can affect the *in vitro* functioning of platelets.(143) When introduced as the chromate ion, this trace metal markedly inhibited the connective tissue, induced aggregation of platelets, and caused other biochemical changes that are normally induced by connective tissue. These effects were observed at concentrations as low as 10 μM (equivalent to 10 μg of Cr/100 ml), i.e., within the physiological range for chromium. Although the *in vitro* results suggest that *in vivo* chromium prevents platelet aggregation, it is possible that the release of chromium from the blood vessel wall not only induces calcification as described above, but also disturbs normal platelet activity in some way that contributes to plaque formation and calcification.

The chemistry, biochemistry, and biological role of chromium have been reviewed by Mertz,(144) and it is the consensus that chromium in biological systems is predominantly present in the trivalent oxidation state. Because of the high oxidation potential between chromium(II) and chromium(III) and between chromium(III) and chromium(IV), a reversible transition between two oxidation states is an unlikely role for chromium in biological systems. Trivalent chromium has a strong tendency to form coordination compounds, complexes, and chelates. The rate of ligand exchange of such compounds is very slow.(145) Chromium(III) commonly has a coordination number of six, with the ligands pointing to the corners of an octahedron. The metal-ligand bonds usually involve oxygen or nitrogen. Free chromium(III) ion is nonexistent in aqueous solution; it is always coordinated with water or other ligands available in solution. At neutral pH, hydrolysis of coordinated water occurs and forms complexes that are polynuclear, with the chromium atoms linked together by oxy and hydroxyl bridges, i.e., Cr—O—Cr and Cr—OH—Cr.(142) This process can continue until large macromolecular complexes are formed [$Cr(OH)_3$, $Cr_2O_3 \cdot xH_2O$] that are chemically inert and insoluble. This formation of polynuclear complexes, together with the inert kinetic behavior of the ligand replacement reactions, is the outstanding feature of the known chemistry of chromium(III). Consequently, the behavior of this ion in biological systems is expected to be dominated by these two factors.

Interactions between chromium(III) and polypeptide chains have been reported in several cases. The best understood of these examples is that of chrome tanning of skin.(144) In this process, skin, with its constituent collagen, is saturated with chromium(III) at low pH. Then the pH is raised. At the low pH the chromium(III) ions bind to water molecules and to the available ligands on the polypeptide chain, namely the free carboxyl groups of the glutamic and aspartic acid residues of the protofibrils of collagen. As the pH is raised, the water ligands hydrolize and become converted into —O— and —OH— linkages, forming polynuclear complexes that create a stable cross-linked aggregate of the protofibrils, i.e., skin is converted into leather. The concentrations of chromium(III) that are required are far greater than those found under physiological conditions, but even at the lower concentrations, the mechanisms of action of chromium(III) are probably similar in all instances.

Cross-linking by several metal ions, including chromium(III), has been investigated by ultracentrifugation with the use of conarachin; a protein of molecular weight 175,000 isolated from ground nuts.[144] Iron(III), aluminum(III), and chromium(III), ions that are known to form polynuclear complexes, all produced cross-linking within the protein, but that resulting from chromium(III) treatment was much more resistant (inert) to competitive replacement of the polynuclear linkages by other ligands, e.g., phosphate. There is also some indication that physiological concentrations of chromium(III) can stabilize the tertiary structure or proteins.[146]

3.8 METALS IN THE ENVIRONMENT

The influence of environmental trace metals on human health is a problem of immediate concern. The presence of "toxic" metals in the environment is a more serious problem than pollution by pesticides or other organic contaminants. No metal is degradable. Once released from their natural sources by mining or industrial processes, metals become a permanent part of our environment. It is also important to note that there are no nonhazardous substances, but only nonhazardous ways of using substances. Even sodium chloride, if ingested in grossly excessive quantity, can be fatal.

Through basic research programs, we need to learn much more about levels of trace metals that are essential for human health, as well as the environmental levels of these metals that can be tolerated over the span of a lifetime without adversely affecting health. Studies are urgently needed on the pathogenesis of chronic metal toxicity and the effects of metals on repair and defense mechanisms at the molecular, cellular, organ, and body levels. Only when the interactions between pollutants and the organism are thoroughly understood can realistic control programs be undertaken.

The immune system provides protection for the body by recognizing and destroying both endogenous and exogenous foreign materials. A malfunction of this system may result in dire circumstances. Evidence has been accumulating which indicates that cadmium and perhaps certain other trace metals may influence the immune system. The current concern over increased levels of cadmium in the environment, and evidence that it has an adverse effect on immunologic phenomena make this an area of immediate concern. In our environment, cadmium is a ubiquitous element. As a result of the rapid increase in the dissipative uses of cadmium over the past 20 years, the concentration of the metal in our environment is reaching critical levels.[147] The effects of low levels of cadmium are amplified by its cumulative deposition in the body. The major sites of cadmium deposition are the kidney and liver; however, there are significant deposits in other soft tissues, including spleen and thymus.[148] Cadmium has been implicated in a number of pathological conditions.[149] These include osteomalacia,[150] hypertension,[151] vascular damage of the testis, as well as testicular atrophy and tumors,[152-154] anemia,[155] severe renal dysfunction, and proteinuria.[156,157] In addition, cadmium has been shown to cross the placenta, which raises the possibility that it could have teratogenic properties.[151] Lifeterm studies of rats[158] and mice[159] exposed to low levels of cadmium have been reported. Median ages of rats (at death) were adversely affected by cadmium. The males that were tested differed from their controls by 156 days and the females by 140 days.

Cadmium-fed animals showed an increased incidence of infection and a concomitant increase in mortality rate. Cadmium-exposed animals also showed an increased incidence of spontaneous tumors. Increased susceptibility to infection and tumors is suggestive of a deficiency in immunologic activity. Cadmium poisoning has been shown to result in defective metabolism of immunoglobulins. It can cause an elevation in serum gamma globulin levels,[155] immunoglobulin light-chain proteinuria,[156,157] amyloidosis,[157-159] and inhibition of macrophage metabolism.[160]

Patients suffering from cadmium-induced renal damage and proteinuria have been found to excrete 2-3 g of protein per day, 30-40% of which is immunoglobulin light chains with additional amounts of smaller fragments shown by immunologic techniques, to be of light-chain origin.[157-165] Light chains are catabolized principally in the renal tubular cells.[161,162] Many of the peptidases involved in the breakdown of light chains require zinc for peptidase activity. Cadmium inhibits at least two of these enzymes (carboxypeptidase A[163] and leucine aminopeptidase[157]) by competing for the zinc in the active site. Even though cadmium can inhibit the breakdown of immunoglobulin light chains in the kidney, it remains difficult to explain the excretion in the urine of 1200 mg or more of light chains per day. The normal daily turnover of immunoglobulins in adult man is on the order of 2 g/day,[157] one third of which is light chains. This would account for, at most, 700 mg of light chains being delivered to the kidney daily for catabolism suggesting that cadmium intoxication results in an excess production of immunoglobulin light chains.

Cadmium adversely affects the ability of an animal to respond to antigenic stimulation. If animals are injected with low levels of cadmium (0.6 mg/kg) for 2 weeks prior to antigenic stimulation, there is an increase in both the primary and secondary responses. If the injections of cadmium are begun 1 week before immunization, primary and secondary responses are decreased. If the injections are begun after the primary immunization, the secondary response is decreased.[164]

Cadmium poisoning in rabbits results in the formation of amyloid deposits in the kidney[165] and spleen.[157] It was established that amyloid tissue consists mainly of fragments of immunoglobulin light chains. Sequence analysis demonstrated that amyloid fibrils[166] consist of the variable part of light chains with portions of the constant region also being present. These fibrils have been produced *in vitro* from Bence-Jones proteins that had been partially digested by enzymes.[167] Studies of immunochemical cross-reactions of human amyloid proteins with human immunoglobulin light polypeptide chains have shown no evidence of the presence of an amyloid specific antigenic determinant. These immunochemical studies support the concept that amyloid fibril proteins are derived from immunoglobulin light chains.[168]

Immunoglobulin light chains are normally synthesized at twice the rate of heavy chains[169] and, after synthesis, they enter a light-chain pool.[170] The light chain then complements the nascent heavy chain to form HL dimers on the heavy-chain polysome. These HL dimers associate to form immunoglobulin molecules $(HL)_2$ either on heavy-chain polysomes or after release into the cisternal space. The majority of the excess uncomplemented light chains that remain free are normally destroyed in the cytoplasm. However, mutants of myelomas have been found that secrete light chains either because they do not complement the heavy chain or because of the lack of heavy-chain synthesis.[166] Although there are many reported cases of light-chain secretion by mouse myelomas, as well as cases of secretion of both light-chain and intact immunoglobulins, there is not a single report of free heavy-chain secretion.[170]

These observations imply that the determinants for secretion of immunoglobulins by the cells are carried on the light chain.[171] Mouse myelomas have been described that produce both free light and heavy chains but secrete only light chains; the heavy chains are destroyed in the cytoplasm. Therefore, the impaired capacity of the cells to form HL dimers, or the inability of these dimers to combine and form immunoglobulin molecules $(HL)_2$, could be responsible for this secretion of light chains only. The high affinity of cadmium for sulfhydryl ligands[151,153,172] suggests that a likely explanation of cadmium-induced excretion of light chains might be the blockage of the formation of the inter-heavy-light-chain disulfide bonds. However, the light chains isolated from the urine of patients with cadmium-induced proteinuria had no associated cadmium. This finding might be explained by postulating that the light chains, after their release from the cells, come in contact with other ligands that have a higher affinity for cadmium. Metallothionein, a cadmium-containing protein found in the kidney, could serve such a function. Cadmium might also block the cytoplasmic catabolism of the excess light chains normally produced. Presumably the peptidases involved would be zinc-containing enzymes similar to those in the kidney.[157]

Cadmium may affect the biosynthesis of antibodies at the cellular level. Other metal ions are known to stimulate blast formation and proliferation of lymphocytes both *in vivo* and *in vitro*. These include nickel,[173] mercury,[174] gold,[175] and zinc.[176] Cadmium induced transformation and proliferation of lymphocytes could account for the excess light-chain production as well as the increased levels of serum immunoglobulins seen in cadmium poisoning.

Other components of the immune system are also affected by cadmium. For example, preliminary experiments in our laboratory have shown that this metal adversely affects the activity of the serum complement system. A concentration of cadmium as low as 5×10^{-6} M is capable of significantly inhibiting guinea pig complement according to assay by hemolysis of the sensitized sheep red blood cells. Concentrations in the range of 10^{-4} M were completely inhibitory in the assay system employed.

Any interaction between the various components of the immune system and cadmium would probably be mediated by cadmium bound to endogenous ligands rather than by the free ionic form of the metal. In many body tissues, cadmium is complexed with the metal-binding protein, metallothionein[177-179] which is probably responsible for absorption, transport, and storage of this metal.[175,180] Ingestion of cadmium results in increased levels of metallothionein[157] and because of its high affinity for metallothionein, as well as for many other ligands, it seems unlikely that cadmium exists as a free ion in body tissues and fluids. The toxicological effects of cadmium are likely to be mediated by a form of the metal complexed to metallothionein or to some other biological ligand.

Metallothionein, initially isolated from equine renal cortex,[177,178] is a protein of low molecular weight (10,500 daltons) that is exceptionally high in cysteine residues. It binds up to 9 mol of cadmium per mole of protein, to a final sulfhydryl to cadmium ratio of 3. These cadmium-binding sites can be occupied by other heavy metals, including zinc, mercury, copper, lead, silver, and gold.[177] Consequently, preparations of the protein are usually heterogeneous in metal content according to the prior history of the organism involved. The apparent relative strengths of metal binding to the protein are, in decreasing order, mercury–cadmium–zinc. The hydrodynamic, spectroscopic, and optical properties of the equine protein have been reported in some detail,[177] but the metallothionein of human origin has been less extensively characterized.[179] Kidneys from patients who died without evidence of primary renal disease and whose ages ranged from 18 to 71 years yielded metallothionein containing about 4 mol of cadmium, 4 mol of zinc, and $\frac{1}{2}$ mol each of mercury and copper. The mercury was traced to prior treatment with therapeutic mercurials.[175]

Cadmium-binding proteins of low molecular weight similar to metallothionein, have been reported recently from a variety of sources.[180-182] Studies of the well-known nutritional antagonism of copper absorption by cadmium led to the identification by gel filtration of a low-molecular-weight protein in the duodenal mucosa that bound administered copper, but which would preferentially bind competitively administered cadmium or zinc.[183] A similar protein was also demonstrated in the cytoplasm from rat duodenum, lever, and kidney[180] and from bovine duodenum and liver.[180,181] In all these studies the protein has only been characterized by molecular weight obtained by gel filtration, cadmium binding obtained from isotope studies, and, in the bovine study, by sulfhydryl group analysis.

A recent study indicates that administered cadmium salts can induce the synthesis of a cadmium-binding protein. The protein isolated from rat liver and kidney homogenates increased with increasing amounts of administered salts. The preparation was shown to be heterogeneous by chromatography and electrophoresis.[184]

Incorporation of the administered cadmium into a cadmium-binding protein has also been demonstrated *in vitro* in human fetal skin and muscle fibroblasts cultured as monolayers. The protein was minimally characterized as a relatively discrete elution peak from chromatography of the supernatant of the cellular lysate on Sephadex G-75.[182]

Review of the discussion above leaves little doubt that cadmium affects the immune system. The data, however, provide very little insight into the mechanism(s) of cadmium involvement in immune processes. It is apparent that cadmium affects both anabolic and catabolic pathways of immunoglobulins. The catabolic effects can account for only a small part of the influence exerted by the cadmium on the immunologic system and it appears likely that this metal also influences the immune system at the cellular level. Among the many other metals that have been shown to affect lymphocytes, nickel, as well as mercury[185] and zinc,[176] acts as a monospecific stimulant for the transformation of peripheral blood lymphocytes *in vitro*.[185] The mechanism of lymphocyte proliferation by these metals appears similar to that of other mitogens.

The immunologic theory of aging purports a central role for the thymus in the development of senescence, including increased tumor incidence, autoimmunity, and marked increase in infectious diseases. An understanding of the role of environmental factors is required before the validity of this hypothesis can be established. Owing to its direct effects on several aspects of immunologic phenomena, cadmium appears to be at least one of the environmental agents that may accelerate immunosenescence. In experimental animals, cadmium poisoning lead to an increased incidence of tumors and infections.[137,158] Accumulation of this metal occurs in the thymus and other tissues and may induce malfunction in the postulated immune surveillance system. Unfortunately, cadmium is a steadily increasing part of our environment. Elucidation of the mechanism(s) of interaction between cadmium and the immune system could provide fundamental knowledge relative to the potential effects of many environmental toxicants and their relationship to a number of chronic diseases.

3.9 PATTERN RECOGNITION

In general, the goal of pattern recognition is to recognize a property or interrelationship among a collection of objects from the results of measurements made on the objects.[186] By virtue of its ability to simultaneously determine the concentrations of many trace metals in a speicmen, emission spectroscopy provides a powerful tool for pattern recognition techniques. Elemental analysis has provided diagnostic patterns useful in geology,[187] criminology, and archeological investigations.[188] Pattern recognition techniques have rarely been applied directly to biomedical problems.[189–193] The judicious application of statistical methods of pattern recognition to the mass of data that can be derived from emission spectrometric analysis should provide a powerful technique in biomedical research. One successful application of the technique of discriminant analysis to a diagnostic problem was described in Section 3.7.1. To further explore the potential of these procedures, several pattern recognition procedures were applied to trace metal analyses performed on tissues derived from the cardiovascular system. In one study[17,194] anatomic sites of the cardiovascular system of 13 pigs were analyzed for their concentrations of 13 trace metals by a modification[195] of the method described in Section 3.2. In addition to a complete multivariate analysis of variance, we applied the statistical techniques of pattern recognition, cluster analysis, and Duncan analysis to ascertain if certain tissues of the cardiovascular system possess characteristic trace metal profiles, and furthermore, to determine if, in fact, trace metal profiles can be used as sufficient descriptions to identify individual tissues with respect to their anatomic origins

Statistical analysis of the data was carried out on a computer using the Statistical Analysis System.[114] As has been observed in other investigations,[196–201] the distributions of the trace metal concentrations were generally skewed toward smaller values. Consequently, all statistical manipulations were carried out after the data had been subjected to a natural logarithmic transformation that provided distributions approximating the normal distribution. Hence, the geometric mean, rather than the arithmetic mean, was used as the central measure of the metal concentrations.

In these determinations the threshold values, i.e., the effective limits of detection, for each of the metals[202] were usually in the range, 1–5 μg%, although copper and strontium could be detected at the 0.5-μg% level. A number of the samples analyzed were computed to contain concentrations of some trace metals below the threshold values. In some cases, particularly zinc, levels in excess of 250 μg%, the upper calibration limit, were found. On the average, about 16% of the values for each trace metal were below threshold. These values, which were below the threshold level, as well as those above 250 μg%, were clearly less accurately determined than the other values, which fell in the range between the threshold value and 250 μg%. In order to better approximate these out-of-range values, they were estimated from the data on specimens that contained concentrations of the trace metals between the threshold values and 250 μg% by a multiple regression procedure,[114] which allowed the values for each metal to regress on the values for all other metals until the total variability, accounted for by the assumed log linear model, was a maximum. The values derived during this

Table 3.15 Coding Abbreviations for Anatomic Sites Sampled[a]

Anatomic site	Abbreviation	Coded letter
Aorta	AO	a
Main pulmonary artery	MPA	b
Right superior vena cava	RSVC	c
Tricuspid valve	TV	d
Mitral valve	MV	e
Pulmonary valve	PV	f
Aortic valve	AV	g
Right atrium	RA	h
Left atrial appendage	LAA	l
Right ventricle (free wall)	RV	m
Left ventricle (free wall)	LV	n
Left ventricle–papillary muscle	LV-PM	o
Interventricular septum	IVS	p
Crista supraventricularis	CR	q
Sinus node	SN	r
Atrioventricular node and His bundle	AVN + B	s
Left bundle branch	LBB	t

[a]From Webb *et al.*,[(194)] by courtesy of American Elsevier Publishing Co.

cyclic iteration procedure were restricted in that those values below threshold could not exceed the threshold, and those about 250 μg% could not be less than 250 μg%. After six iterations, the data set accounted for about 90% of the total variability. The mean trace metal concentrations were altered slightly in some cases by this iterative procedure.

Table 3.16 lists the geometric mean concentration of each metal calculated from analytical determinations on 11 specimens of each tissue (denoted by the

Table 3.16 Concentrations[a] (μg/100 g of Wet Tissue) of 13 Trace Metals in 17 Anatomic Sites of Pig Heart Tissue

	Cu	Mn	Mo	Zn	Cr	Ni	Cs	Ba	Sr	Cd	Al	Sn	Pb
	45.7	8.4	2.7	706	4.1	3.2	7.7	14.1	6.8	12.6	74.2	19.7	20.0
AO	*38.4*	*5.1*	*1.9*	*300*	*2.3*	*1.8*	*3.9*	*9.9*	*5.4*	*7.1*	*31.0*	*15.2*	*8.8*
	32.2	3.1	1.3	127	1.2	1.0	2.0	6.9	4.2	4.0	13.0	11.7	3.9
	58.6	6.8	3.5	1323	2.5	3.0	5.8	8.4	8.2	19.3	45.4	18.7	16.7
MPA	*47.6*	*3.8*	*2.1*	*774*	*1.2*	*1.7*	*2.3*	*4.9*	*3.7*	*9.7*	*13.7*	*15.9*	*8.8*
	38.7	2.1	1.3	453	0.6	1.0	0.9	2.9	1.7	4.8	4.1	13.4	4.6
	150	15.2	6.4	834	5.8	6.9	12.3	13.1	9.7	25.0	106.7	43.9	38.3
RSVC	*112*	*9.8*	*4.6*	*521*	*3.0*	*3.9*	*4.9*	*7.3*	*4.4*	*11.8*	*35.9*	*32.8*	*20.1*
	83.6	6.3	3.3	325	1.6	2.2	1.9	4.0	2.0	5.6	12.1	24.5	10.6

(*Continued*)

Table 3.16 (*Continued*)

	Cu	Mn	Mo	Zn	Cr	Ni	Cs	Ba	Sr	Cd	Al	Sn	Pb
	66.6	7.9	5.0	557	7.9	4.1	10.9	15.3	9.1	40.4	83.7	31.0	55.8
TV	*52.7*	*3.7*	*3.2*	*358*	*3.0*	*2.7*	*3.7*	*6.6*	*3.6*	*17.6*	*28.6*	*18.5*	*25.2*
	41.7	1.8	2.1	230	1.1	1.8	1.2	2.9	1.4	7.6	9.8	11.1	11.4
	55.6	6.6	4.5	489	4.3	5.5	7.4	9.5	7.6	24.0	63.9	19.9	27.0
MV	*42.0*	*3.6*	*3.2*	*327*	*2.1*	*3.5*	*3.1*	*5.1*	*3.7*	*9.7*	*22.0*	*15.4*	*15.1*
	31.7	2.0	2.3	218	1.1	2.2	1.3	2.7	1.8	3.9	7.6	11.9	8.5
	133	32.0	12.9	1165	18.2	19.3	83.4	82.0	39.0	74.6	599	47.3	106
PV	*102*	*21.9*	*5.9*	*744*	*7.6*	*8.4*	*47.3*	*54.4*	*16.4*	*33.9*	*280*	*36.9*	*55.0*
	78.6	15.0	2.7	475	3.1	3.7	26.9	36.1	6.9	15.4	131	28.9	28.4
	81.7	19.3	7.0	460	11.3	9.6	37.5	51.2	18.1	23.6	425	34.0	63.2
AV	*61.2*	*12.8*	*4.2*	*224*	*5.7*	*5.2*	*21.9*	*31.7*	*7.8*	*13.8*	*126*	*25.2*	*32.5*
	45.8	8.4	2.5	109	2.8	2.9	12.8	19.6	3.4	8.1	37.5	18.6	16.8
	169	9.3	4.5	660	4.3	3.9	5.0	5.4	5.8	13.7	32.9	40.7	22.0
RA	*140*	*6.4*	*3.3*	*429*	*2.2*	*2.8*	*2.0*	*2.3*	*2.3*	*7.5*	*5.0*	*33.9*	*12.2*
	117	4.4	2.5	278	1.2	2.0	0.8	1.0	0.9	4.1	0.8	28.2	6.7
	158	11.0	4.0	687	1.6	3.6	2.5	1.2	2.3	16.9	31.1	38.5	16.6
LAA	*137*	*7.9*	*3.2*	*353*	*0.7*	*2.8*	*1.1*	*0.7*	*1.2*	*7.9*	*6.5*	*33.5*	*9.5*
	118	5.6	2.5	181	0.3	2.2	0.5	0.4	0.6	3.7	1.3	29.1	5.4
	208	9.2	5.0	419	3.8	2.9	2.0	1.5	3.9	13.2	39.1	44.5	13.2
RV	*163*	*6.7*	*3.8*	*233*	*2.0*	*2.2*	*0.8*	*0.7*	*1.6*	*7.5*	*6.3*	*38.9*	*6.7*
	128	4.9	2.9	130	1.0	1.6	0.3	0.3	0.7	4.2	1.0	34.0	3.4
	207	9.8	6.7	722	3.6	3.7	1.4	1.7	3.6	12.0	42.0	53.0	13.8
LV	*171*	*7.2*	*4.6*	*396*	*1.7*	*2.8*	*0.7*	*0.8*	*1.5*	*7.3*	*6.4*	*44.2*	*7.6*
	141	5.3	3.2	217	0.8	2.1	0.3	0.4	0.7	4.4	1.0	36.8	4.2
	204	10.1	6.7	702	3.4	3.9	1.8	1.5	4.0	11.7	43.5	47.6	17.9
LV-PM	*170*	*7.8*	*5.3*	*330*	*1.8*	*2.6*	*0.9*	*0.8*	*2.1*	*6.8*	*7.6*	*41.6*	*9.3*
	141	6.0	4.2	156	0.9	1.7	0.4	0.4	1.1	4.0	1.3	36.3	4.9
	215	9.8	6.3	381	2.1	3.7	3.3	2.0	3.7	12.5	46.4	46.7	18.3
IVS	*171*	*6.8*	*4.6*	*248*	*1.1*	*2.8*	*1.4*	*0.9*	*1.5*	*7.0*	*9.5*	*39.9*	*10.3*
	137	4.8	3.4	162	0.5	2.1	0.6	0.4	0.6	3.9	1.9	34.1	5.8
	193	9.2	6.1	748	3.3	4.3	3.8	3.7	5.3	12.3	103	42.5	24.0
CR	*160*	*6.9*	*4.1*	*493*	*1.5*	*2.9*	*1.3*	*1.6*	*1.9*	*7.7*	*17.8*	*35.6*	*12.5*
	132	5.2	2.8	325	0.6	1.9	0.5	0.7	0.7	4.8	3.1	29.7	6.5
	180	8.7	4.9	776	3.3	3.7	2.6	3.1	4.6	13.4	43.1	36.4	20.2
SN	*145*	*6.6*	*4.2*	*548*	*1.7*	*2.9*	*1.1*	*1.5*	*2.2*	*7.7*	*8.6*	*31.5*	*12.4*
	117	5.0	3.6	386	0.9	2.3	0.5	0.8	1.0	4.4	1.7	27.2	7.6
	178	9.0	5.7	466	2.4	3.5	2.6	2.4	5.5	12.0	27.5	41.4	11.6
AVN + B	*144*	*6.5*	*4.2*	*284*	*1.3*	*2.5*	*1.3*	*1.3*	*3.3*	*6.5*	*5.4*	*32.8*	*7.5*
	116	4.6	3.0	174	0.7	1.8	0.6	0.7	2.0	3.5	1.1	26.1	4.9
	203	10.1	6.5	644	3.0	3.4	3.6	2.5	4.2	11.1	62.7	48.7	22.6
LBB	*164*	*7.7*	*5.4*	*449*	*1.5*	*2.8*	*1.5*	*1.2*	*2.1*	*7.6*	*12.8*	*39.7*	*12.6*
	133	5.8	4.5	313	0.7	2.3	0.6	0.6	1.0	5.2	2.6	32.4	7.0

[a]The geometric mean of 11 determinations is shown in italic type, with the upper and lower 95% confidence limits above and below.

abbreviations of Table 3.15) with the upper and lower 95% confidence limits shown above and below the geometric mean. Tissues are grouped, at least in part, according to the physiological function, progressing from arterial (aorta, main pulmonary artery) and venous (right superior vena cava) tissue, to valves (tricuspid, mitral, pulmonary, and aortic), to ordinary (e.g., ventricular free walls) and specialized (e.g., sinus node, left bundle branch) myocardium.

The data in Table 3.16 indicate that the trace metal composition of cardiac tissues derived from different anatomic sites varies widely. For example, mean copper levels are about 100 μg% or less for the first seven anatomic sites listed in the table, namely, the blood vessels and heart valves, but greater than about 140 μg% for the 10 remaining sites of ordinary and specialized myocardium. Particularly high concentrations of cesium, barium, strontium, cadmium, aluminum, and lead were noted in the pulmonary and aortic valves. The high concentrations of trace metals in the pulmonary valve were particularly striking. Thus, the aluminum concentration of the pulmonary valve was twice as great as that of the aortic valve and an order of magnitude greater than that of any of the other 15 anatomic sites sampled.

3.9.1 Multivariate Analysis of Variance

The statistical significance of the differences in trace metal concentrations among the 17 tissues was tested by a multivariate analysis of variance, using the Wilks criterion,[203] or likelihood-ratio test (F value). For every metal the variations from tissue to tissue are highly significant ($p < 0.0001$).

The F values and probability levels for the variability among the tissues for each metal, calculated with the pig-to-pig variability removed (as a blocking effect), are listed in Table 3.17. The table also lists the mean-square error (MSE), which is a measure of the spread among the determinations of each metal, and R_1 and R_2, the correlation coefficients between the trace metal concentrations and the first and second canonical variables, respectively. The canonical variables are linear indices constructed from the data to best discriminate among the tissues.[114] The first canonical variable accounts for 61.7% of the total variability, and the second accounts for 23.8%. The magnitude of R gives a measure of the contribution of each metal to the canonical variable that is independent of the relative magnitude of the different metal levels determined. The R_1 values shown in Table 3.17 indicate large positive contributions from copper (0.49) and tin (0.27) with much smaller positive contributions from manganese (0.06), molybdenum (0.08), and chromium (0.09). Large negative contributions are made by barium (-0.32), cesium (-0.21), aluminum (-0.14), lead (-0.12), and strontium (-0.12). For the second canonical variable, almost all contributions are positive, and significant contributions are made by several metals that contributed to the first canonical variable.

Table 3.17 Statistical Analysis of ln (metal concentration) (μg%) from Tissue to Tissue for 13 Trace Metals in 17 Anatomic Sites[a]

	F^b	Prob. > F	MSE[c]	$R_1{}^d$	$R_2{}^d$
Cu	56.7	0.0001	0.059	0.49	-0.02
Mn	12.4	0.0001	0.174	0.06	0.18
Mo	4.15	0.0001	0.247	0.08	0.09
Zn	4.05	0.0001	0.372	-0.03	-0.10
Cr	6.07	0.0001	0.598	0.09	0.14
Ni	4.91	0.0001	0.314	-0.01	0.18
Cs	21.4	0.0001	0.677	-0.21	0.28
Ba	33.4	0.0001	0.595	-0.32	0.25
Sr	5.47	0.0001	0.902	-0.12	0.08
Cd	4.28	0.0001	0.477	-0.07	0.11
Al	10.2	0.0001	1.37	-0.14	0.17
Sn	20.0	0.0001	0.075	0.27	0.06
Pb	14.7	0.0001	0.242	-0.12	0.28

[a]From Webb *et al.*,[194] by courtesy of American Elsevier Publishing Co.
[b]Wilks parameter.[18]
[c]Mean square error.
[d]Correlation coefficients between ln (metal concentration) (μg%) and the first (R_1) and second (R_2) canonical variables.[16]

The extent of the contributions by the various metals to these linear indices can be seen from Figs. 3.14 and 3.15, in which the analytical data for the 13 metals determined in the 17 tissues are plotted, using the coded lowercase letters of Table 3.15 to signify the various tissues. The figures emphasize the asymmetry of the distribution of the metals among the tissues. Copper, the metal that makes the largest positive contribution to the first canonical variable, illustrates in Figs. 3.14 and 3.15 the dominant contrast within the data. The seven heart valves and blood vessels all have lower concentrations of copper than do the other 10 tissues derived from myocardium. This distinction persists for tin, but to a lesser extent. Only five of the seven valves and vessels are below the other 10 tissues in the concentrations of tin. Consequently, the correlation of tin with the first canonical variable is less (0.27) than that of copper (0.49). Since the variability is in the same direction, the correlations are of the same algebraic sign. By contrast, the variability observed for aluminum, lead, strontium, barium, and cesium is opposite from that of copper. The concentrations of these metals are higher in the blood vessels and heart valves than in the 10 anatomic sites containing myocardium, leading to negative correlation coefficients with the first canonical variable. Zinc concentrations show little apparent trend across the tissues, which is reflected in the small R_1 value for this metal (-0.03).

3.9.2 Duncan Analysis

The statistical significance of the differences in the metal concentrations from tissue to tissue was tested using Duncan's procedure.[204] The results of the Duncan analysis are shown in Table 3.18 for zinc, copper, tin, aluminum,

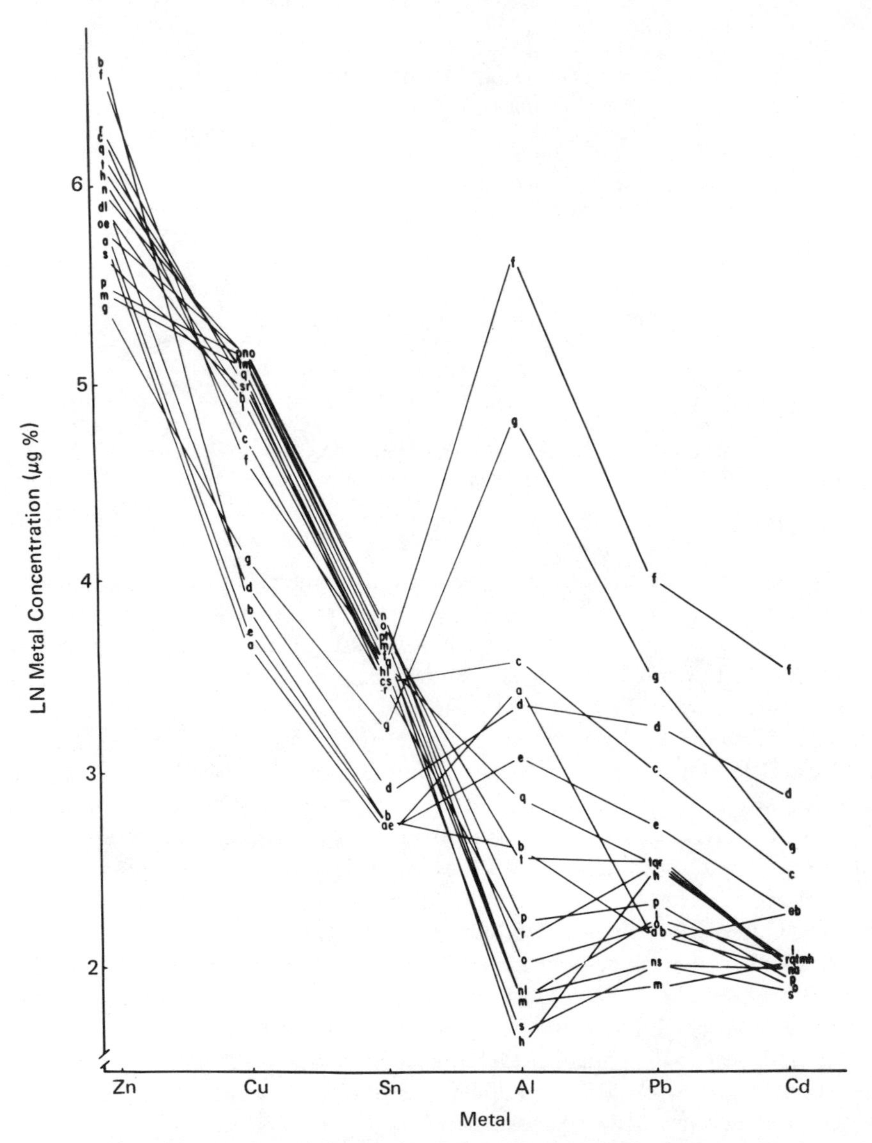

Fig. 3.14 Plot of ln (metal concentration), μg% of 13 trace metals present in 17 anatomic sites. Tissues are denoted by coded letters of Table 1. [From Webb *et al.*,[194] p. 68, by courtesy of the American Elsevier Publishing Co., Inc.]

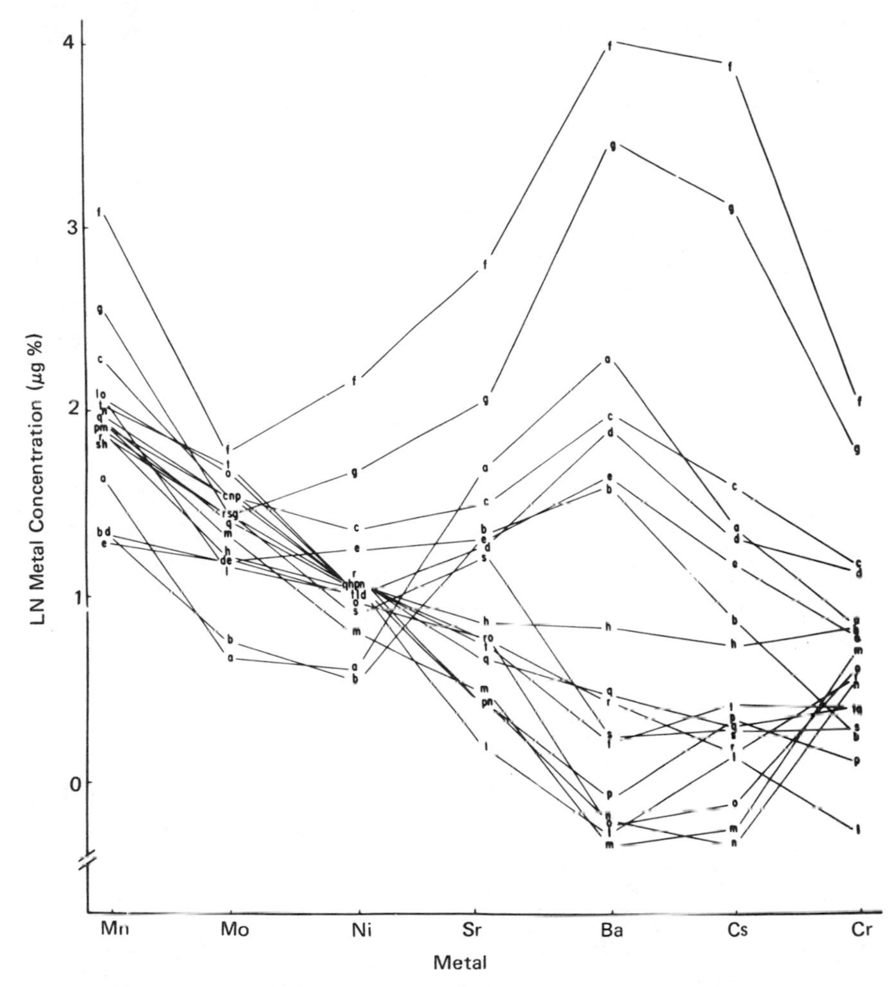

Fig. 3.15 Plot of ln (metal concentration), μg% of 13 trace metals present in 17 anatomic sites. Tissues are denoted by coded letters of Table 3.15. [From Webb *et al.*,(194) p. 69, by courtesy of the American Elsevier Publishing Co., Inc.]

lead, cadmium, manganese, molybdenum, nickel, strontium, barium, cesium, and chromium. The tissues are arranged in these tables from left to right in the order of decreasing metal concentrations, corresponding to the vertical sequences of Figs. 3.14 and 3.15. The horizontal lines are used to summarize the results of the statistical test. Any tissues that are joined together by a single line are not significantly different from each other, but they are different from all other tissues. The statistical significance of the differences observed in the figures can be readily seen from the tables. The pulmonary valve is significantly different from all other tissues in lead, cadmium, manganese, nickel, and cesium.

Table 3.18 Results[a] of Duncan Analysis of Concentrations of 13 Trace Metals in 17 Anatomic Sites[b]

Metal	Anatomic sites
Zn	MPA PV SN RSVC CR LBB RA LV TV LAA LV-PM MV AO AVN+B IVS RV AV
Cu	IVS LV LV-PM LBB RV CR SN AVN+B RA LAA RSVC PV AV TV MPA MV AO
Sn	LV LV-PM IVS LBB RV PV CR RA LAA AVN+B RSVC SN AV TV MPA MV AO
Al	PV AV RSVC AO TV MV CR MPA LBB IVS SN LV-PM LAA LV RV AVN+B RA
Pb	PV AV TV RSVC MV LBB CR SN RA IVS LAA LV-PM AO MPA LV AVN+B RV
Cd	PV TV AV RSVC MV MPA LAA SN CR LBB RV RA LV AO IVS LV-PM AVN+B
Mn	PV AV RSVC LAA LV-PM LBB LV CR IVS RV SN AVN+B RA AO MPA TV MV
Mo	PV LBB LV-PM LV RSVC IVS SN AVN+B AV CR RV RA TV MV LAA MPA AO
Ni	PV AV RSVC MV SN CR RA IVS LV LBB LAA TV LV-PM AVN+B RV AO MPA
Sr	PV AV AO RSVC MPA MV TV AVN+B RA SN LV-PM LBB CR RV LV IVS LAA
Ba	PV AV AO RSVC TV MV MPA RA CR SN AVN+B LBB IVS LV LV-PM LAA RV

(Continued)

Table 3.18 (*Continued*)

Cs	PV	AV	RSVC	AO	TV	MV	MPA	RA	LBB	IVS	CR	AVN+B	SN	LAA	LV–PM	RV	LV
Cr	PV	AV	RSVC	TV	AO	RA	MV	RV	LV–PM	SN	LV	LBB	CR	AVN+B	MPA	IVS	LAA

[a] For any metal, all tissues joined by a single line are not significantly different ($p > 0.05$).
[b] Tissues are denoted by abbreviations from Table 3.15.

The pulmonary and aortic valves are significantly different from all other tissues in aluminum and barium. The uniqueness of the trace metal concentration in the pulmonary valve can be seen from Table 3.18, which indicates that the pulmonary valve is significantly different from the aortic valve in the levels of 8 of the 13 trace metals. In contrast, these latter two valves differ from each other only in two metal levels (Cu, Pb). The aortic valve differs from the mitral and tricuspid valves in the levels of about half of the metals determined.

There is also a similar contrast among the three blood vessels. While the aorta and main pulmonary artery are significantly different from each other in only two trace metal concentrations (Zn, Cu), they differ from the right superior vena cava in the concentrations of six and eight trace metals, respectively.

Most of the tissues of ordinary and specialized myocardium are characterized by extensive similarity in the levels of many metals, and frequently are joined together in Table 3.18 by horizontal lines.

3.9.3 Cluster Analysis

In the cluster analysis, linear combinations of the means of the 13 trace metal concentrations of each tissue were formed that best separated the tissues into groups, or clusters.[205] The criterion for the best partition was the maximum value of a transformed Wilks parameter, i.e., the likelihood-ratio criterion.[203] The cluster pattern created in this way for two through seven groups of tissues was determined. The cluster patterns were displayed as a plot in two dimensions created by the linear indices that best separate the tissue into the required number of groups. Each tissue site was plotted according to its coefficients in these two indices, EV-1 and EV-2. Tissues in the same group will have similar values for these two coefficients, while those outside that group will have quite different values. Tissues included in any given group are not significantly different from each other, but are significantly different from all other tissues outside the group.

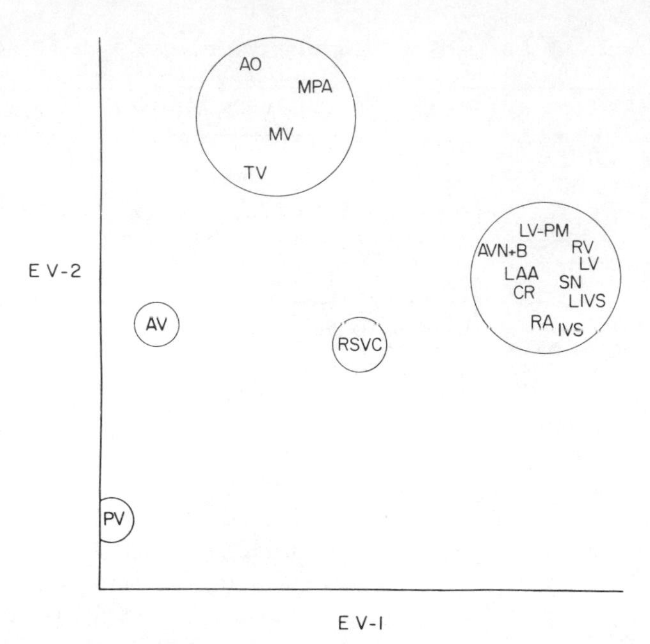

Fig. 3.16 Tissue groups created by the five-cluster analysis plotted according to their coefficients (EV-1, EV-2) in the linear indices constructed to create the cluster. [From Webb *et al.*,(206) p. 390, by courtesy of Academic Press, Inc.]

The results of a cluster analysis for five groups of tissues are shown in Fig. 3.16. The analysis for two through seven groups of tissues is generally consistent with this grouping. However, for a total group containing only 17 members, the number of subgroups chosen as the final clustering pattern, namely five, was chosen to be neither unreasonably small nor unreasonably large. The coordinates, EV-1 and EV-2, are the coefficients of the linear indices that best separate the tissues into five clusters. The values of these coefficients are quite different for each group, giving a well-differentiated clustering pattern. In this analysis, three tissues, the pulmonary valve, aortic valve, and right superior vena cava, constitute unique clusters of one tissue each that are well distinguished from each other, as measured by their different coordinates on the EV-1, EV-2 plot. The aorta, main pulmonary artery, mitral, and tricuspid valves constitute a group of four tissues that are significantly different from the other tissues, but are not significantly different among themselves. The final group of tissues are the samples derived from the myocardium. All 10 sample sites appear indistinguishable from each other on the basis of their profile of 13 trace metal concentrations.

3.9.4 Discriminant Analysis

In the discriminant analysis procedure, each of the 187 tissues was "plotted" in 13-dimensional space using as coordinates the 13 trace metal levels for each specimen. Since 11 samples of each anatomic site were studied, the centroid for each of the 17 types of tissue was calculated from its 11 "points" in the 13-dimensional array. The pairwise square generalized distance between each pair of tissues, i.e., the distance between each pair of centroids in this multidimensional space, was then calculated. Tissues that are alike have centroids close together while the centroids of unlike tissues are farther apart. This procedure gives a measure of the degree of "likeness" and "unlikeness" among the tissues. Having defined these centroids, the location of each of the 187 individual tissue specimens analyzed was considered in turn. The probability of identifying each specimen as each of the 17 types of cardiovascular tissue studied was then determined on the basis of its location in the 13-dimensional space relative to the previously calculated centroids for each tissue. The individual tissue was classified as one of the 17 types of cardiovascular tissue according to the largest calculated probability.

The results are shown in Fig. 3.17 as a 17-X-17 matrix of actual and inferred identities for the tissues. Since 11 pig hearts were dissected, the sum of the entries in any row of the matrix must be eleven. Diagonal entries correspond to correct inferences. Off-diagonal entries represent misclassifications, or errors. The rows and columns of the classification matrix have been arranged to maximize the diagonalization, thus bringing the most alike tissues closest together. The five entries in the matrix that have a superscript (solid dot) are those for which the probability of correct classification is less than 0.05. All other off-diagonal entries have a greater than 5% probability of being correctly classified.

For example, 10 samples of aorta were classified correctly as aorta, and one was improperly classified as mitral valve. This improperly classified sample had, however, a 10% probability of being classified as aorta. For the two samples of mitral valve that were incorrectly classified as tricuspid valve the probability of correct classification was 0.29 and 0.19. In summary, of the 187 specimens analyzed, 121, or 65%, were correctly classified. The probability of properly identifying 49 of 66 specimens that are listed in off-diagonal positions in Fig. 3.4 was 0.25 or less.

The groups of tissues that correspond to the clusters shown in Fig. 3.16 are blocked out in the discriminant matrix (Fig. 3.17). These groups are generally consistent with the diagonalization pattern within the matrix. It should be noted that while classification errors are few in number for the blood vessels and heart valves, they were comparatively common in the myocardium samples. No errors occurred in classifying the 11 samples of both the main pulmonary artery and the pulmonary valve. Misclassifications within the blocked out groups usually involved only other members of that group. Thus, the single misclassified sample of aorta was listed as mitral valve, and two mitral valve samples were considered to be tricuspid valves. Some uncertainty exists in regard to the most suitable grouping for the RSVC/RA pair of tissues. The right superior vena cava and the right atrium were both classified as the other tissue in three instances. These six specimens represent the only examples of errors in discriminating between tissues derived from the myocardium and those from the blood vessels and heart valves.

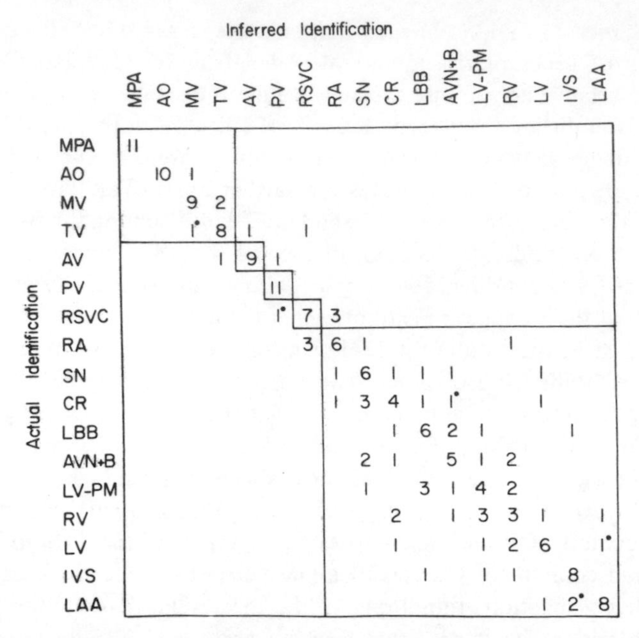

Fig. 3.17 Classification by discriminant analysis of the 187 tissue specimens on the basis of their concentrations of 13 trace metals. [From Webb *et al.*,(206) p. 390, by courtesy of Academic Press, Inc.]

The results of this study show that certain tissues of the cardiovascular system have distinctive trace metal contents. Statistical treatment of the data revealed a number of similarities and also differences among the trace metal compositions of the different tissues studied. Several of these observations may be of special significance:

1. The levels of many trace metals are much higher in the aortic and the pulmonary valves than in any other tissue analyzed. In addition, these two valves are significantly different from each other and from both the mitral and tricuspid valves, which have similar trace metal contents.

2. The two samples of arterial tissue (the aorta and main pulmonary artery), which are quite similar in their trace metal content, are very different from the single venous tissue studied (the right superior vena cava).

3. All three blood vessels are significantly different from the four heart valves in several trace metals.

4. There is little difference among the tissues composed largely of ordinary and/or specialized myocardium. These tissues are markedly different from the blood vessels and heart valves.

This reproducible asymmetry in the distribution of many trace metals among the various functionally different anatomic regions of the heart raises the possibility that some of these metals may be directly related to physiological function. Such speculations appear premature at this stage of these investigations, particularly for those trace metals with no known biological function. Metabolic differences, including varying requirements for the essential trace metals copper, manganese, molybdenum, and zinc, are well known, for example, between connective tissues and tissues consisting predominantly of muscle.[206–210]

It is clear that certain tissues of the cardiovascular system have distinctive trace metal profiles. Most of the significant patterns recognized in these analyses concern the blood vessels and heart valves, not the myocardium samples. The less striking discrimination among the several samples of ordinary and specialized myocardium is undoubtedly due in part to the difficulty of discretely and reproducibly sampling these anatomic sites. For example, specimens of left bundle branch will inevitably contain some tissue closely resembling free ventricular wall. Discrete sampling of the vessels and valves is not impeded as seriously by such technical difficulties.

These examples of the application of pattern recognition techniques to the interpretation of trace metal data suggest that similar applications may provide new approaches to the evaluation of trace metals in biological processes. Extension of the technique described here to include certain organic metabolites of known biological significance might provide further insight into the role of trace metals in metabolic processes. Studies directed toward this end have been initiated[211] using a group of semiisolated dairy farmers of Sliab Luachra, a 20-square-mile highland bog on the border of counties Cork and Kerry in southwest Ireland. This population was selected because of its isolation and remote location from urban and industrial sources of environmental pollution. The vast amount of clinical, genetic, historical, and metabolic information that has been gathered on this group during the last several decades[212, 213] was another factor in the choice.

Blood serum samples from these people were analyzed for 14 trace metals and 26 organic constituents. Although statistical analysis of the data is not yet complete, an interesting correlation involving molybdenum has been observed. As shown in Table 3.19 a correlation of high statistical significance was observed

Table 3.19 Correlation of the Concentrations[a] of Thyroxine in Blood Serum[b]

Thyroxine concentration	Mo 0–127[a]		Mo 128–131[a]		Mo 132+[a]	
	Number of subjects		Number of subjects		Number of subjects	
	Found	Expected[c]	Found	Expected[c]	Found	Expected[c]
0.0–3.4	28	18.3	9	16.9	13	14.8
3.5–4.2	44	43.9	46	40.5	30	35.6
4.3+	18	27.8	28	25.2	30	22.6

[a] Concentrations are given in terms of spectrometer response only since relative values, rather than absolute quantities, are of interest.
[b] $\chi^2 = 16.12, n = 4, p = 0.001$.
[c] Number expected on basis of random distribution.

between molybdenum and thyroxine (T4) ($p = 0.001$). This observation suggests that molybdenum may in some way be related to the function of the thyroid gland, and thus points to a hitherto unexplored approach to a more complete understanding of the normal function of this gland and the diseases with which it is associated.

In further studies on this group of Irish dairy farmers, certain correlations were observed between anthropologic features and the concentrations of trace metals in blood serum.[(214)] Among 204 dairymen born on (20–86 years prior; median 48 years) and still residents on family farms (which served as their principal source of food), serum molybdenum and manganese were significantly related to each other (χ^2-44.5, $p = 10^{-10}$) and to stature (χ^2-8.2, $p = 0.004$). Dairymen with both Mo and Mn low tended to be short, and their family farms were clustered in a discrete area; those with both Mo and Mn high tended to be tall and their family farms were clustered in other discrete areas. Serum copper, zinc, and iron intercorrelated with each other and with Mo and Mn but not with stature among the dairymen.

Among the 54 controls, also of "Old Irish" stock but born in (20–83 years prior; median 49 years) and still resident in area villages and towns (food sources varied since birth), there were no significant correlations or even tendencies between the five trace elements or between the trace elements and stature. These observations probably reflect the role of trace metals in the diet on growth and metabolism.

Perhaps the greatest potential biomedical application of pattern recognition techniques is in the laboratory diagnosis of disease. Application of the method to diagnosis of myocardial infarction was described in Section 3.7. Trace metal profiles are probably abnormal in a number of chronic diseases, as was shown for rheumatoid arthritis is Section 3.4. The disease specificity of these abnormal profiles is not known. By expanding the data base to include a battery of organic metabolites that can be determined by automated clinical laboratory methods, it seems likely that specific profiles can be identified for a number of diseases. Thus, pattern recognition techniques may provide a powerful new technique for the laboratory diagnosis of disease.

3.10 REFERENCES

1. I. H. Tipton, M. J. Cook, R. L. Steiner, C. A. Boye, H. M. Perry, Jr., and H. A. Schroeder, *Health Phys. 9*, 89 (1963).
2. I. H. Tipton and M. J. Cook, *Health Phys. 9*, 103 (1963).
3. I. H. Tipton, H. A. Schroeder, H. M. Perry, Jr., and M. J. Cook, *Health Phys. 11*, 403 (1965).
4. H. A. Schroeder and J. J. Balassa, *J. Chronic Diseases 14*, 408 (1961).
5. H. A. Schroeder, J. J. Balassa, and I. H. Tipton, *J. Chronic Diseases 15*, 51 (1962).
6. H. A. Schroeder and W. H. Vinton, Jr., *Amer. J. Physiol. 202*, 515 (1962).
7. H. A. Schroeder, W. H. Vinton, Jr., and J. J. Balassa, *Proc. Soc. Exp. Biol. Med. 109*, 859 (1962).

8. H. A. Schroeder, J. J. Balassa, and I. H. Tipton, *J. Chronic Diseases 16*, 55 (1963).
9. H. A. Schroeder, J. J. Balassa, and I. H. Tipton, *J. Chronic Diseases 16*, 1047 (1963).
10. H. A. Schroeder, J. J. Balassa, and I. H. Tipton, *J. Chronic Diseases 17*, 483 (1964).
11. H. A. Schroeder and J. J. Balassa, *J. Chronic Diseases 18*, 229 (1965).
12. H. A. Schroeder, *J. Chronic Diseases 18*, 217 (1965).
13. H. A. Schroeder and J. J. Balassa, *J. Chronic Diseases 19*, 85 (1966).
14. H. A. Schroeder, J. J. Balassa, and I. H. Tipton, *J. Chronic Diseases 19*, 545 (1966).
15. H. A. Schroeder and J. J. Balassa, *J. Chronic Diseases 19*, 573 (1966).
16. H. A. Schroeder, A. P. Nason, I. H. Tipton, and J. J. Balassa, *J. Chronic Diseases 19*, 1007 (1966).
17. H. A. Schroeder, J. Buckman, and J. J. Balassa, *J. Chronic Diseases 20*, 147 (1967).
18. H. A. Schroeder, A. P. Nason, I. H. Tipton, and J. J. Balassa, *J. Chronic Diseases 20*, 179 (1967).
19. H. A. Schroeder and J. J. Balassa, *J. Chronic Diseases 20*, 211 (1967).
20. H. A. Schroeder, A. P. Nason, and I. H. Tipton, *J. Chronic Diseases 20*, 869 (1968).
21. H. A. Schroeder, A. P. Nason, and I. H. Tipton, *J. Chronic Diseases 21*, 815 (1969).
22. H. A. Schroeder, D. V. Frost, and J. J. Balassa, *J. Chronic Diseases 23*, 227 (1970).
23. W. Neidermeier, J. H. Griggs, and J. Webb, *Appl. Spectrosc. 28*, 1 (1974).
24. E. M. Butt, R. E. Nusbaum, T. C. Gilmour, and S. L. Didio, *Arch. Environ. Health 6*, 286 (1963).
25. H. R. Imbus, J. Cholak, L. H. Miller, and T. Sterling, *Arch. Environ. Health 6*, 286 (1963).
26. R. E. Nusbaum, E. M. Butt, T. C. Gilmour, and S. L. Didio, *Amer. J. Clin. Pathol. 35*, 44 (1961).
27. L. M. Paixao and J. H. Yoe, *Clin. Chim. Acta 4*, 507 (1959).
28. R. Monacelli, H. Tanaka, and J. H. Yoe, *Clin. Chim. Acta 1*, 577 (1956).
29. A. J. Bedrosian, R. K. Skogerboe, and G. H. Morrison, *Anal. Chem. 40*, 854 (1968).
30. K. M. Hambidge, in *Trace Substances in Environmental Health* (D. D. Hemphill, ed.), University of Missouri Press, Columbia, Mo.
31. K. M. Hambidge, *Anal. Chem. 43*, 103 (1971).
32. G. H. Morrison, R. K. Skogerboe, A. J. Bedrasian, and A. M. Rothenberg, *Appl. Spectrosc. 23*, 349 (1969).
33. W. Niedermeier, J. H. Griggs, and R. S. Johnson, *Appl. Spectrosc. 25*, 53 (1971).
34. F. Twyman and F. Simeon, *Trans. Opt. Soc. (London) 31*, 169 (1929).
35. R. S. Johnson, W. Niedermeier, J. H. Griggs, and J. F. Lewis, *Appl. Spectrosc. 22*, 552 (1968).
36. C. D. West and D. N. Hume, *Anal. Chem. 36*, 412 (1964).
37. R. Mavrodineanu and R. C. Hughes, *Spectrochim. Acta 19*, 1309 (1963).
38. L. May, *Spectroscopic Tricks*, Plenum Press, New York (1967), pp. 120ff.
39. R. E. Weekley, and J. A. Norris, *Appl. Spectrosc. 18*, 21 (1964).
40. J. B. Scarborough, *Numerical Mathematical Analysis*, The Johns Hopkins Press, Baltimore (1966), pp. 528–533.
41. J. Ramirez-Munoz, J. L. Malakoff, and C. P. Aime, *Anal. Chim. Acta 36*, 328 (1966).
42. F. A. Pohl, *Anal. Chem. 209*, 19 (1965).
43. M. F. Hasler, *Anal. Chem. 39*(6), 26A (1967).
44. Anon., *Chem. Eng. News 46*(5), 38 (1968).
45. J. J. Lowe and R. J. Martin, *J. Appl. Spectrosc. 23*, 587 (1969).
46. M. Margoshes and S. D. Rasberry, *Anal. Chem. 41*, 1163 (1969).
47. B. W. Glover, R. D. Orwell, and P. J. Adams, *Hilger J. 10*, 7 (1967).
48. B. T. Cronhjort, *Appl. Spectrosc. 21*, 232 (1967).
49. J. Schubert, in *Metal-Binding in Medicine* (M. J. Seren, ed.), J. B. Lippincott, Philadelphia, Pa.(1960), pp. 325–328.
50. L. L. Wiesel, *Metabolism 8*, 256 (1959).

51. Empire Rheumatism Council, *Bull. Rheumatic Diseases 11*, 255 (1960).
52. A. S. Prasd, *Federation Proc. 26*, 172 (1967).
53. H. C. Robinson, A. Telser, and A. Dorfman, *Proc. Natl. Acad. Sci. U.S. 56*, 1859 (1966).
54. E. E. Grebner, C. W. Hall, and E. F. Nuefeld, *Arch. Biochem. Biophys. 116*, 391 (1966).
55. R. M. Leach, Jr., *Federation Proc. 26*, 118 (1967).
56. G. C. Cotzias and A. C. Foradori, in *The Biological Basis of Medicine* (E. E. Bittar and N. Bittar, eds.), Vol. 1, Academic Press, Inc., New York (1968), p. 105.
57. C. H. Hill, B. Starcher, and C. Kin, *Federation Proc. 26*, 129 (1967).
58. M. W. Ropes and W. Bauer, *Synovial Fluid Changes in Joint Disease*, Harvard University Press, Cambridge, Mass. (1953).
59. C. T. Stafford, W. Niedermeir, H. L. Holley, and W. Pigman, *Ann. Rheumatic Diseases 23*, 152 (1964).
60. W. Pigman and S. Rizvi, *Biochem. Biophys. Res. Commun. 1*, 39 (1959).
61. W. Niedermeier, *Ann. Rheumatic Diseases 24*, 544 (1965).
62. W. Niedermeier, *Alabama J. Med. Sci. 2*, 196 (1965).
63. W. Niedermeier, R. P. Laney, and C. Dobson, *Biochim. Biophys. Acta 148*, 400 (1967).
64. W. Niedermeier, E. E. Creitz, and H. L. Holley, *Arthritis Rheumat. 5*, 439 (1962).
65. A. Lorber, L. S. Cutler, and C. C. Chang, *Arthritis Rheumat. 11*, 65 (1968).
66. I. Sternlieb, J. I. Sandson, A. G. Morell, E. Korotkin, and I. H. Scheinberg, *Arthritis Rheumat. 12*, 458 (1969).
67. A. Lorber, *Arthritis Rheumat. 12*, 459 (1969).
68. K. D. Muirden and G. B. Senator, *Ann. Rheumatic Diseases 27*, 38 (1968).
69. G. B. Senator and K. D. Muirden, *Ann. Rheumatic Diseases 27*, 49 (1968).
70. L. O. Plantin and P. O. Strandberg, *Acta Rheumatol. Scand. 11*, 30 (1965).
71. W. J. Pories, J. H. Henzel, and C. G. Rob, *Lancet 1*, 121 (1967).
72. W. J. Pories, C. G. Rob, J. L. Smith *et al.*, paper presented at 7th International Congress of Gerontology, Vienna, Austria, June 28, 1966.
73. G. C. Cotzias, P. S. Papvasilious, and E. R. Hughes, *J. Clin. Invest. 47*, 992 (1968).
74. E. Nettelbladt and L. Sundblad, *Opuscula Med. 12*, 224 (1967).
75. W. Niedermeier, R. Cross, and W. P. Beetham, Jr., *Arthritis Rheumat. 8*, 335 (1965).
76. J. Forestier, *Arch. Interamer. Rheumatol. 6*, 310 (1963).
77. R. H. Freyberg, W. D. Block, and S. Levey, *J. Clin. Invest. 20*, 401 (1941).
78. R. H. Freyberg, *Clinics 1*, 537 (1940).
79. J. J. Bertrand, H. Waine, and C. A. Toblas, *J. Lab. Clin. Med. 33*, 1133 (1948).
80. H. A. Swartz, J. E. Christian, and F. N. Andrews, *Amer. J. Physiol. 199*, 67 (1960).
81. J. C. Caygill and F. R. Jevons, *Clin. Chim. Acta 11*, 233 (1965).
82. R. H. Persellin, J. D. Smiley, and M. Ziff, *Arthritis Rheumat. 6*, 787 (1963).
83. G. Ungar, *Lancet 2*, 742 (1952).
84. M. W. Whitehouse, *J. Pharm. 15*, 556 (1963).
85. M. Adam, P. Bartl, Z. Deyl, and J. Rosmus, *Ann. Rheumatic Diseases 24*, 378 (1965).
86. W. Niedermeier and J. H. Griggs, *J. Chronic Diseases 23*, 527 (1971).
87. S. Segal, *Nonparametric Statistics*, McGraw Hill Book Company, New York (1956), p. 184.
88. J. L. Hardwick and C. J. Martin, *Helv. Odontol. Acta 11*, 62 (1967).
89. F. Brudevold, L. T. Steadman, and F. A. Smith, *Ann. N.Y. Acad. Sci. 85*, 110 (1960).
90. S. Wah Leung, in: *Caries-Resistant Teeth* (G. E. W. Wolstenholme and M. O'Connor, eds.) Little, Brown, and Company, Boston (1965), pp. 266–283.
91. J. L. Hardwick (G. E. W. Wolstenholme and M. O'Connor, eds.), in *Caries-Resistant Teeth*, Little, Brown and Company, Boston (1965), pp. 222–237.

92. S. Dreizen, H. A. Spies, Jr., and T. D. Spies, *J. Dental Res. 31*, 137 (1952).
93. D. Afonsky, *Saliva and Its Relation to Oral Health*, University of Alabama Press, University, Ala. (1961).
94. T. Arwill, N. Myrberg, and R. Soremark, *Odontol. Rev. 18*, 1 (1967).
95. H. Spencer and J. Samachson, in *Secretory Mechanisms of Salivary Glands* (L. H. Schneyer and C. A. Schneyer, eds.), Academic Press, Inc., New York (1967), pp. 101–112.
96. S. Dreizen, B. M. Levy, W. Niedermeier, and J. H. Griggs, *Arch. Oral Biol. 15*, 179 (1970).
97. A. N. Zengo, I. D. Mandel, A. Solomon, and P. Block, *J. Oral Therap. Pharm. 4*, 359 (1968).
98. C. A. Schneyer and H. D. Hall, *Proc. Soc. Exp. Biol. Med. 121*, 96 (1966).
99. E. Frieden, *Sci. Amer. 227*, 52 (1972).
100. B. L. Vallee, *Metabolism 1*, 420 (1952).
101. W. E. C. Wacker, D. D. Ulmer, and B. L. Vallee, *New Engl. J. Med. 255*, 449 (1956).
102. C. A. D'Alonzo and S. Pell, *Arch. Environ. Health 6*, 381 (1963).
103. F. W. Sunderman, Jr., S. Nomoto, A. M. Pradhan, H. Levine, S. H. Bernstein, and R. Hirsch, *New Engl. J. Med. 283*, 896 (1970).
104. F. W. Sunderman, Jr., M. I. Decsy, and M. D. McNeely, *Ann. N.Y. Acad. Sci. 199*, 300 (1972).
105. S. Nomoto, M. D. McNeely, and F. W. Sunderman, Jr., *Biochemistry 10*, 1647 (1971).
106. A. F. Parisi and B. L. Vallee, *Biochemistry 9*, 2421 (1970).
107. E. J. Underwood, *Trace Elements in Human and Animal Nutrition*, 3rd ed., Academic Press, Inc., New York, (1971), p. 170.
108. M. Tal, *Biochim. Biophys. Acta 195*, 76 (1969).
109. S. P. Bessman and N. J. Doorenbos, *Ann. Internal Med. 47*, 1036 (1957).
110. F. H. Nielsen and H. E. Sauberlich, *Proc. Soc. Exp. Biol. Med. 134*, 845 (1970).
111. F. H. Nielsen and D. J. Higgs, in *Trace Substances in Environmental Health* (D. D. Hemphill, ed.), Vol. 4, University of Missouri Press, Columbia, Mo. (1971), p. 241.
112. C. S. Nomoto, M. D. McNeely, and F. W. Sunderman, Jr., *Biochem. 10*, 1647 (1971).
113. J. Webb, K. A. Kirk, W. H. Jackson, W. Niedermeier, M. E. Turner, C. E. Rackley and R. O. Russell, Jr., *Exp. Mol. Pathol. 25*, 322 (1976).
114. A. J. Barr and J. H. Goodnight, *Statistical Analysis System*, North Carolina State University, Raleigh, N.C. (1971).
115. P. O. Webster, *Scand. J. Clin. Lab. Invest. 17*, 357 (1965).
116. S. S. Wilks, *Biometrika 24*, 471 (1932).
117. D. M. Gobler and D. A. Winfield, *Brit. Heart J. 34*, 597 (1972).
118. G. L. Curran and D. L. Azarnoff, *A.M.A. Arch. Internal Med. 101*, 685 (1958).
119. F. Snyder and W. E. Cornatzer, *Nature 182*, 462 (1958).
120. G. L. Curran, *J. Biol. Chem. 210*, 765 (1974).
121. F. Bernhein and M. L. C. Bernheim, *J. Biol. Chem. 210*, 765 (1954).
122. H. A. Schroeder, *J. Nutr. 94*, 465 (1968).
123. A. W. Voors, *Lancet 2*, 1337 (1969).
124. N. F. Volkov, *Federation Proc. 22*, T897 (1963).
125. D. Harman, *J. Gerontol. 20*, 151 (1965).
126. H. A. Schroeder, A. P. Nason, and I. H. Tipton, *J. Chronic Diseases 23*, 123 (1970).
127. I. H. Tipton, in *Metal-Binding in Medicine* (M. J. Seven, ed.), J. B. Lippincott, Philadelphia (1960), p. 27.
128. H. A. Schroeder and J. J. Balassa, *Amer. J. Physiol. 209*, 433 (1965).
129. H. A. Schroeder, J. J. Balassa, and W. H. Vinton, Jr., *J. Nutr. 86*, 51 (1965).
130. W. J. Campbell and W. Mertz, *Amer. J. Physiol. 204*, 1028 (1963).
131. W. Mertz, E. E. Roginski, and H. A. Schroeder, *J. Nutr. 86*, 107 (1965).

132. H. A. Schroeder, J. J. Balassa, and I. H. Tipton, *J. Chronic Diseases 15*, 941 (1962).
133. I. H. Tipton and M. J. Cook, *Health Phys. 9*, 103 (1963).
134. I. H. Tipton, H. A. Schroeder, H. M. Perry, Jr., and M. J. Cook, *Health Phys. 11*, 403 (1965).
135. H. A. Schroeder, *Circulation 35*, 570 (1967).
136. W. H. Glinsmann and W. Mertz, *Metabolism 15*, 510 (1966).
137. H. W. Staub, G. Reussner, and R. Thiessen, Jr., *Science 166*, 746 (1969).
138. T. G. Farkas and W. C. Weese, *Exp. Eye Res. 9*, 132 (1970).
139. H. A. Schroeder, *J. Nutr. 88*, 439 (1966).
140. E. E. Roginski and W. Mertz, *J. Nutr. 97*, 525 (1969).
141. H. A. Schroeder, *J. Nutr. 97*, 237 (1969).
142. B. N. Bachra and G. A. van Harskamp. *Calcif. Tissue Res. 4*, 359 (1970).
143. E. E. Kattlove and T. H. Spaet, *Blood 35*, 659 (1970).
144. W. Mertz, *Physiol. Rev. 49*, 163 (1969).
145. F. Basolo and R. G. Pearson, *Mechanisms of Inorganic Reactions*, 2nd ed. John Wiley & Sons, Inc., (1966), p. 818.
146. V. A. Pchelin, N. V. Grigor'eva and V. N. Izmailova, *Dokl. Akad. Nauk SSSR 151*, 134 (1963); *Chem. Abstr. 59*, 10229 (1963).
147. D. F. Flick, H. F. Kraybill, and J. M. Dimitroff, *Envrion. Res. 4*, 71 (1971).
148. H. A. Schroeder and J. J. Balassa, *J. Chronic Diseases 14*, 236 (1961).
149. L. Friberg, M. Piscator, and G. Nordberg, *Cadmium in the Environment*, C.R.C. Press, Cleveland, (1971).
150. B. T. Emmerson, *Ann. Intern. Med. 73*, 854 (1970).
151. H. A. Schroeder and J. Buckman, *Arch. Environ. Health 14*, 693 (1967).
152. S. A. Gunn, T. C. Gould, and W. A. D. Anderson, *J. Natl. Cancer Inst. 31*, 745 (1963).
153. S. A. Gunn, T. C. Gould, and W. A. D. Anderson, *Proc. Soc. Exp. Biol. Med. 122*, 1036 (1966).
154. S. A. Gunn, T. C. Gould, and W. A. D. Anderson, *J. Reprod. Fertil. 15*, 65 (1968).
155. D. J. Lawford, *Biochem. Pharm. 7*, 109 (1961).
156. B. Azelsson and M. Piscator, *Arch. Environ. Health 12*, 360 (1966).
157. E. C. Vigliani, *Amer. Ind. Hyg. Assoc. J. 30*, 329 (1969).
158. H. A. Schroeder, J. J. Balassa, and W. H. Vinton, *J. Nutr. 83*, 239 (1964).
159. J. Baum and H. G. Worthen, *Nature 213*, 1040 (1967).
160. M. G. Mustafa, C. E. Cross, and W. S. Tyler, *Arch. Intern. Med. 127*, 1050 (1971).
161. R. D. Wochner, W. Strober, and T. A. Waldman, *J. Exp. Med. 126*, 207 (1967).
162. W. V. Epstein, P. F. Gulyassy, M. Tan, and A. I. Rae, *Ann. Intern. Med. 68*, 48 (1968).
163. B. L. Vallee, J. F. Riordan, and J. E. Coleman, *Proc. Natl. Acad. Sci. U.S. 49*, 109 (1963).
164. R. H. Jones, R. L. Williams, and A. M. Jones, *Proc. Soc. Expl. Biol. Med. 137*, 1231 (1971).
165. J. Baum and H. G. Worthen, *Nature 213*, 1040 (1967).
166. G. G. Glenner, W. Terry, M. Harada, C. Isersky, and D. L. Page, *Science 172*, 1150 (1971).
167. G. G. Glenner, D. Elin, E. D. Eaner, H. A. Bladen, W. Terry, and D. L. Page, *Science 174*, 712 (1971).
168. L. J. Gathercole and L. Klein, *Biochim. Biophys. Acta 257*, 531 (1972).
169. A. L. Shapiro, M. D. Scharff, J. V. Maizel, and J. W. Uhr, *Nature 211*, 243 (1966).
170. M. Cohn, *Cold Spring Harbor Symp. Quant. Biol. 32*, 211 (1967).
171. E. S. Lennox and M. Cohn, *Ann. Rev. Biochem. 36*, 365 (1967).
172. D. D. Perrin and A. E. Watt, *Biochim. Biophys. Acta 230*, 96 (1971).
173. N. Aspergren and H. Rorsman, *Acta Dermato-Venereol. 42*, 412 (1962).

174. E. Schopf, K. H. Schulz, and I. Isensee, *Arch. Klin. Exp. Derm. 234*, 420 (1969).
175. F. Scheifarth, H. W. Baenkler, and S. Pfister, *Int. Arch. Allergy 40*, 117 (1971).
176. H. Kirchner and H. Ruhl, *Exp. Cell. Res. 61*, 229 (1970).
177. J. H. R. Kagi and B. L. Vallee, *J. Biol. Chem. 236*, 2435 (1961).
178. J. H. R. Kagi and B. L. Vallee, *J. Biol. Chem. 235*, 3460 (1960).
179. P. Pulido, J. H. R. Kagi, and B. L. Vallee, *Biochemistry 5*, 1768 (1965).
180 G. W. Evans, P. F. Majors, and W. E. Cornatzer, *Biochem. Biophys. Res. Commun. 40*, 1142 (1970).
181. G. W. Evans, P. F. Majors, and W. E. Cornatzer, *Biochem. Biophys. Res. Commun. 41*, 1244 (1970).
182. O. J. Lucis, Z. A. Shaikh, and J. A. Embil, Jr., *Experientia 26*, 1109 (1970).
183. B. C. Starcher, *J. Nutr. 97*, 321 (1969).
184. Z. H. Shaikh and O. J. Lucis, *Federation Proc. 29*, 298 (1970).
185. A. Pappas, C. E. Orfanos, and R. Bertram, *J. Invest. Dermatol. 55*, 198 (1970).
186. R. B. Kowalski and C. F. Bender, *J. Amer. Chem. Soc. 94*, 5632 (1972).
187. J. M. Rhodes, *Lithos 2*, 223 (1969).
188. R. B. Kowalski, T. F. Schatze, and F. H. Strass, *Anal. Chem. 44*, 2176 (1972).
189. J. Webb, K. A. Kirk, W. Niedermeier, J. H. Griggs, M. E. Turner, and T. N. James, *Alabama J. Med. Sci. 10*, 377 (1973).
190. J. Webb, K. A. Kirk, W. Niedermeier, J. H. Griggs, M. E. Turner, and T. N. James, *Bioinorg. Chem. 5*, 261 (1976).
191. J. Webb, K. A. Kirk, W. Niedermeier, J. H. Griggs, M. E. Turner, and T. N. James, *Bioinorg. Chem. 5*, 261 (1976).
192. J. Webb, K. A. Kirk, W. Niedermeier, J. H. Griggs, K. Kirk, M. E. Turner, and T. N. James, *Anal. Chim. Acta 81*, 143 (1976).
193. J. Webb, K. A. Kirk, W. Niedermeier, J. H. Griggs, M. E. Turner, and T. N. James, *Bioinorg. Chem.*, in press (1977).
194. J. Webb, K. A. Kirk, W. Niedermeier, J. H. Griggs, M. E. Turner, and T. N. James, *Bioinorg. Chem. 3*, 39 (1973).
195. J. Webb, W. Niedermeier, J. H. Griggs, and T. N. James, *Appl. Spectrosc. 27*, 342 (1973).
196. I. H. Tipton and M. J. Cook, *Health Phys. 9*, 103 (1963).
197. I. H. Tipton, H. A. Schroeder, H. M. Perry, and M. J. Cook, *Health Phys. 11*, 403 (1965).
198. P. O. Wester, *Acta Med. Scand. Suppl. 17*, 439 (1965).
199. P. O. Wester, *Scand. J. Clin. Lab. Invest. 17*, 357 (1965).
200. P. O. Wester, *Biochim. Biophys. Acta 109*, 268 (1965).
201. P. O. Wester, *Acta Med. Scand. 178*, 789 (1965).
202. W. Niedermeier, J. H. Griggs, and R. S. Johnson, *Appl. Spectrosc. 25*, 53 (1971).
203. S. S. Wilks, *Biometrika 24*, 471 (1932).
204. C. W. Duncan, *Biometrics 11*, 1 (1955).
205. J. Rubin and H. P. Friedman, *A Cluster Analysis and Taxonomy System for Grouping and Classifying Data*, IBM Corporation, New York (1967).
206. J. Webb, K. A. Kirk, W. Niedermeier, J. H. Griggs, M. E. Turner, and T. N. James, *J. Mol. Cell. Biol. 6*, 383 (1974).
207. D. J. Prockop, *Federation Proc. 30*, 984 (1971).
208. W. H. Carnes, *Federation Proc. 30*, 995 (1971).
209. R. M. Leach, Jr., *Federation Proc. 30*, 991 (1971).
210. N. Westmoreland, *Federation Proc. 30*, 1001 (1971).
211. Sr. C. Kennedy, R. Rauner, and O. Gawron, *Biochem. Biophys. Res. Commun. 47*, 285 (1972).

212. A. E. Casey, W. Niedermeier, F. Gravelee, L. Hall, E. Downey, and C. Lupton, *Alabama J. Med. Sci. 9*, 418 (1972).
213. C. W. Dupertuis, A. E. Casey, E. I. McGowan, W. W. Barham, and C. Holland, *Alabama J. Med. Sci. 9*, 69 (1972).
214. A. E. Casey, W. Niedermeier, C. W. Dupertuis, and E. L. Downey, *Alabama J. Med. Sci. 12*, 388 (1975).

Application of Spectroscopy to Toxicology and Clinical Chemistry 4

Eleanor Berman

4.1 INTRODUCTION

The modern science of spectroscopy marked its beginning before the eighteenth century. By 1700, L'Abbé Marie[1] had presented the first ideas about photometry. Then in 1732, Geoffrey, and later Melville and Marygraf[2] in 1752 and 1758, respectively, described colors imparted to flames by metallic salts. About 100 years later, an American physician, David Alter,[3] attempted to use the flame as a source for qualitative spectrochemical analysis. Subsequently, Bunsen and Kirchhoff[3] in Germany presented similar findings.

A visual flame photometer, named the "spectronatrometre" by Champion, Pellet, and Grenier in 1873,[4] probably represents the first instrument for quantitative flame analysis. Three years later Gouy[5] discussed an atomizer method for introducing samples into a flame.

From our vantage point more than a century later, the works of application spectroscopists such as Valentin, Bence-Jones, Emil Rosenberg, and others seem admirable and exciting. In 1863 Valentin[6] published a book on the application of the spectroscope to physiology and medicine. Besides employing this instrument in the diagnosis of metal poisoning, he also examined narcotic poisons and remedies spectroscopically.

H. Bence-Jones described his investigations of chemicals circulating in the blood at a meeting of the Royal Institute in 1865. Among other things, he described the characteristics of a peculiar protein[7] excreted into the urine of subjects afflicted with a condition now recognized as multiple myeloma, a type

Eleanor Berman • Division of Biochemistry, Cook County Hospital, Chicago, Illinois

of bone marrow tumor. Unlike ordinary albumin and globulin, which begins to precipitate out when their solution is heated to between 60 and 70°C, the Bence-Jones protein comes down at a lower temperature (40–60°C), then redissolves as the temperature reaches 100°C. It then reappears as the solution cools. To this date, establishing the presence of the protein described by Bence-Jones is considered useful in the diagnosis of multiple myeloma.

Quinke,[(8)] along with others in 1872, compared spectroscopically blood obtained from disease states such as anemia and leukemia. They also examined spectra resulting when blood had been exposed to carbon monoxide, cyanide, carbon dioxide, stibine, and arsine.

In 1876 Emil Rosenberg,[(8)] a Philadelphia physician, published a monograph entitled *The Use of the Spectroscope: Its Application to Scientific and Practical Medicine*. Considering his crude instrumentation and limited analytical techniques, he managed to acquire an amazing amount of information concerning the distribution of potassium, sodium, calcium, and lithium in tissues and biological materials. Rosenberg also investigated hemoglobin pigments and derivatives in blood and other body fluids in both normal and disease states.

Around this period of time, spectroscopy also began being applied in the identification of "suspicious stains" for forensic purposes.

In 1885 Hartley showed that principal alkaloids gave highly characteristic ultraviolet absorption spectra which could be used for identification and for ascertaining their purity. He further demonstrated that alkaloids which are closely related in structure, for example quinine and quinidine, morphine and codeine, cinchonine and cinchonidine, generated similar spectra.[(9–13)] It is interesting to note that Harley's work, when first presented at a meeting of the Royal Society, was greeted with less than the expected enthusiasm by certain of the august gentlemen present. One, in particular, foresaw the dreadful day when the "expert" would be supplanted by ignorant girls with machines. Dobbie and associates, from 1903 until 1914, did further studies on the relationship between the chemical nature of alkaloids and their ultraviolet spectra.[(14–18)]

In 1908 Coblentz[(19)] developed and studied the spectra for 135 compounds in the infrared region.

From 1900 to approximately 1940, some fine studies concerning trace metal deposits in tissues in health and disease were made with arc-spark spectrography. Since many of the preparative techniques are used with slight modifications today when applying emission and atomic absorption instruments, these will be discussed in a subsequent section on current applications of spectroscopy to trace metal analysis of biological materials.

Despite the inspired activity of Sheldon and Ramage, the Gerlachs, Benoit, Policard, and others, chemical spectroscopy in general developed slowly. Few selected highly specialized areas had access to the required instrumentation. It was necessary, initially, to develop an appreciation of the potential applications of such instrumentation and their actual importance as aids in solving routine diagnostic problems.

Until about 20 years ago, analytical toxicology was still concerned, primarily, with establishing the cause of a possible chemical death, just as it had been in the early nineteenth century. Furthermore, compared to the present, there were relatively few toxic substances available. Once a poison was isolated from a generous amount of biological material, identification would be made, more or less, on the basis of some chemical procedures, various color reactions, and microcrystalline tests. It was not unusual for an "expert" to taste an extract in order to obtain a presumptive "diagnosis."

Animal assays would be employed in confirming a possible chemical identification. For example, the production of tonic-clonic convulsions in a frog following the injection of suspected material was considered positive confirmation for the presence of strychnine. Should an extract instilled into the eye of a cat or rabbit dilate the animal's pupil, the identification of a substance as being atropine or a belladonna alkaloid was then considered substantiated. Morphine injected into a mouse caused the tail to curl in such a manner as to form an S with its body. In addition, a morphinized mouse did not jump off a heated hot plate. And so, the usual toxicologist of the preinstrumentation era practiced his art.

A comparator block and Duboscq visual colorimeter seemed adequate for the needs of the usual clinical chemistry laboratory of the pre and early post-World War II period. The author recalls the time when a Duboscq colorimeter with a built-in light source was considered an elegant instrument indeed.

Following World War II, it became apparent that the industry of the chemist had created a deluge of compounds for use or misuse, plus a variety of economic poisons and household aids, all of which could become toxicological problems.

At about this time, the emphasis in the practice of toxicology began shifting from the forensic to more immediate clinical situations involving a still-living patient. Sample size available for analysis was less, and more substances had to be considered as posible toxic agents.

In addition, the physician began to realize the value of routine clinical chemical determinations of blood glucose, urea, electrolytes, hemoglobin, and so on, as an aid in more adequately establishing diagnoses and guiding the treatment of patients.

This great demand for better means of analyses has resulted in a plethora of instrumental developments and their applications to toxicology and clinical chemistry in general. Consequently, today, one is often hard pressed to choose the instrumentation and analytical techniques of the proper degree of sophistication for a specific job. It is important that the analyst understand the basic chemistry involved, as well as the instrument's function, since all methods, instrumental and chemical, have their limitations.

4.2 SPECTROSCOPIC APPLICATIONS TO ANALYSES OF ORGANIC SUBSTANCES

The evolution of spectroscopic applications to the identification and quantitation of normal and abnormal substances in biological materials represents man's emergence from the comparator block. Many clinical chemistry analyses involve the measurement of color intensities produced by chemical reactions. In the laboratory of the not-too-distant past, when the Duboscq colorimeter represented sophistication, concentrations of unknown and standard substances were compared visually, really much like matching thread to a fabric. Given a proper light source, a skilled technician, who was neither color-blind nor too fatigued, could make these judgments with a fair degree of accuracy.

Application of a spectrophotometer in the visual range eliminates some of the human error and permits more precise measurements on comparatively smaller quantities of sample, provided that the instrument is properly calibrated and standardized for a specific analysis. An ignorant and inept "analyst" can generate beautifully inaccurate, meaningless, and misleading results with great precision on very "sophisticated" instrumentation. It must be remembered, however, that not all substances of interest produce easily measurable, fairly specific color reactions.

4.2.1 Ultraviolet Spectrophotometry

Since Hartley first published his findings concerning ultraviolet absorption spectra of certain alkaloids, the spectra of thousands of compounds have been documented. Innumerable applications to clinical- and forensic-type analyses, among others, have been published. The compendium of R. G. White represents but a fraction of this effort.[20]

A major application of use in forensic toxicological analysis was made by Heinz Schiller in 1937.[21] After isolating such compounds as atropine, cocaine, strychnine, and veratrine from tissues, he identified them by ultraviolet spectrophotometry. It was with Leo Goldbaum's work on a rapid method for identifying and quantitating barbiturates in blood by ultraviolet spectrophotometry[22] that the era of practical applications of this instrumentation to toxicological investigations of living subjects really began. In 1948, this represented a quite specific, yet rapid, technique. Even then, the cobalt nitrate colorimetric procedure for barbiturate analysis was considered less than reliable because of its reaction with amide and imide groups. For example, at one time the tap water in a specific city consistently gave a positive reaction because the chlorine used in treating drinking water was being neutralized with ammonia.

Ultraviolet spectrophotometry does possess certain limitations. First, the compound to be identified and quantitated must be in the pure state, free of constituents in its vehicle as well as other drug groups. An extract or solution containing more than one organic substance yields a spectrum representative of none. For example, a trace of salicylate can blot out the spectrum of a significant amount of a barbiturate. The spectrum merely indicates that "something" is present. Furthermore, not every compound of possible interest yields a spectrum sufficiently characteristic to enable its use as a parameter for identification. A rather common drug, meprobamate, does not contain a multiple bond in its structure and so does not generate a usable spectrum. Nevertheless, meprobamate does absorb ultraviolet light and can occlude spectral characteristics of other substances which may be present.

As Hartley pointed out in 1885, members of a drug family possess similar absorption characteristics. Therefore, ultraviolet absorption spectra alone cannot be used to distinguish the individual members of a drug family, for example the barbiturates. On the other hand, various compounds with structural similarities, but with widely divergent actions, cannot be differentiated. The amphetamines, propoxyphene, meperidine, phencyclidine, and methoxyanisole, to name a few, yield similar absorption spectra in the ultraviolet region.

The practical value of ultraviolet spectrophotometry in toxicological analyses is noteworthy despite "the interferences." Even though an essentially meaningless spectrum is generated, it does indicate that a substance of interest may be present. Further investigation is then in order. Separation techniques such as thin-layer, paper, gas–liquid, or high-pressure liquid chromatography can be combined with ultraviolet spectrophotometry for purposes of identifying compounds.

Incidentally, no single analytical technique is now considered adequate as the sole means of identifying a substance for a toxicological investigation. It is now the routine that more than one method be employed.

4.2.2 Infrared Spectroscopy

Infrared spectroscopy is a superb tool for the identification of chemical substances. Its applications to biochemical and toxicological investigations have been rather varied. For example, Freeman[23] was able to distinguish strains of tubercle bacilli by their infrared spectra. Morton[24] studied chemical characteristics of cell mitochondria by this means. Klein *et al.*[25] identified the constituents in renal and biliary calculi with infrared spectroscopy.

Stewart[26] employed this instrumentation to determine ethanol and carbon tetrachloride in blood. Along with numerous other investigators, Alha and Tamminen[27] and Rieders[28] applied infrared techniques in the identification of organic materials isolated during the course of toxicological analyses.

As with ultraviolet spectrometry, meaningful infrared spectra can be obtained only from pure compounds. However, infrared, can also be combined with different chromatographic techniques. For example, combining an existing infrared spectrometer with thin-layer chromatography provides a comparatively inexpensive, yet effective means of confirming the identification of an isolate. Areas of interest on a thin-layer chromatogram can be scraped off, eluted with a solvent, and then combined into the discs for subsequent infrared analysis. Infrared spectrometers can also be coupled with gas–liquid or high-pressure liquid chromatographs.

4.2.3. Mass Spectrometry

As early as 1957, Fowler used mass spectrometry for the study of gas exchange in the lungs.[29] Biemann[30] later applied mass spectrometry in elucidating the structure of certain organic molecules and studying their metabolism. Currently, mass spectrometers, coupled with gas chromatographs, are being employed rather extensively for the more definitive identification of compounds isolated during the course of toxicological investigations. Other applications in biochemistry have been numerous, but a discussion is beyond the intended scope of this review.

One must point out, however, that in a gas chromatograph/mass spectrometer system the heart of the system is the gas chromatographic column. Success in an analysis is dependent upon the skill and expertise of the chromatographer. The mass spectrometer is, in truth, merely another detector, although a very sensitive one. But if the material does not elute from the gas chromatograph because of improper technique or an inadequate column, the very sensitive mass spectrometer will not detect it.

4.2.4 Fluorimetry and Phosphorimetry

Numerous applications of fluorimetry and phosphorimetry have been made in biology and medicine for analyses of aromatic and heterocyclic compounds of interest. As with the other methods, accuracy and success with these procedures is dependent upon the purity of the chemical reagents used in the analytical process.

Fluorimetry and phosphorimetry are complementary rather than competitive techniques. For example, salicylic acid can be determined at low concentrations by fluorimetry but is rather insensitive to phosphorimetry.[31] Aspirin (acetyl salicylic acid), on the other hand, can be detected in low concentrations by phosphorimetry but is very insensitive to fluorimetry.

With the exception of tryptophan, phosphorimetry is of little use in the analysis of constituents normally present in blood. This is probably due to their low solubility in chloroform and to their weak phosphorescence. However, phosphorimetry probably could have limited application as a rapid means of characterizing a blood serum or plasma sample as being drug-free. For example, the normally low phosphorescence background of serum permits the application of phosphorimetry in the analysis of trace quantities of certain drugs. Caffeine, phenacetin, and aspirin are sensitive to phosphoresence. Sulfa drugs, codeine, atropine, procaine, cocaine, chlorpromazine, and phenobarbital are among the compounds exhibiting poor sensitivity to phosphorescence.

Phosphorimetry and fluorimetry can be applied to the localization of organic compounds on paper and thin-layer chromatograms. Fluorimetry has been applied to the analyses of amino acids and to those compounds with metal complexes, such as porphyrins and chlorophylls. The technique has been employed in assaying enzyme activity and vitamins.[32,33] Since most transition metals do not form fluorescent metal chelates, fluorimetry applications to trace metal analyses have been limited to the determination of calcium and magnesium.

4.3 APPLICATION OF SPECTROSCOPY TO TRACE METAL ANALYSES OF BIOLOGICAL MATERIALS

Prior to 1940, spectroscopic applications in investigations of trace metal contents of biological materials were limited primarily to flame and arc spectrographic methods. Sheldon and Ramage in 1931 attempted to ascertain what chemical elements, in addition to those already recognized as being universal constituents of protoplasm, were constantly present in tissues.[34] Earlier investigators had already established that sodium, potassium, magnesium, calcium, iron, and copper were present in all tissue. Other objectives of Sheldon and Ramage were to discover variations that might exist among the different organs, as well as possible variations between healthy and diseased tissues. Using an oxyacetylene flame and a Hilger quartz spectrograph, these investigators were able to qualitate and quantitate lead, cadmium, nickel, manganese, and other elements in small pieces of tissue and 0.1-ml quantities of blood. In preparation for spectrographic analyses, tissues were washed, dried to constant weight, and ground into powder. Blood specimens were soaked into filter paper and dried to constant weight, then burned in the oxyacetylene flame.

A majority of other early investigations were carried out with arc-spark methods. In the early 1930s such investigators as the Gerlachs,[35,36] Benoit,[37] and Policard[38,39] studied mineral deposits in tissues by histospectrography, a method which consists of passing a high-frequency spark through a section of tissue (fixed, fresh, or dried) and then analyzing the rays emitted by means of a quartz spectrograph. Firm tissue such as glands, muscle, and skin were burned in the spark directly. Fragile tissue from organs and various fluids were introduced into the spark in a carrier of ashless filter paper. Some investigators ashed organs at low temperatures, dissolved the residues in hydrochloric acid, and absorbed the solution on filter paper prior to spectrographic analysis. Potassium, sodium, magnesium, gold, boron, manganese, zinc, iron, silver, phosphorus, silicon, and carbon were among the elements identified at this time.

Later, Gerlach *et al.*[40] attempted analyses for lead and thallium in blood and tissue. The blood was wet-ashed before spectrographic analysis. Scott and Williams[41,42] further explored the inorganic constituents of protoplasm. They listed a number of points to consider when attempting interpretations of spectrograms of biological materials.

Boyd and De,[43] working in India, combined chemical means of separation with emission spectrographic analysis in their examination of tissues and various Indian vegetables for trace metal content.

Gaul and Staud[44-46] analyzed biopsy material from a series of patients receiving gold thiosulfate or silver arsphenamine therapy.

Faced by an epidemic of childhood lead poisoning in Baltimore, Shipley, with coworkers Scott and Blumberg,[47] developed in 1932, a semiquantitative spectrographic method for detection of lead in blood. Five milliliters of blood and 2 ml of sulfuric acid were charred over an open flame, then placed in a muffle furnace not exceeding 500°C. Ashing was completed in 12–24 h. The resultant ash was powdered and examined spectrographically.

From 1935 to 1944, Cholak[48] and coworkers contributed various spectroscopic methods for the analyses of lead, bismuth, cadmium, antimony, manganese, tin, copper, and silver in biological materials. Techniques employed ranged from arc-spark spectrography to colorimetry following chelate formation with dithizone. Cholak was a versatile investigator.

During this period Kehoe[49] attempted to establish "normal" values for certain trace element concentrations in biological materials.

An outstanding contribution of the 1950s is the work of Isabel Tipton. As a health physicist interested in learning the concentrations and distribution of stable elements in the normal body, she employed the spectrograph in the investigation and distribution of 18 heavy metals in tissues obtained from accident victims throughout the United States.[50]

Emission spectrography has been utilized fairly extensively in toxicological studies. Its major advantage lies in the breadth of coverage—the ability to obtain information on many metals simultaneously.

4.3.1 X-Ray Spectroscopy

The principles of x-ray spectroscopy developed from the work of Baikla in 1909 and from that of Moseley.[51] X-ray emission spectroscopy is a highly useful tool for the direct, nondestructive measurements of elements in biological materials. Sample-size requirements are small. Both anions and cations have been determined by this technique.

Grebe and Esser investigated strontium, lead, thallium, nickel, copper, cadmium, chromium, zinc, manganese, magnesium, and cobalt in tissues.[52] However, Samuel Natelson is considered to be the pioneer in the application of x-ray spectroscopy to the clinical laboratory. He determined potassium, calcium, sodium, thallium, lead, gold, mercury, bismuth, zinc, sulfur, and chloride on microliter samples of serum.[52–56]

Lund and Mathies used x-ray spectroscopy to determine protein-bound iodine in only 50 μl of serum.[57] In addition, they measured potassium, zinc, iron, phosphorus, magnesium, and other elements in minute tissue samples of approximately 1-mg size.[58] Lund and Mathies also measured nitrogen, carbon, and oxygen indirectly.[59]

F. J. Brooks and colleagues published a method for determining mercury in urine by x-ray spectroscopy.[60]

Although x-ray spectroscopy is an elegant means for determining various elements in biological materials, its use has been restricted to a few specialty areas. Flame emission and atomic absorption have proved to be more adaptable to the specific needs of the clinically oriented laboratory, i.e., possessing the ability for multiple trace metal analyses in volume and with good specificity, precision, and accuracy.

4.3.2 Emission and Atomic Absorption Spectroscopy

It can be said that the availability of good, reasonably priced flame emission and atomic absorption instruments has revolutionized various activities in clinical laboratories. Trace metal analyses, formerly done only "in extremis" and then not very well, are now routine. As a result, philosophies for managing patient care more adequately have changed.

Prior to the introduction of such instruments, the role of such elements as sodium, potassium, calcium, and magnesium in maintaining normal bodily function and the malfunction resulting from their imbalance were not fully appreciated. This lack of awareness can be attributed, in part, to analytical techniques then extant. These were cumbersome, imprecise, time consuming, and of questionable accuracy by comparison with current procedures.

The author can recall the time when a busy clinical laboratory serving a 400-bed hospital performed only one or two analyses per year for the sodium or potassium content of blood serum. The analytical method, a gravimetric procedure, required an entire day to complete and the findings were, of course, academic at best. The patient in question either departed from this earth or else improved despite the inadequacies.

With the instrumentation available today, simultaneous sodium and potassium analyses can be completed at the rate of 30 s per sample. Consequently, serum electrolytes are an established part of every patient's initial laboratory "workup." In addition, electrolytes of serum and other bodily fluids can be monitored as often as deemed necessary on all hospitalized patients receiving intravenous fluids.

This ability to supply an attending physician with accurate and rapid determination of sodium, potassium, calcium, and magnesium often proves to be lifesaving.

Analyses for the toxic metals lead, cadmium, mercury, and thallium, among others, can be completed within relatively few minutes by atomic absorption spectrometry, flame and flameless. There is little need to belabor the clinician's dependence upon these various analytical findings in establishing diagnoses and following therapeutic responses of patients.

Publications dealing with applications of spectroscopic instrumentation number into the thousands. Emission and atomic absorption are complementary techniques. Each is superior for certain elements. Each suffers from interferences. The nature and composition of the matrix must be considered when developing applications. Concentrations of a substance in relation to other elements present should also be taken into account. For example, a normal blood level is 20 μg% (0.2 μg/ml). A level of 60 μg% is considered diagnostic of lead poisoning. Normal blood sodium content is about 216 mg% (2160 μg/ml), a concentration 10,000 times as great. Normal potassium content in the blood is approximately 16% (160 μg/ml). Sodium and potassium concentrations in urine are still greater.

When these elements and lead are present in solution in equivalent concentrations, they exert no apparent effect upon light absorption at the analytical line for lead. However, at the concentrations normally present in biological materials, sodium and potassium will absorb light at any wavelength. The phenomenon is termed "light-scattering effect" or nonspecific molecular absorption." If such effects are not eliminated either instrumentally, i.e., by deuterium background correction and/or chemical pretreatment of samples analyzed, the concentrations reported will be many times greater than the actual value present. This phenomenon can be eliminated to some degree by subtracting absorbance readings obtained from a nonabsorbing line from the readings obtained at the resonance line for the specific element.

Flameless atomic absorption devices, whether they be the graphite furnace, carbon rod analyzer, or the tantalum strip, do augment the analytical capabilities of atomic absorption spectrometers. While only about 5% of a solution aspirated into a flame is atomized, the entire aliquot that is introduced into a flameless sampling system is atomized. Excellent!

Nevertheless, one must be aware that these flameless systems have limitations. It is true that the practical limits of detection for elements of interest are extended. However, the spectral interferences that emanate from the constituents in a biological matrix are still present. For example, the 10,000:1 sodium/lead ratio in a normal blood specimen remains unaltered and can contribute to an additional decrease in the light intensity from the cathode. The deuterium continuum for background correction is effective, but only in part, in the elimination of the nonspecific molecular absorption.

The viscosity of solutions being aspirated into a flame or measured into a flameless atomizer (graphite furnace, carbon rod, or tantalum strip) is another factor for consideration. The more viscous the solution, the slower the flow, and so lesser quantities of elements in solution are aspirated and atomized per unit of time. To minimize error, test and reference solutions should be of comparable viscosity.

Viscous solutions tend to cling to the sampling device. Consequently, a specific pipet will actually deliver a lesser volume of blood to the carbon furnace, for example, than it will of water or of a solvent such as methyl isobutyl ketone.

Solid materials and tissue specimens should be solubilized before atomic absorption analysis. Organic matter can be destroyed by wet- or dry-ashing techniques. Often precipitation of blood or tissue proteins by trichloroacetic acid is sufficient.

Although it is true that a few milligrams of tissue, hair, or whatever can be introduced into a flameless sampler without prior chemical treatment, results obtained from such atomic absorption analyses can be considered to be qualitative at best. Instrumental background correction does not eliminate the entire background "spewed" forth by a piece of tissue or aliquot of blood. If precise and accurate analyses are desired, organic matter must first be removed.

Although developments and advances in instrumentation have extended the chemist's analytical capabilities, the necessity for practicing the principles of "chemistry" has not been eliminated, by any means. Adequate acceptable analytical methodology can evolve only from the judicious application of one's understanding of the physics of a specific instrument in use and the chemistry of the materials subjected to analysis.

4.4 SPECIFIC APPLICATIONS OF EMISSION AND ATOMIC ABSORPTION SPECTROPHOTOMETRY

Representative applications of flame and flameless methods to clinical and toxicological analyses will be discussed in order of frequency of use. See Table 4.1 for a summary of the information given below.

4.4.1 Sodium and Potassium

Flame emission spectrometry is the preferred means for the determination of sodium and potassium. In most clinical laboratories the two elements are measured simultaneously. Some analytical systems employ internal standardization with lithium to achieve better control of results.

Sera are analyzed at a 1:250 dilution in water against aqueous reference standards containing both elements. At this dilution, protein in serum does not

Table 4.1 Trace Elements in Biological Materials, Concentrations, and Detection Limits: Summary

Element	Class	Normal concentration (μg/ml)	Material	Reference	Detection limit (μg/ml)	Procedure
Na	Essential	3100–3560	Serum	65	1×10^{-4}	FE
Na					5×10^{-3}	AA
K	Essential	140–190	Serum	65	2×10^{-4}	FE
K					5×10^{-3}	AA
Ca	Essential	90–110	Serum	65	9×10^{-4}	FE
Ca					3×10^{-3}	AA
Mg	Essential	10–20	Serum	65	2×10^{-2}	FE
Mg					5×10^{-4}	AA
Cu	Essential	0.85–1.5	Serum	65	0.1	FE
Cu		<0.05	Urine	65	5×10^{-3}	AA
Fe	Essential	0.85–1.5	Serum	65	1×10^{-2}	AA
Zn	Essential	0.8–1.1	Serum	66	2×10^{-3}	AA
Al	Unknown	0.2–0.94	Blood	67	1	AA
Co	Essential	1×10^{-2}	Blood	68	1×10^{-2}	AA
Cr	Essential	$1.1\text{–}6.5 \times 10^{-2}$	Blood	69	0.1	FE
Cr					5×10^{-3}	AA
Mn	Essential	$6\text{-}11 \times 10^{-3}$	Whole blood	70	6×10^{-3}	AA
Mn		$1.2\text{–}1.6 \times 10^{-3}$	Plasma	71		
Mn		$2.2\text{–}2.4 \times 10^{-2}$	Red cells	71		
Ni	Essential	0.02–0.1	Serum	72, 73	1×10^{-2}	AA
Au	Therapeutic	0	Blood, etc.	74–76	5×10^{-2}	AA/ES
Li	Therapeutic					
	Normal	0	Blood, etc.	77	3×10^{-6}	FE
	Thera	10–20			5×10^{-3}	AA
	Toxic	>25				
As	Toxic	$0\text{–}4 \times 10^{-3}$	Blood, urine	78	0.2	AA
Be	Toxic	0	Blood	79	2×10^{-2}	AA
Be		0	Urine		3×10^{-5}	AA-furnace
Bi	Toxic	0	Blood, etc.	80	0.4	AA
Cd	Toxic	$<5 \times 10^{-2}$	Blood, urine	81-82	4.2	FE
Cd					1×10^{-3}	AA
Pb	Toxic	About 0.2	Blood	83	3×10^{-2}	AA
Pb		to 6.5×10^{-2}	Urine	84		
Hg	Toxic	$<3 \times 10^{-2}$	Blood	85	2.5	FE
Hg		$<2.5 \times 10^{-2}$	Urine		0.2	AA
Tl	Toxic	0	Blood, etc.	86	0.2	AA
Sb	Toxic	0	Blood, etc.	87	0.1	AA
Se	Toxic	0	Blood, etc.	87	0.5	AA

contribute significantly to the viscosity of the solution. Other elem
do not interfere at the emission lines commonly used for these analys

At its 5890-Å resonance line, sodium is detected in a conce
0.0001 μg/ml by flame emission spectrometry while potassium is
a concentration of 0.002 μg/ml at the 7665-Å resonance line. Both ... ents
have detection limits of 0.005 μg/ml with atomic absorption spectrometry.

4.4.2 Calcium

Calcium has a detection limit of 0.0009 μg/ml by flame emission and 0.003 μg/ml by atomic absorption spectrometry at the 4227-Å resonance line. Although flame emission is more sensitive, atomic absorption spectrometry is the method of choice for the analysis of calcium in biological materials, since interferences from constituents in the biological matrix are more easily eliminated. To perform accurate, reproducible analyses with atomic absorption, one needs only to dilute both standards and samples with a 1% solution of a lanthanum salt. On the other hand, to do accurate analyses by flame emission requires that standards be prepared in a matrix containing concentrations of protein, sodium, and potassium equivalent to those present in normal sera.

Phosphate has been found to suppress the absorbance signal of calcium. Normal calcium levels in serum are 9–11 mg%, while the phosphate content is around 4 mg%. Since lanthanum salts are more competitive than calcium for phosphate ions, biological matrices are usually prepared in a lanthanum solution in order to eliminate the "phosphate effect."

A suggested method for the analysis of calcium in sera involves only the preparation of a 1:25 dilution of standards and sera with an aqueous solution containing 1% lanthanum and 5% trichloroacetic acid. Although the serum protein content at this dilution has little effect upon the absorbance signal and viscosity of solution, we prefer to precipitate proteins present by trichloroacetic acid and eliminate particles by centrifugation. Thus, problems emanating from clogging of the burner head and nebulizer are minimized.

Dilutions employed for other biological fluids depend upon the calcium content.

One of the criticisms of the atomic absorption method for calcium determination has been the apparent instrumental instability (drift) and the consequent variation of analyses. This drift can be eliminated by allowing sufficient "warmup" time for both the hollow cathode lamp and the burner. Conditions are stable in about $\frac{1}{2}$ h. Instrumental stability should be monitored further, by rereading a standard solution after every five determinations.

Calcium is best determined with a reducing flame, one that is fuel (acetylene)-rich. Calcium oxide formation is minimized.

4.4.3 Magnesium

At its resonance line, 2852 Å, the detection limits for magnesium are 2×10^{-2} and 5×10^{-4} μg/ml by flame emission and atomic absorption, respectively.

Prior to the appearance of atomic absorption instruments in the clinical laboratory, magnesium analyses were considered somewhat problematical. Spectrophotometric techniques were time consuming and lacking in precision. Fluorimetric methods, with proper techniques, represented a slight improvement. Since the normal magnesium content of serum is 1–2 mg%, flame emission applications were not practical. The background created by the high concentration of sodium overwhelmed the weak signal from magnesium.

With atomic absorption spectrometry, however, magnesium does not appear to be subjected to major interferences. Some investigators claim that protein in solution slightly suppresses the absorbance signal from magnesium. This can be eliminated by the addition of a competitive ion, such as lanthanum or strontium.

The dilution at which serum magnesium is determined is dependent upon the sensitivity of a specific instrument. With a unit equivalent in sensitivity to a Perkin-Elmer 303, serum magnesium can be determined in a 1:50 dilution. Both reference and test solutions are prepared with 0.5% strontium or lanthanum nitrate. Dilutions necessary for other biological fluids will vary.

Magnesium is determined in a reducing flame, one less rich than for calcium, however.

4.4.4 Lithium

Lithium is used clinically in the treatment of manic–depressive states. Since the metal is potentially toxic, its dosage must be controlled. Serum lithium levels provide a reliable index of impending toxicity. The therapeutic range is 1–2 mg%.

Limits of detection at the 6708-Å lithium line are 3×10^{-6} μg/ml for flame emission and 5×10^{-3} μg/ml for atomic absorption. Both instrumental techniques are used routinely. A lean oxidizing flame is used with the atomic absorption method.

The interferences due to sodium and potassium observed with flame emission can be eliminated by adding these elements, in amounts normally present in serum, to the reference standards used for calibration. Reference standards may also be prepared with lithium-free normal serum, then diluting both standards and test samples with water.

Since there are no apparent interferences with atomic absorption, lithium concentrations in serum can be determined following a simple 1:25 dilution with water. Viscosity is not a problem at this dilution. Adequate flushing of the burner with distilled water between samples will minimize the possibility of burner clogging from protein "buildup."

4.4.5 Copper

With the 3248-Å resonance line, the detection limits for copper in aqueous solutions are 0.1 μg/ml by flame emission and 0.005 μg/ml by atomic absorption.

Copper levels in normal blood serum range between 85 and 150 μg% (0.8–1.5 μg/ml). Levels are less than 0.6 μg/ml in Wilson's disease, a disturbance in copper metabolism that is genetically linked. Normally, urinary copper excretion is less than 50 μg/liter (0.05 μg/ml). In Wilson's disease, urinary copper excretions are markedly increased.

From an examination of the detection limits and concentrations normally found in serum and urine, it would seem that with atomic absorption spectrometry, copper in these materials could be determined by direct analysis following approximately a 1:5 dilution. There appear to be no significant interferences with copper in aqueous solutions. However, the particulates in serum and urine do generate sufficient background to considerably affect the absorption signal. When accurate, precise analyses are the objective, this effect must be removed.

Prior to analyzing sera by flame atomic absorption, the serum proteins are first precipitated by trichloroacetic acid. Then after the adjustment of pH to 5.5–6.5, the copper is chelated by dithiocarbamate and extracted into methyl isobutyl ketone. Addition of EDTA (versenate) increases the chelation and extraction of copper by blocking the chelation and extraction of other metallic ions present. Urine specimens with normal copper content must be concentrated about 4:1 prior to analyses by a flame instrument. The aliquot is adjusted to pH, dithiocarbamate is added, then the aliquot is extracted into a lesser volume of methyl isobutyl ketone, which is then aspirated.

A specific concentration of copper in methyl isobutyl ketone will produce an absorption signal about four times greater than that for the same concentration in aqueous solution. This is probably the result of the cooling effect of water. A lean oxidizing flame is used for copper analysis.

When determining the copper content of sera with the graphite furnace, standards and unknown sera are diluted 1:5 with 5% trichloroacetic acid. The atomization temperature is 2500°C. Reference standards and unknowns need not be read against background correction, because the signal generated by the nonspecific molecular absorption due to serum proteins and sodium occurs a few seconds before that of the atomization of copper itself.

Urine specimens must be chelated and extracted prior to graphite furnace analyses for two reasons. First, the signal generated by the normal constituents in urine completely overwhelms that of copper. Background correction does not eliminate the very large sodium signal. Second, the copper content in normal urine samples must be concentrated to be adequately measured.

Tissues and other solid materials must be solubilized prior to attempting copper determinations. It is understood, of course, that a series of standards treated in an identical manner to that of the test samples will be carried simultaneously through the analytical procedures.

4.4.6 Iron

A detection limit of 0.01 μg/ml by flame atomic absorption spectrometry is claimed for iron. Limits by flame emission are generally higher than the biological iron concentration being determined.

Normal serum iron levels range between 85 and 150 μg% (0.85–1.5 μg/ml). Since there is little significant interference with iron, it is possible to determine normal serum iron levels by aspirating serum samples diluted 1:5 with trichloroacetic acid, provided that the instrument employed possesses sufficient sensitivity.

The concentration technique developed by Zettner[61] which consists of protein precipitation followed by chelation with bathophenanthroline and extraction into methyl isobutyl ketone is suggested for analyses of sera from anemic patients whose iron levels may be far below 0.5 μg/ml.

It is preferred to concentrate and extract urine specimens with normal iron content (about 100 μg/liter to 0.1 μg/ml) before analysis. The nonspecific light-scattering effect interferes at this level. On the other hand, urinary iron levels from cases of iron poisoning can be determined with minimal error by direct analyses. Levels of 2–8 mg/liter (2–8 μg/ml) are usual in such cases.

The iron content of tissues can be analyzed directly following digestion. A lean (fuel poor) flame is used for iron analyses. The graphite furnace lends itself readily to iron analyses. Serum iron levels can be determined on one tenth the volume required for analysis by the flame aspirator–burner or colorimetric procedure. Unknowns and standards are diluted 1:5 with trichloroacetic acid.

Background correction for serum iron analyses is not necessary, since the background signal appears well before the iron signal and is clearly separated from it. However, because of the greater sodium and potassium content present in urine and tissue digestants, background correction is necessary when determining the iron content of these matrices.

A word of caution concerning iron contaminants on plastic ware should be interjected at this point. Polyethylene and some other plastic test tubes, tubing, and pipette tips are made, usually, on stainless steel molds. As a consequence, these contain quantities of iron, cadmium, zinc, and lead. All plastic ware should be precleaned with 10% nitric acid and rinsed copiously in deionized water. Each precleaned tip is then rinsed twice with the diluent used for sample preparation. An aliquot is dispensed into the graphite furnace. Should the resulting signal be too great, the tip is discarded.

4.4.7 Zinc

Since the introduction of atomic absorption spectrometry provided simple, yet precise and accurate analyses of zinc in biological materials, interest in the role of zinc in health and disease states has been accelerating.

Zinc and copper levels are monitored during chemotherapy for leukemia and lymphoma. The zinc levels in sera initially are lower than the normal 100–150 μg%, while copper levels are elevated. With a favorable response to therapy, zinc levels rise as the copper levels decrease. Among its other roles, zinc is an activator of the enzyme carbonic anhydrase. It is interesting that following a stroke, carbonic anhydrase and zinc content in the affected area of the brain are markedly decreased. Serum zinc is decreased in stress situations. A relationship between zinc content and wound healing has also been postulated.

Detection limits for zinc at its resonance line 2139 Å, is 0.002 μg/ml by atomic absorption. An oxidizing flame is suggested.

The zinc content of sera can be determined in a 1:5 dilution. When a large number of analyses are to be done, protein precipitation followed by centrifugation is desirable to prevent burner clogging problems.

While zinc is not subject to significant interference in the serum determination described, the molecular absorption and scattering effects due to urinary constituents are considerable. This can be eliminated by chelation and extraction plus the use of the deuterium background corrector. Zinc chelates with dithiocarbamate and can be quantitatively extracted over a pH range of 4–11.

The zinc content of tissue is determined on a dilution of the digestant.

Were it not for the massive contamination of plastic tips and tubing with zinc, analyses on microliter amounts of sample in the graphite furnace would be practical. It should be emphasized that zinc is incorporated into rubber tubing and stoppers. Thus, these can be a source of contamination of specimens stored in glass bottles with rubber stoppers. However, polypropylene tubes with Teflon-lined caps are essentially zinc-free.

4.4.8 Gold

There is a limited clinical interest in gold. Early in this century it was used in the treatment of tuberculosis. At present, some rheumatologists consider the element efficacious for the relief of arthritis. Since gold is a toxic metal, affecting the kidneys, its plasma levels and urinary excretions should be monitored when it is being administered.

Detection limits for gold at the 2430-Å resonance line is said to be about 0.05 μg/ml by atomic absorption spectrometry. Supposedly, no significant interferences for gold have been reported. However, certain components in biological materials, primarily potassium, phosphate, and proteins, markedly enhance the absorbance signal for gold. Consequently, despite the favorable sensitivity of detection, gold must be isolated from its biological matrix.

To accomplish this, proteins in both tissues and plasma are destroyed by wet or dry ashing. Only a small fraction of the gold circulating in the blood (about 5%) is free. The major portion is tightly bound to the fibrinogen and $\alpha 1$ globulin fractions of plasma proteins. The trichloroacetic acid precipitation will not release this bound gold. Proteins must be totally destroyed by wet or dry ashing.

This binding of gold to plasma or serum occurs *in vitro* as well. When TCA precipitation is employed, only 5% of the gold added is recovered. Following wet ashing, however, recovery of added gold is total. Urine specimens must also be chelated and extracted prior to analysis.

The diethyldithiocarbamate chelate of gold can be extracted with methyl isobutylketone at pH 5-6. A lean flame is used for the analyses.

4.4.9 Manganese

Various enzymes, among them phosphatases, peptidases, and arginases, are activated by manganese. Thus, manganese is considered to be an essential trace element. In excess amounts, however, manganese can exert toxic effects.

The entire body has been reported to contain a total of 12-20 mg/70-kg man.[62] Whole blood manganese levels of 6-10 μg/liter and serum levels of 1-1.5 μg/liter have been reported in "normal" subjects. Significantly elevated manganese levels of red cells in rheumatoid arthritis patients have been observed while the serum levels remained unchanged.

Manganese poisoning produces a neurological disorder which resembles Parkinson's disease. It is interesting to note that upon examining brain tissues of a victim of Parkinson's disease, the substantia nigra contained a manganese content of 10 mg/100 g. In similar sections of "normal" brains, any manganese present was below the detection limits by the flame atomic absorption spectrometer in the author's laboratory.

Detection limits for manganese by flame atomic absorption is in the order of 0.005 μg/ml at the 2795-Å resonance line.

Manganese analysis in the graphite furnace following chelation and extraction is practical. The diethyldithiocarbamate chelate is quantitatively extracted at pH 6-9 with methyl isobutyl ketone. Versenate completely masks, i.e., interferes with, the extraction of this element. Manganese is also reported to form an extractable complex with 8-hydroxyquinoline.

4.4.10 Nickel

Interest in the probable role of nickel has been growing. Serum nickel is elevated following a coronary occlusion.

With a sensitive atomic absorption spectrometer, detection limits for nickel at its 2320-Å resonance line are claimed to be 0.01 μg/ml. Serum levels of 0.02–0.1 μg/ml are considered as normal. Proteins and various other constituents in biological materials enhance the absorption signal. Therefore, nickel present in biological materials must be isolated from its matrix and concentrated prior to analysis.

Nickel is extractable into methyl isobutyl ketone following chelation by sodium diethyldithiocarbamate at pH 5-11. Versenate and cyanide are interferences. Dimethylglyoxime also forms an extractable nickel chelate at pH 2.5–8.

Copper and palladium also form extractable chelates. A lean oxidizing flame is employed for this analysis.

As might be expected, the application of the graphite furnace to nickel analysis in biological material extends the practical detection limits.

4.4.11 Cobalt

Cobalt, one of the essential metals, is a component of the vitamin B_{12} molecule. About 0.01 μg/ml of cobalt is present in human blood. Tipton[50] found cobalt in all tissues examined.

Detection limits of 0.01 μg/ml at the 2407-Å line have been attained with flame atomic absorption spectrometry. The cobalt content in blood samples must be concentrated prior to analysis. Although no significant interferences for cobalt have been reported, the light-scattering effect produced by constituents of biological fluid must be eliminated.

Cobalt (valence 3) as the diethyldithiocarbamate chelate is extractable into an organic solvent over the pH range 4–11. The cobalt acetylacetone chelate is extractable at pH 1. A lean flame is used for cobalt analyses.

Cobalt can also be analyzed in the graphite furnace, which does extend its detection limits.

4.4.12 Chromium

For many years, interest in chromium was confined primarily to its toxic properties for plants and man. However, information being accumulated concerning its metabolic functions indicate that chromium can probably be classed as an essential element. For example, it appears to increase hepatic cholesterol and fatty acid synthesis. It is also involved in glucose tolerance and protein synthesis. Increased serum chromium levels have been observed following myocardial infarction. Normal blood chromium levels are of the magnitude of 0.03 μg/ml.

Chromium at its 3579-Å resonance line can be detected at levels of 0.1 μg/ml and 0.005 μg/ml by flame emission and atomic absorption, respectively. Interferences from iron and nickel are reduced by adding ammonium chloride. A fuel-rich (reducing) flame is used.

Detection limits in the graphite furnace are more favorable. However, constituents in biological matrices (plasma, urine, wet-ashed tissues) generate interference signals too chose to or superimposed upon the chromium signal. To eliminate these interferences, Davidson and Secrest[62] adapted a double-atomization procedure. After the initial drying and charring cycle, the sample was atomized at 1400°C and then subjected to an additional drying and charring cycle and finally to atomization at 2400°C. There were no losses of chromium, since this metal volatilizes above 1400°C; only the interfering salts are volatilized at the initial atomization.

Chromium forms extractable chelates with 5-methyl-8-hydroxyquinoline at pH 5.5-9.5. The metal at valence 3 is inert to chelate formation with either acetylacetone or diethyldithiocarbamate.

4.5 THE TOXIC METALS

Health problems resulting from exposure to the toxic heavy metals, lead, mercury, and arsenic are not a modern entity. Their effects were probably recognized far earlier than the first century B.C. A description of lead colic is found in Hippocrates. Dioscorides described the toxic effects of swallowing mercury metal and inhaling its fumes. It should be pointed out, however, that while mercury vapors prove toxic when inhaled, cold metallic mercury is essentially nontoxic when swallowed. Metallic mercury as such is poorly absorbed from the gastrointestinal tract. Through the ages, arsenic has been a medicine, as well as a favorite poison. Recognition of the hazards of cadmium and thallium are somewhat more recent.

An appreciation of the biochemistry of these toxic metals and the various interrelationships among them and the "essential" trace elements is rapidly developing. There are now analytical tools with which to study these elements. Prior to the availability of atomic absorption spectrophotometric instrumentation, most heavy metal analyses, especially those done in clinical laboratory areas, were performed by some sort of colorimetric procedure. Instrumentation for spectrographic analyses were beyond their financial and technical capabilities.

Classical colorimetric methods, while sensitive, do lack the desired specificity. For example, dithizone, the chelating agent most commonly employed, forms colored complexes with approximately 17 metals. A visual spectrophotometer, unfortunately, does not possess the capability of distinguishing among the dithizonates or dithiocarbamates of cadmium, lead, and thallium.

4.5.1 Lead

Lead is a general protoplasmic poison. Various enzyme systems regulating metabolic functions are inhibited by lead. Those enzyme systems containing sulfhydryl groups are primarily involved. Lead is a cumulative poison, subtle in its action, often mimicking other disease states. All bodily systems are affected.

Lead is a common contaminant in our environment. The major route of lead absorption is through the gastrointestinal tract. According to Kehoe, the average adult ingests between 0.3 and 0.6 mg of lead in his diet.[63] Approximately 10% is retained in the body.

Except in certain lead-associated industries, for example foundries, smelters, or factories producing lead compounds, exposure to lead by inhalation of fumes is minimal. Occasionally, a nonindustrial situation such as exposure to fumes from burning leaded battery casings for fuel, results in a serious and acute exposure to lead fumes.

The lead particulates in the air of metropolitan areas such as Chicago amounts to approximately 1.5 $\mu g/m^3$ of air. We breath about 22 m^3 in a 24-h period. Only a fraction of the particulates inhaled are absorbed. Even if absorption were total, the degrees of lead exposure from such air is equivalent to a mere fraction of that lead which is ingested with our food.

Normal blood lead levels are about 0.2 μg/ml or less. Normal urinary lead excretion in children and adults are stated as being 0.08 μg/ml and 0.15 μg/ml, respectively. A blood lead level of 0.6 μg/ml is considered suggestive of lead poisoning and worthy of further investigation. Ninety-five percent of the lead circulating in blood is loosely bound to the red cell (erythrocyte) membrane.

At the 2833-Å resonance line, lead has a detection limit of 0.03 μg/ml by flame atomic absorption. A detection limit of 0.01 μg/ml is claimed for the 2170-Å line. However, the electronic "noise" encountered at the latter with some atomic absorption instruments may make that a less desirable analytical line.

Other metals interfere little with lead. On the other hand, constituents in a biological matrix (blood, urine, tissue) can affect the absorption signal considerably. To attain more accurate, precise results, lead should be isolated from the matrix. Lead in red blood cells can be released by hemolyzing the cells with water. Lead in solution can then be chelated by a dithiocarbamate at pH 5.5–6.5 and extracted into methyl isobutyl ketone. A minute amount of a surfactant such as Triton X-100 is added to minimize emulsion formation.

Urines are adjusted to pH 5.5–6.5 prior to the chelate and extraction into methyl isobutyl ketone. Tissues must be wet- or dry-ashed. An aliquot of the diluted digestant is then adjusted to pH, chelated, and extracted. Some investigators also prefer to wet-ash blood samples.

A lean oxidizing flame extending no more than 1 inch or so above the burner head is the preferred flame for lead analyses. Flow rates of 3.5–4 liters/min for acetylene and 5–5.5 liters/min for air are suggested. The methyl isobutyl ketone extract should be aspirated slowly, about 1–1.5 ml/min. A single-slot burner head is more practical than the three-slot Boling burner for this application. The latter back flashes when used at the suggested lower-gas-flow rates.

Delves[(64)] proposed a micro technique for blood lead determinations using 20 μl of sample. Blood proteins are denatured by heat and peroxide prior to atomizing the sample in the flame of a Boling burner. There have been various modifications of the technique.

With proper correction of background absorption due to matrix and with a suitable recorder, adequate blood lead analyses can be performed by the Delves technique. A skilled, knowledgable spectroscopic technician is essential. This technique is not too practical for "screening" programs where hundreds of analyses must be completed daily by a few people who are often less than expert technically.

Accurate and precise lead analyses in the graphite furnace are feasible, provided that the biological matrix is removed prior to analysis. Background correction must also be employed.

4.5.2 Cadmium

Cadmium, like lead, is a common contaminant in our environment. It has been implicated as a contributing factor in the production of heart disease and severe hypertension. Cadmium inactivates the SH enzymes. Interest in this trace element's possible role in health and disease is increasing. The cadmium contents of blood and urine are normally quite low. Levels above 0.05 μg/ml in blood and urine are worthy of further investigation.

The detection limits for cadmium at its 2288-Å resonance line are 4.2 μg/ml by flame emission and 0.001 μg/ml by atomic absorbance. Cadmium appears not to be subject to interferences from other metals; however, the constituents of biological material enhance the absorption signal manyfold. Therefore, cadmium should be isolated from its biological matrix. A solvent extractable chelate with sodium diethyldithiocarbamate forms at pH 5.5–6.5. A lean oxidizing flame, as described for lead analysis, is used.

Cadmium analyses can be performed with the graphite furnace. However, the analyst must be aware of the rich contamination of polyethylene pipet tips and tubing with cadmium, which must be removed.

4.5.3 Thallium

In the past, thallium salts were employed in depilatory preparations, both prescribed and proprietary. Currently, their major use is in the preparation of rodenticides. Thallium inactivates sulfhydryl-containing enzymes and is a potent neurological poison. Levels of 0.1 μg/ml in blood and 0.05 μg/ml in urine are considered significant.

Detection limits for thallium at the 2768-Å resonance line are listed as 0.2 μg/ml for atomic absorption spectrometry. Blood and urine contents must be concentrated two- to fivefold prior to analysis. While other metals do not interfere, constituents of biological materials enhance the absorption signal.

Thallium can be chelated by sodium diethyldithiocarbamate at pH 5.5–6.5 and extracted into a lesser volume of MIBK. A lean flame is preferred for analysis. Following isolation from its biological matrix, thallium can be determined in the graphite furnace. Background correction is employed to eliminate the nonspecific signal which is due primarily to sodium.

Since thallium, lead, and cadmium as well as antimony are extracted simultaneously by MIBK following chelation with sodium diethyldithiocarbamate at pH 5.5–6.5, a single extract of a specimen can be employed for the determination of these toxic elements. Only the hollow cathode lamps have to be changed.

4.5.4 Antimony

The detection limits for antimony at its 2069-Å resonance line is 0.1 μg/ml by atomic absorption spectrometry. While the 2178-Å is more sensitive, it is not practical for clinical or toxicological analyses because of its proximity to a strong lead line. Antimony can be removed from biological materials by the technique described for lead. It is not a normal constituent of such materials.

Antimony, as well as lead, is a major contaminant in the environment of police firing ranges. Significant increases in blood and urinary antimony levels have been encountered among firing-range personnel.

4.5.5 Arsenic

By atomic absorption spectrometry, the detection limits for arsenic at the 1937-Å resonance line are 0.2 μg/ml in aqueous solution. However, in an air–acetylene flame with organic solvents, any sodium present cause high absorption signals. Substituting an argon flame improved matters somewhat. Use of the tantalum boat extended the detection limits. Nevertheless, until the introduction of flameless techniques and electrodeless discharge lamps, the analyses of biological materials for arsenic by atomic absorption were not very practical. The large sample size necessary made the old Gutzeit procedure the more preferred method.

Arsenic is not a normal constituent of biological material. A level of 0.05 μg/ml in urine is significant.

By utilizing chelation-extraction techniques, the graphite furnace, and an electrodeless discharge lamp, we have been able to detect arsenic at a level of 0.001 μg/ml in blood and urine. Trivalent arsenic as the dithiocarbamate chelate can be extracted by MIBK at pH 4–6.

4.5.6 Beryllium

Detection limits claimed for beryllium at the 2348-Å resonance line by flame atomic absorption are about 0.002 μg/ml. Nitrous oxide is the suggested oxidant. A reducing flame is used.

Beryllium is not a normal constituent of biological materials. Even in pathological situations, beryllium concentrations are less than the detection limits claimed for flame atomic absorption. Practical limits are attained when the graphite furnace is utilized. We have been able to detect concentrations in the range 0.01 ng/ml by first chelating the beryllium at pH 6 and extracting the complex with MIBK, then atomizing the sample in the graphite furnace.

Cupferron, 2% aqueous solution, was found to be the most efficient complexing agent. The dithiocarbamates do not form chelates with this metal.

4.5.7 Mercury

Mercury is not a normal constituent of biological materials. A level of 0.05 μg/ml in urine is considered of sufficient significance to warrant further investigation. Mercury inactivates the sulfhydryl-containing enzymes.

At the 2537-Å resonance line, detection limits for mercury are 2.5 μg/ml by flame emission and 0.2 μg/ml by flame atomic absorption. Phosphates, sodium, potassium, and proteins in biological matrices markedly enhance the absorption signal from mercury. Proteins in blood and tissue specimens are precipitated with trichloroacetic acid, then centrifuged. The mercury, in salt form, is chelated by dithiocarbamate at pH 3–4 and extracted into MIBK. Following pH adjustment, urines can be chelated and extracted. Red cells contain twice as much mercury as does plasma.

Since mercury is such a volatile substance, tissues cannot be wet- or dry-ashed in the usual manner. Trichloroacetic acid precipitation of a tissue puree is a satisfactory means for the removal of proteins. Mercury in the supernatant can then be chelated and extracted after pH adjustment. Some investigators prefer an alkaline digestion technique. The lean oxidizing flame, such as that used for lead, is used.

Applying flameless atomic absorption techniques extends the practical detection limits. Specimens are treated as described above. Background correction is necessary because of the nonspecific molecular absorption primarily from sodium. Extracts are dried and charred at 75°C to prevent volatilization of mercury during these cycles. About 1 ng/ml of mercury can be detected in an extract from a 5-ml specimen.

4.5.8 Selenium

Detection limits for selenium by flame atomic absorption are 0.5 ng/ml at the 1960-Å resonance line. Limits are extended to less than the ng/ml range when both flameless atomization techniques and electrodeless discharge lamps are employed.

Analyses of selenium in simple matrices, water for example, are being performed quite successfully. Biological materials, however, present an exquisite, but not impossible, challenge. The challenge lies with the chemistry, not the instrumentation.

Possibly the best approach, at present, for all matrices, i.e., blood, urine, tissue, etc., is complete destruction of organic materials by wet or dry ashing. Chelation techniques are grossly inefficient. Sodium diethyldithiocarbamate and ammonium pyrollinidine dithiocarbamate chelate selenium over the pH range

4-9. Cupferron chelates selenium at pH 5-5. We have found both the dithiocarbamate and cupferron complexes of selenium to be very poorly extracted by carbon tetrachloride and chloroform. Ethyl acetate and methyl isobutyl ketone proved somewhat more effective.

4.5.9 Miscellaneous Metals

Practical procedures, from the clinical laboratory viewpoint, for analyses of tin, boron, tellurium, platinum, aluminum, and other trace metals now considered esoteric are yet to be derived. The instrumentation available has the potential capability, once the matter of interferences due to the matrices are resolved.

4.6 PLASMA EMISSION SPECTROMETRY

Plasma emission spectrometry may become another practical analytical tool in the future clinical laboratory. Fine precision, accuracy, and detection limits are now attainable when measuring various metallic elements in simple aqueous solutions. However, the challenges generated by biological matrices must also be resolved with the plasma emission.

4.7 REFERENCES

1. L'Abbé Marie, *Nouvelle découverte sur la lumière pour en mesurer et compter les degrés*, Paris (1700).
2. R. Herrmann and C. T. J. Alkemade, *Chemical Analysis by Flame Photometry*, Wiley-Interscience (1963).
3. C. A. MacMunn, *Spectrum Analysis Applied to Biology and Medicine*, Longmans, Green & Company Ltd., London (1914).
4. P. Champion, H. Pellet, and M. Grenier, *Compt Rend. 76*, 707 (1873).
5. A. Gouy, *Compt. Rend. 83*, 269 (1876).
6. G. Valentin, *Der Gebrauch des Spektroscopes zu physiologischen und ärztlichen Zwecken*, Winter'sche Buchhandlung, Leipzig (1863).
7. H. Bence-Jones, *On Chemical Circulation in the Body*, Proceedings of the Royal Institute, May 26, 1865.
8. E. Rosenberg, *The Use of the Spectroscope in Its Applications to Scientific and Practical Medicine*, G. P. Putnam's Sons, New York (1876).
9. W. N. Hartley, *Chem. News 5*, 135 (1885).
10. W. N. Hartley, *Phil. Trans. Pt. II*, 9 (1885).
11. W. N. Hartley, *Proc. Roy. Soc. 38*, 1 (1885).
12. W. N. Hartley, *Proc. Roy. Soc. 38*, 191 (1885).
13. W. N. Hartley, *J. Chem. Soc. 87*, 1796 (1905).
14. J. J. Dobbie and J. J. Fox, *J. Chem. Soc. 103*, 1193 (1913).
15. J. J. Dobbie and J. J. Fox, *J. Chem. Soc. 105*, 1639 (1914).

16. J. J. Dobbie and A. Lauder, *J. Chem. Soc. 83*, 605 (1903).
17. J. J. Dobbie and A. Lauder, *J. Chem. Soc. 97*, 1546 (1910).
18. J. J. Dobbie and A. Lauder, *J. Chem. Soc. 99*, 1254 (1911).
19. W. W. Coblentz, *Investigations of Infra Red Spectra*, Carnegie Institution, Washington, D.C. (1908).
20. R. G. White, *Handbook of Ultraviolet Methods*, Plenum Press, New York (1965).
21. H. Schiller, *Deut. Z. Ges. Gerichtl. Med. 29*, 104 (1937–38).
22. L. R. Goldbaum, *J. Pharm. Exp. Therap. 94*, 68 (1948).
23. N. K. Freeman, *Advan. Biol. Med. Phys. 4*, 167 (1956).
24. R. A. Morton, *Nature 193*, 314 (1962).
25. B. Klein, M. Weissman, and J. Berkowitz, *Clin. Chem. 6*, 453 (1960).
26. R. D. Stewart, T. R. Torkelson, C. L. Hake, and D. S. Erley, *J. Lab. Clin. Med. 56*, 148 (1960).
27. R. Alha and V. Tamminen, *Ann. Med. Exp. Biol., Fenniae (Helsinki) 37*, 157 (1959).
28. F. Rieders, *J. Forensic Sci. 6*, 401 (1961).
29. K. T. Fowler and P. Hugh-Jones, *Brit. Med. J. 5029*, 1205 (1957).
30. K. Biemann, *Ann. Rev. Biochem. 32*, 755 (1963).
31. D. M. Hercules, ed., *Fluorescence and Phosphorescence Analysis: Principles and Applications*, Wiley–Interscience (1966).
32. S. Undenfriend, *Fluorescence Assay in Biology and Medicine*, Vol. II, Academic Press, Inc., New York (1969).
33. S. Undenfriend, D. F. Duggan, B. M. Vasta, and B. B. Brodie, *J. Pharm. Exp. Therap. 120*, 26, (1957).
34. J. H. Sheldon and H. Ramage, *Biochem. J. 25*, 1608 (1931).
35. W. Gerlach and W. Gerlach, *(Virchows) Arch. Pathol. Anat. Physiol. 28*, 209 (1931).
36. W. Gerlach and W. Gerlach, *Die Chemische Emissions-Spektralanalyse. II. Anwendung in Medizin, Chemie, und Mineralogie*, Leopold Voss, Leipzig (1933).
37. W. Benoit, *Z. Ges. Exp. Med. Biol. 90*, S.421 (1933).
38. A. Policard, *Protoplasma 19*, 602 (1933).
39. A. Policard and A. Morel, *Bull. Histol. Appl. I, 9*, 57 (1932).
40. W. Gerlach, W. Rollwaggen, and R. Intonti, *(Virchows) Arch. Path. Anat. Physiol 301*, 588 (1938).
41. G. H. Scott and P. S. Williams, *Proc. Soc. Exp. Biol. Med. 32*, 505 (1934).
42. G. H. Scott and P. S. Williams, *Anat. Record 64*, 107 (1935).
43. T. C. Boyd and N. K. De, *Indian J. Med. Res.* 20 (3), 789 (1933).
44. L. E. Gaul and A. H. Staud, *Arch. Dermatol. Syph. 28*, 790 (1933).
45. L. E. Gaul and A. H. Staud, *Arch. Dermatol. Syph. 30*, 433 (1934).
46. L. E. Gaul and A. H. Staud, *J. Amer. Med. Assoc. 104*, 1387 (1935).
47. P. G. Shipley, F. T. M. Scott, and H. Blumberg, *Bull. Johns Hopkins Hosp. 51*, 327 (1932).
48. J. Cholak, *Ind. Eng. Chem. 7*, 287 (1935a).
49. R. A. Kehoe, *J. Nutri. 19*, 579 (1940).
50. I. H. Tipton, in *Metal-Binding in Medicine* (M. J. Seven, ed.), J. B. Lippincott Company, Philadelphia (1960), p. 27.
51. S. Natelson, M. R. Richelson, B. Shield, and S. L. Bender, *Clin. Chem. 5*, 579 (1959).
52. V. L. Grebe and F. Esser, *Fortschr. Gebiete Roentgenstrahlen 54*, 185 (1936).
53. S. Natelson, *Clin. Chem. Suppl. 11*, 290 (1965).
54. S. Natelson, M. R. Richelson, B. Shield, and S. L. Bender, *Clin. Chem. 5*, 579 (1959).
55. S. Natelson and B. Shield, *Clin. Chem.* 6, 299 (1960).
56. S. Natelson and B. Shield, *Clin. Chem. 8*, 17 (1962).
57. P. K. Lund and J. C. Mathies, *Amer. J. Clin. Pathol. 40*, 132 (1963).

58. P. K. Lund, D. A. Morningstar, and J. C. Mathies, *Biochem. Biophys. Res. Commun. 14*, 177 (1964).
59. P. K. Lund and J. C. Mathies, *Amer. J. Clin. Pathol. 44*, 398 (1965).
60. E. J. Brooks, O. R. Gates, and M. Nottingham, *Amer. J. Clin. Pathol. 41*, 154 (1964).
61. A. Zettner, L. C. Sylvia, and L. Capacho-Delgado, *Amer. J. Clin. Pathol. 45*, 533 (1966).
62. I. W. F. Davidson and W. L. Secrest, *Anal. Chem. 44*, 1808 (1972).
63. R. A. Kehoe, J. Cholak, and R. V. Story, *J. Nutr. 19*, 579 (1940).
64. H. T. Delves, *Analyst 95*, 431 (1970).
65. R. J. Henry, *Clinical Chemistry: Principles and Techniques*, Harper & Row, Publishers, New York (1969).
66. J. A. Halsted, B. Hackley, and J. C. Smith, *Lancet 2*, 278 (1968).
67. H. Wolff, *Biochem. Z. 319*, 1 (1948).
68. A. Heyrovsky, *Casopis Lekaru Ceskych 91*, 680 (1952).
69. F. J. Feldman, E. C. Knoblock, and W. C. Purdy, *Anal. Chim. Acta 38*, 489 (1967).
70. G. C. Cotzias, S. T. Miller, and J. Edwards, *J. Lab. Clin. Med. 67*, 836 (1966).
71. P. S. Papavasilion, S. T. Miller, and G. C. Cotzias, *Amer. J. Physiol. 211*, 211 (1966).
72. F. W. Sunderman, *Amer. J. Clin. Pathol. 44*, 182 (1965).
73. S. Nomoto and F. W. Sunderman, Jr., *Clin. Chem. 16*, 477 (1970).
74. K. Liebscher and H. Smith, *Arch. Environ. Health 17*, 881 (1968).
75. H. M. Perry, Jr., I. H. Tipton, H. A. Schroeder, and M. J. Cook, *J. Lab. Clin. Med. 60*, 245 (1962).
76. I. H. Tipton and M. J. Cook, *Health Phys. 9*, 103 (1963).
77. R. Robertson, K. Fritze, and P. Grof, *Clin. Chim. Acta 45*, 25 (1973).
78. D. Brune, K. Samsahl, and P. O. Wester, *Clin. Chim. Acta 13*, 285 (1966).
79. J. Lieben and F. Metzner, *Amer. Ind. Hyg. Assoc. Abstr.*, Ind. Health Conference, 1959.
80. H. A. Schroeder, in *Metal-Binding in Medicine* (M. J. Seven, ed.), J. B. Lippincott, Philadelphia (1960).
81. H. R. Imbrus, J. Cholak, I. H. Miller, and T. Sterling, *Arch. Environ. Health 6*, 286, (1963).
82. J. Kubata, A. Lazar, and F. L. Losee, *Arch. Environ. Health 16*, 788 (1968).
83. R. A. Kehoe, J. Cholak, and R. V. Story, *J. Nutr. 19*, 579 (1940).
84. C. C. Patterson, *Arch. Environ. Health 11*, 344 (1965).
85. L. J. Goldwater, *Mercury: A History of Quicksilver*, York Press, Baltimore (1972).
86. E. M. Richeson, *Ind. Med. 27*, 607 (1958).
87. R. Mavrodineanu, ed., *Analytical Flame Spectroscopy*, Springer-Verlag, New York (1970).

Author Index

Subject Index